The Science of Bureaucracy

Inside Technology Series

Edited by Wiebe E. Bijker, W. Bernard Carlson, and Trevor Pinch

A list of the series appears at the back of the book.

The Science of Bureaucracy

Risk Decision-Making and the US Environmental Protection Agency

David Demortain

The MIT Press
Cambridge, Massachusetts
London, England

This book was set in Stone Serif and Stone Sans by Westchester Publishing Services. Printed and bound in the United States of America.

Library of Congress Cataloging-in-Publication Data

Names: Demortain, David, author.
Title: The science of bureaucracy : risk decision-making and the US Environmental Protection Agency / David Demortain.
Description: Cambridge, MA : The MIT Press, [2019] | Series: Inside technology series | Includes bibliographical references and index.
Identifiers: LCCN 2019010651 | ISBN 9780262537940 (pbk. : alk. paper)
Subjects: LCSH: United States. Environmental Protection Agency--Management--History. | Risk management--Government policy--United States--History. | Environmental policy--United States--Decision making. | Environmental policy--United States--History.
Classification: LCC TD171 .D46 2019 | DDC 363.700973--dc23
LC record available at https://lccn.loc.gov/2019010651

10 9 8 7 6 5 4 3 2 1

For Charline

Contents

Acknowledgments ix

Introduction 1

1 Risk Sciences: Expertise for Decision-Making and Dispute 31

2 The Invention of Quantitative Risk Assessment 51

3 Prioritizing Toxics: The Prehistory of Risk Management 79

4 Codifying the Risk Assessment–Risk Management Framework 105

5 Ruckelshaus and Risk: Representing the EPA 137

6 Risk Management: The EPA as a Decision-Making System 161

7 Designing a Science for Decisions 185

8 The Rise and Fall of Comparative Risk Assessment 211

9 Scientization and the Reform of the Risk Assessment–Risk Management Framework 231

10 Beyond the Risk Paradigm 273

Conclusion 305

Archival Sources and Documents 321

Acronyms 323

Notes 325

References 371

Index 421

Acknowledgments

This book is nothing like I imagined it would be when I embarked on a study of the transnational history of the risk assessment-risk management framework. Like many others, I knew I needed to study in-depth what had happened at the National Research Council, which published *Risk Assessment in the Federal Government*, or the "Red Book," in 1983, often presented as the unique point-source of the framework. I also knew that the Red Book earned an aura because William Ruckelshaus gave it visibility, in a moment when it was working to restore the authority of the Environmental Protection Agency. As I made progress through archives and interviews, though, it very quickly became clear that the institutionalization of risk decision-making and of the EPA were inseparable. They have emerged together, gave form and credibility to one another, and changed jointly over time. Risk decision-making is the bureaucratic translation of the ambition to govern uncertain hazards in a rational, optimal, and scientific manner. It is an instance of Max Weber's ideal type of the bureaucracy as administration legitimized by expertise, except that in this case science does not represent a historical force progressing in our societies in a linear fashion. It gives shape to bureaucratic institutions from within, responding to their difficulties to govern risks and the public controversies they give rise to.

Many people helped me along the way, and were indeed essential to make the most out of this moment of serendipity: Pierre-Benoit Joly and Mike Power saw this work in its early stages and helped me bring it to fruition. The conversations with them were essential to gear up to perform this research. Two persons made the journey with me: I worked for a couple of years with Soraya Boudia on the subject, which was a great pleasure and an important source of inspiration, and method. I have had recurrent discussions with Jean-Louis Jouve, one of the earliest French readers of the Red Book and also one of my first interviewees on the subject. Sheila Jasanoff enabled me to develop this book in many essential ways: through her own seminal work on agencies and risk expertise, from which I learned so much; through her recollections, observations, and analyses of the various people, organizations, and moments that comprise the history I write about here; through her close and critical reading of a first version of this book. Three more people tremendously helped me improve this book thanks to their critical and inspiring comments: Philippe Bézès, Marie-Laure Salles-Djelic, and Christine Noiville. Genuine thanks to all three of them.

Thanks to the many colleagues and friends who helped me, reading parts of this text and forcing me to consolidate my ideas: Jean-Pierre Galland, Martin Giraudeau, Ashveen Peerbaye, and Jean-Claude Thoenig. I am also grateful for the many conversations, in-depth or in passing, which seeded my thoughts over time, with Marc Barbier, Daniel Benamouzig, Olivier Borraz, Henri Boullier, François Busquet, Daniel Carpenter, Jean-Philippe Cointet, Steve Collier, Amy Dahan, François Dedieu, Jim Dratwa, John Downer, Liz Fisher, Scott Frickel, Jean-Paul Gaudillière, James Hammitt, Boris Hauray, Steve Hilgartner, Bridget Hutter, Nathalie Jas, Jean-Noel Jouzel, Nils Kessel, Andrew Lakoff, Brice Laurent, Javier Lezaun, Yves Lichtenberger, Allison Loconto, Martin Lodge, Tom McGarity, Marianne Noel, Craig Oren, Catherine Paradeise, Sylvain Parasie, Brian Pentland, Lorène Prost, Carsten Reinhardt, Arie Rip, Akos Rona-Tas, Henry Rothstein, Yves Schemeil, Syd Shapiro, Vincent Spenlehauer, David Vogel, and David Winickoff.

I have interviewed more than fifty people from the EPA or NRC panels over the years, without whom this history would not exist, so thanks to all of them for giving their time and sharing their stories and insights. I would mention just a few people with whom I have had fundamental discussions, and great pleasure conversing: Terry Davies, Adam Finkel, Dan Fiorino, Ted Greenwood, Larry McCray, Richard Morgenstern, Peter Preuss, Milton Russell, and Maurice Zeeman. Archivists of the NAS, the Baker Center, and the EPA National Library were immensely helpful, so thanks to Daniel Barbiero, Erin Boest, Kris Bronstad, Janice Goldblum, and Kezee Procita.

Particular thanks go to Katie Helke, the editor for science and technology studies at MIT Press, for expressing her interest in this book and fostering its publication. Thanks to Laura Keeler, too, for making the review and production process as smooth as it can be. Three women edited the text and helped me turn it into a proper book: Liz Libbrecht, Susan McClung, and Melody Negron. The Institution Francilien Recherche Innovation Société (IFRIS) and the Laboratoire Interdisciplinaire Sciences Innovations Sociétés (LISIS), my research center, have provided financial support for the publication of this book, thanks for this to their members and respective directors, Rigas Arvanitis and Pierre-Benoit Joly.

Finally, thank you to Charline, Gaspard, and Margot for their patience and support, for containing my onetime obsession with EPA and risk, and for making life besides book writing so worthy and fulfilling.

Introduction

Since Donald Trump became president of the United States in January 2017 and two successive proponents of environmental deregulation were placed at the head of the US Environmental Protection Agency (EPA)—Scott Pruitt first and then Andrew Wheeler—this agency has undergone a massive disruption, both in the orientation of its policies and in the traditional way in which it develops standards and regulations for environmental protection.

Protecting the public against environmental and health threats, one of the agency's key goals for decades, has been replaced by a job protection mantra, as if the agency's mission were to promote the economy and as if regulatory measures to clean up the environment and protection from health risks linked to environmental degradations, in and of themselves, were detrimental to the economy. This reorientation motivated the reversal of many standards advanced by President Barack Obama's EPA, from greenhouse gas emissions standards for cars and light trucks, to the contemplated ban of the pesticide chlorpyrifos, to the obligation to declare methane emissions from drilling operations. These efforts look like an unprecedented attack on environmental rules adopted in the United States and the fundamental vocation of the agency (Dillon et al. 2018).

In keeping with this reorientation, the new leadership of the agency has radically altered the way that the agency uses science. The budget cuts inflicted on the EPA have severely reduced the capacities of its Office of Research and Development (ORD). References to the agency's scientific mission in its official website presentation have been obliterated and, very recently, the use of the term *science-based* has been banned in official agency communication. The membership of the Science Advisory Board (SAB), a panel of external scientists who review key policies and proposals of the agency, has been completely renewed, based on a rule that excludes researchers who benefited from research grants by the agency—thus promoting scientists funded by other sources, notably by the industries that the agency regulates. In 2018, the EPA administrator unilaterally decided to apply at the agency a bill that had repeatedly failed to pass in Congress, according to which unpublished scientific studies may not be used as evidence by the agency (Hakim and Lipton 2018). This provision de facto excludes epidemiological studies of the health effects of pollution on populations (such as biomonitoring studies) that are generally not published in order to preserve the confidentiality of the people included in the cohorts. It severely deconstructs what is considered acceptable evidence, and restricts the use of the regulatory knowledge that is most conducive to protective standards (Mayo and Hollander 1991; Wagner and Steinzor 2018; Wagner et al. 2018).

With these measures, these two administrators reversed the traditional policy focus of the agency on the reduction of risk, as well as the pattern of using science to advance this policy. They are ignoring what had become a rule in the EPA: the formal separation between scientific assessment of environmental issues and policy development—a central tenet of the risk paradigm that the agency embraced in the early 1980s. For past administrators of the agency, whether Republican or Democrat, this is a complete and dangerous reversal of the rules that made the credibility of the EPA (Ruckelshaus 2017; McCarthy and McGabe 2018).

This attack on science and the traditional process for regulating risks is due, first, to the belief among Republicans that science is a bastion of politically biased environmentalism. Over the past two decades, Republicans have become fierce opponents of environmental policy and regulation (Layzer 2012; Sellers 2018). Second, the attack reflects a more fundamental and gradual decline in trust in science among conservatives. Since the 1990s, political conservatism compounded with religious conservatism to create a rise in the proportion of Republicans who do not believe in evolution (nearly 70 percent) and who maintain that the effects of global warming have been exaggerated (50 percent) (Gauchat 2012). During the two terms of President George W. Bush, this stance translated into aggressive editing or twisting of the science produced inside the agency to support the Republican's agenda (UCS 2008; see also Rich and Merrick 2006, Freeman and Vermeule 2007, Rest and Halpern 2007, Shulman 2008, and Shapiro 2009). The attack on the EPA's science in the current presidency is not the first one that the agency's science has had to endure (McGarity 2001; Fredrickson et al. 2018).

But conservatives attack the use of science in policy for a third, perhaps even more fundamental reason. Risk regulation is rooted in science, and their wish is precisely to roll back or eliminate altogether environmental constraints on business development. Science is, to use Mark Brown's apt phrase, a "proxy battleground for politics" (Brown 2009, 3) and, further, a medium for the assertion of legitimate policies and institutions. Conservatives are not attacking science as much as science-backed federal administration (Miller 2017). The radicalism of these attacks reflects the deep, historical institutionalization of the recourse to science for policymaking at the EPA, as well as the way in which science has authorized the EPA to address environmental and health hazards. The EPA is an agency that regulates thanks to scientific work and large programs of measurement, experimentation, and modeling, and thanks to the contribution of multiple constituencies of scientists in various parts of the agency. Science-based decision-making is an essential part of its identity, legitimacy, and autonomy.

Science, Risk, and the Design of the EPA

The US EPA was established by the National Environmental Policy Act, adopted by President Richard Nixon in late 1970, which placed the services that administered statutes related to the environment and variegated kinds of pollution under the same institutional roof. It is in charge of the implementation of an immense variety of legislation and programs (Weinberg and Reilly 2013), among which the best known are probably the program for criteria air pollutants [National Ambient Air Quality Standards (NAASQ)] and hazardous air pollutants [National Emissions Standards for Hazardous Air Pollutants (NESHAPS)], for pesticides [Federal Insecticide, Fungicide, and Rodenticide Act (FIFRA)] and toxic substances [Toxic Substances Control Act (TSCA)], for Superfund sites [Comprehensive Environmental Response, Compensation, and Liability Act (CERCLA)], for hazardous waste [Resource Conservation and Recovery Act (RCRA)], and for drinking water [Safe Drinking Water Act (SDWA)] and surface water [Clean Water Act (CWA)]. These statutes are risk-based: They rest on the principle that the hazards due to technologies (notably chemicals) can be anticipated and should be addressed before they materialize (Shapiro and Glicksman 2003, 31). And even though not all the programs administered by the agency are risk based (many statutes command actions of measurement and monitoring, depollution, decontamination, and technology assessment and substitution based on established, deterministic knowledge, from the Emergency Planning and Community Right-to-Know Act to the Endangered Species Act, through the Energy Independence and Security Act, the Marine Protection, Research, and Sanctuaries Act, the Noise Control Act, or the Energy Policy Act), the notion of risk has consistently influenced the image of the EPA's bureaucracy and its handling of environmental and human health problems.

William Ruckelshaus inaugurated this practice of deriving the identity of the EPA, as well as an image of how it is acting, from this technical, rationalistic language of risk. In a speech he gave at the National Academy of Sciences (NAS) at the end of June 1983, soon after being reappointed by President Ronald Reagan to end the crisis to which Ann Gorsuch had brought the agency (Ruckelshaus had been the first administrator of the EPA between 1970 and 1973), he articulated the public mission of his agency and the ways of fulfilling it in just this way (Ruckelshaus 1983c, 1026):

> EPA is an instrument of public policy, whose mission is to protect the public health and the environment laid down by its statutes. That manner is to set standards and enforce them, and our enforcement powers are strong and pervasive. But the standards we set, whether technology- or health-related, must have a sound scientific base ... Scientists assess a risk to find out what the problems are. The process of deciding what to do about the problems is risk management. The National Academy of Science report recommends that these two functions—risk assessment and risk management—be separated as much as possible within a regulatory agency. This is what we now do at EPA and it makes sense.

To today's reader, this might not sound like a very original argument. At the time that this speech was pronounced, however, the rubrics of *risk assessment* and *risk management* were hardly used in law, in public discourse, or in the actual processes of the EPA to describe what it was doing or had to do as a whole. It was probably the first time that an EPA administrator could formulate what the identity and goal of the agency were in such an integrated manner. At this moment, *risk* became used to refer to the class of things that the EPA, across the many laws it applies and the various missions of its offices, could identify with, above and beyond cancer-causing chemicals, oil spills, species extinction, emission-reducing technologies, or other elements. Along with risk emerged a legitimate description of this

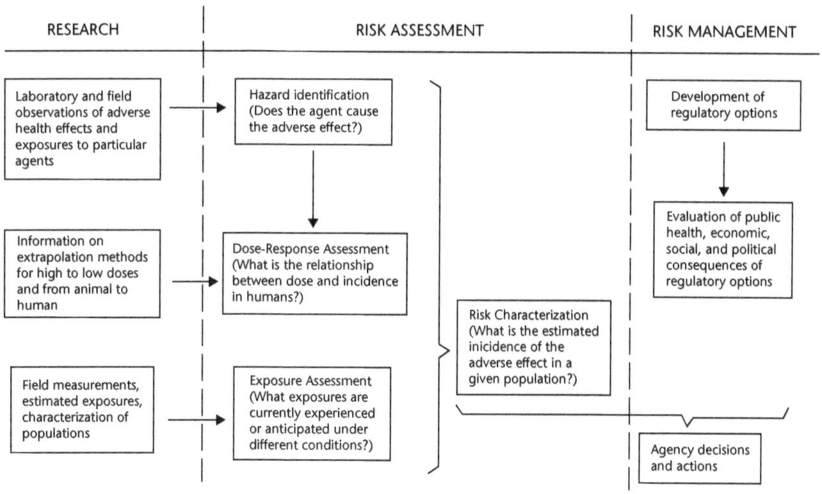

Figure 0.1
Elements of risk assessment and risk management (adapted from NRC 1983).

bureaucracy's own way of producing effective and legitimate rules. *Hazard identification, hazard characterization, exposure assessment, risk characterization, and risk management*: each of these rubrics referred to disciplinary exercises that are necessary to the agency and embody its expertise and competence. They helped rationalize and represent the way that the EPA was using science, which science it chose to use, and how it used this science to legitimately regulate the environment and the behavior of industries and citizens through the making of *decisions*.

This mixed scientific and administrative vocabulary acquired its authority through a report by the National Research Council (NRC),[1] with a vivid red cover, later called the Red Book (Johnson and Reisa 2003; Rodricks 2007). In this report, titled *Risk Assessment in the Federal Government: Managing the Process* (RAFG) (NRC 1983),[2] regulatory practices on uncertain environmental health hazards were codified as consisting of risk assessment (comprising several scientific exercises, as described previously, and preceded by a research stage) and risk management. An oft-reproduced graph included in the NRC report shows the logical organization of these processes to produce a decision (see figure 0.1).

It comprises what is known as the *risk paradigm* (Barnes 1994),[3] a set of conventional commandments, which may be summarized as follows:

- Analyze risk quantitatively, using data or studies enabling the description of the hazard (hazard identification), its seriousness under various circumstances (hazard characterization), and the extent to which people face the hazard in the actual environment (exposure assessment).
- Characterize the risk and ensure that all judgments and assumptions used to calculate the risk are made explicit.
- In the adoption of the final standard, or the risk management stage, take into account a wide variety of elements, from the results of the risk assessment to the costs and benefits of the potential standard; consideration of fairness or other values is legitimate at this stage only, and it should always be distinguished from the risk assessment.
- Make sure that those responsible for making choices, either analytical or political, are identified, and distinguish (at least formally) the risk assessment from the risk management.

Since the 1980s, this paradigm has evolved a great deal. In the initial design presented in figure 0.1, the regulatory process is presented as making a frank decision, based on an initial act of knowledge production. Risk management is separate and downstream in the process. Over time, it has appeared more legitimate to place risk assessment within the sphere of risk management (Power and McCarthy 1998; Jardine et al. 2003; Power 2007; Renn 2008; NRC 2009; Assmuth et al. 2010). Simultaneously, the need to involve the public and to create opportunities for deliberating about the balance between economic development and environmental goals has been increasingly recognized. The public should not be at the receiving end of a linear decision-making process; rather, it should participate throughout (Stern 2009). These transformations toward more risk communication and public participation were promoted by landmark reports of the NRC that, after RAFG, renovated the paradigm of science-based decision-making for environment and health hazards (NRC 1994, 1996, 2009). More rules were articulated in these reports and complemented the initial paradigm (e.g., Van Zwanenberg and Millstone 2005; Renn 2008; NRC 2009; Pellizzoni 2009; Abt et al. 2010; Hultman et al. 2010). Key new prescriptions include:

- Monitor and research the variety of potential hazards facing the public.
- Measure and rank problems according to their gravity, but also according to the costs and benefits of addressing them.
- Once you select the problem, formulate it clearly.

- Do not solely minimize individual risks, but also aim to improve health and environmental conditions.
- Engage with the public during all stages of the decision-making process.
- Evaluate and review the consequences of the adopted standard.

This evolving paradigm has structured the image and identity of the agency, inside and out. The paradigm has value for the internal integration of the agency. It has been used as a way to assemble an organization with perennial problems of internal cohesion, split as it is between multiple program offices. Functional offices for research and development, for policy management or for legal affairs as well as the agency administrator and deputy administrator also, frequently have difficulties controlling the rules and decisions that are shaped within program offices. Processes for joint assessment or analysis of risks and environmental conditions, in this context, have been conceived as an instrument to re-assemble the organization. And indeed, nearly all the guidance documents adopted since the early 1980s, be it by the offices governing air pollution, water, pesticides or other topics, build on the four standard exercises of hazard identification, dose-response calculations, exposure assessment, and risk characterization. Most, if not all cross-agency programmatic documents are founded on this process and division of responsibility, referring back to RAFG or to the agency's flagship reports in which this paradigm is frequently presented (e.g., EPA 1984b, 1995b, 2012). This paradigm is used to justify the role of the ORD vis à vis program offices. It has given rise to the creation of risk assessment and risk management services inside regulatory offices. It has also translated into the creation of a host of high-level transversal bodies in the agency, such as the Risk Assessment Forum and Risk Management Council—now called the Science Policy Council. And in the nearly five decades of the agency's existence, almost its entire staff has been trained in the technical exercises and procedures of risk assessment and risk management.

But the risk framework is essential for the outside, too. The model underpins the public representation of the agency, of its unity, credibility and rationality. There have been fourteen confirmed administrators of the agency since its creation in 1970.[4] William Ruckelshaus was the first to embrace the paradigm publicly, but his immediate successors basically took the same course, publicly emphasizing the risk-based nature of the agency. Bill Reilly, EPA administrator during the term of President George H. W.

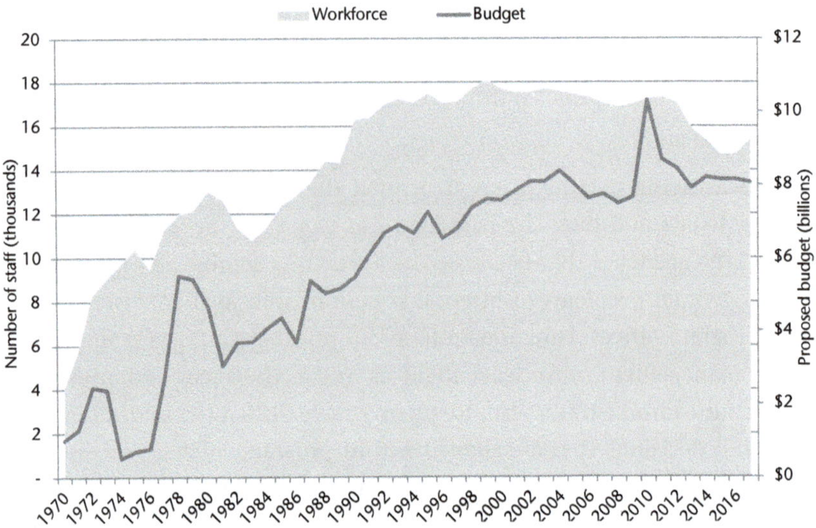

Figure 0.2
EPA budget and workforce, 1970–2016.
Source: EPA's Budget and Spending (https://www.epa.gov/planandbudget/budget, last consulted December 13, 2018).

Bush, also promoted risk analysis. Carol Browner, appointed by Bill Clinton, initially seemed to distance herself from science and risk analysis, but she ended up defending them as an important tradition of the agency. Christine Todd Whitman, appointed in 2001 by George W. Bush, praised the "long-standing traditions" of "precaution, science-based risk analysis, and sound risk management, including consideration of benefit/cost" (cited in Gibb 2001). Interestingly, risk assessment and risk management continued to be central pillars of the EPA in terms of defining its generic objects of action, its typical knowledge, and its way of making decisions, even during times when the president chooses to restrict its budget and staff. The statistics in this regard from 1970 through 2016 are given in figure 0.2.

Most of what Daniel Carpenter would call the audiences of the EPA (Carpenter 2010)—including other agencies, principals in the executive or in Congress, the regulated industries, scientists, environmental nongovernmental organizations (NGOs), and courts, among others—have to come to accept this identity of the agency as an entity that assesses and manages risk. They describe its operations and goals in these terms too, and hold agency administrators accountable related to this vision. Most assume and expect that some pattern corresponding to these criteria and codes of decision-making will be apparent in what the agency does. When the Government Accountability Office (GAO) of Congress reviews the EPA's performance under a given program, it assigns the risk assessment/risk management model to the agency's actions (GAO 2008). When an EPA official needs to inform a congressperson in a hearing about how the EPA works, she starts by explaining what a risk assessment is, using the four operations defined in RAFG and in agency guidelines, and then discusses how risk assessment informs risk management measures. The courts review the EPA's proposed rules against the benchmark of risk assessment methods and the due separation between risk assessment and risk management processes. NGOs describe (and also attack) the EPA using these terms to refer to what it does and should do.

The Puzzle

The notion of risk and procedures for the assessment and management of uncertain hazards has thus been very central in the institutionalization of the EPA, as well as in the assertion of a public image of how it acts and why.

There is a seldom noted paradox in this centrality of risk for this agency: Risk evokes models of coherent and rational administrative action—models of decision-making based on analytical knowledge and adjusted to defined probabilities. The paradigm works as an integrated and coherent representation of knowledge and action (Power 1997). It looks like a mini-regime of risk regulation, comprised of a delimited set of processes to address risks systematically (Hood et al. 2001). But *risk* is also the generic name for extremely controversial issues that the US polity and institutional system are struggling to frame and solve. The governance of environmental and health hazards is an area of extreme political polarization and constant disputes about what the real risks are, the appreciation of experts of these risks, and hence the legitimacy of administrations to manage them. How, then, are integrated designs for administrative knowledge and action invented and reinvented, projecting an impression of rationality, credibility, and control in societies that constantly question these very attributes of administrations? How does one forge an accepted, legitimate model for making decisions amid intense controversy, doubt, and actual difficulties to establish shared, frank decisions about environmental problems?

Risk is synonymous with calculation and control, but also with controversy (Nelkin 1985; Beck 1992; Leiss 2000). Hazards constitute "zones of ignorance" (Callon et al. 2009). To make claims about these hazards (and, in particular, to anticipate these hazards), one needs to make a number of assumptions about what information is needed in order to know about these hazards, what kind of calculations can reveal their true, future extent, and the parameters to use in making these calculations. Risk assessment thus involves choices that are determined by viewpoints, moral assumptions, and visions of nature and technology. Given that these are diverse across the polity (Douglas and Wildavsky 1983; Beck 1986; Lash et al. 1996; Collingridge and Reeve 1986), technical uncertainty soon turns into a political or structural kind of uncertainty (Wynne 1987; Schwarz and Thompson 1990; Jamieson 1996; Borraz 2008, 2011). Structural uncertainty translates into the flexible, irreconcilable interpretation of information (Collins 1981). Injecting more science into the debate makes things worse, not better, because its diversity and disunity mean that anyone will find facts, information, or theories to support his or her existing opinion (Sarewitz 2004).

Risk invites controversy all the more easily, as it is a subject of pressing institutional action and urgent social demands. With environmental risks, we are in a regime of "post-normal science" (Funtowicz and Ravetz 1993, 739). We stand outside the realm of normal or research science, into what has been called *regulatory science* (see Salter 1988; Jasanoff 1990): a science that aims to fill knowledge gaps and make difficult predictions of human health risks in the context of multiple, contradictory, and pressing institutional demands. In postnormal or regulatory science, the deconstruction of scientific claims and the apparent impossibility to decide controversy are the rule rather than the exception (Jasanoff 1995; Irwin et al. 1997). The assertion of truth and scientific authority is a constant and daunting (if not impossible) challenge in such a context. Again, risks involve complicated interpretations of information, but visions of what information is needed and why and for what scientific information is used also matter. Scientists may be accused by policymakers of making decisions in their stead, while pretending not to. Policymakers, in turn, are regularly criticized for not being able to face the facts and for making decisions without taking into account the available information and evidence.

In matters of risk, we see constant disagreements, miscommunications, and opposition of interests between scientists and policymakers, making this boundary a site of dispute as much as a site of mutual construction of authority (Gieryn 1983; Jasanoff 1987, 1990; Halffman 2005). And so, risk assessment itself, the cornerstone of its model of science-based decision-making, is an object of dispute (Clarke and Short 1993; Wartenberg and Chess 1993; Slovic 1999). The "ideals of good governance" based on rational risk analysis have increasingly been publicly challenged in contemporary societies (Power 2007, 20). Even though they sometimes initially supported the application of risk assessment and cost-benefit analysis,[5] public interest groups and environmental activists have gradually distanced themselves from these methods, which they found to be "slow, to be insensitive, to be costly for agencies and society as a whole, and to frustrate the health-protective goal of environmental health legislation" (Cranor 1997, 103). They are accused of delaying the adoption of regulatory measures on the issue—which is called "paralysis by analysis" (Mirer 2003, 1129)—and of serving the interests of polluters (Shabecoff 2003). Its underlying utilitarian meaning is frequently indicted as well (Shapiro and Glicksman 2003). Alternatives to risk assessment, for this reason, have been actively explored (Silbergeld 1993; Frosdick 1997; O'Brien 2000), even as scientists and the regulated industries were busy defending and advancing it (Michaels 2008).

Last but not least, risk invites controversy because of its implications about the distribution of power in society (Nelkin 1995). Risk implies measurement and comparison of situations based on observable levels of hazard, probabilities, but also and relatedly, costs and benefits. It is the subject of a large quantification of the natural world and life. It grants power to those who quantify and compute, as well as to new experts, who impose prescriptions about what the true risks are and judge whether citizens are right to worry about something. Under notions of risk perception and communication, risk thus establishes a certain relationship between the state and society, or between knowledgeable experts and the individual. This rationality of government is at the very least worthy of discussion, given the way that it treats the individual citizen as someone who does not understand risk and that should make utilitarian decisions. It is discussible, and in fact it is often contested for this very reason. Risks are objects of intense conflict, in which the scientific measurement of risk gets embroiled in broader disputes about the "meaning and morality of protecting the environment, about the distribution of resources and about the locus of power" (Nelkin 1995, 445).

All of this accentuates the paradox. There are, so to speak, two facets to risk. *Risk* is the name of an apparently rational way of defining and addressing environmental issues, underpinned by the reduction of scientific uncertainty, the calculation of odds, and the adjustment of public decisions to what knowledge establishes (Bernstein 1996). And the EPA has been at the forefront of the codification of this way of handling environmental issues, organizing and representing itself along the lines of such a model of risk-based administration of the environment.

On the other hand, *risk* is also the name of a state of controversy—mixing scientific, moral, and policy dimensions—that has become pervasive in industrial societies and leads to radically questioning the institutional and administrative system by which the environment and public health are protected (Beck 1992). This level of controversy is something that the EPA also embodies. Bill Reilly, EPA administrator between 1988 and 1992, once noted that 80 percent of all the standards he signed ended up being challenged in court (O'Leary 1995). Each piece of data advanced in support of its standards, its calculations, or models gets forensically deconstructed. Most of its choices for the assessment of risk and its determinations of how to reduce these risks are attacked in courts by corporations and environmental groups alike. The levels of risk that it declares acceptable are constantly debated. These divisions are reflected inside the agency.

The EPA is an agency that suffers from problems of internal cohesion. Its offices enact different pieces of legislation that, in their diversity, reflect the structural uncertainty existing in society, and among audiences, concerning what a risk is and how to establish it. Offices of the agency, therefore, make policies in different ways. Its professionals frequently disagree about when an issue implies a joint determination of the risk and of environmental standards. In matters of environment, where something affecting water also tends to affect the air, and ultimately the health of populations, it quite often does so. The EPA is required to act in an integrated manner, and yet it cannot.

So, what we have with the risk paradigm is the unlikely coexistence of a harmonious, standard design for governing the environment on the one hand, projecting the smooth coordination of all involved—scientists, economists, layers, political managers, industries, communities, and environmental groups—to make collective decisions on seemingly accepted criteria; and on the other, continuous public disputes about uncertain environmental and health issues, their nature, their importance, and the legitimacy of those who are charged with addressing them. The fashioning of an accepted model of rationality in a context of controversy and conflict of rationalities is the main puzzle that this book is interested in addressing.

Folded within this paradox, one finds a number of other intriguing questions about risk-based decision-making and its institutionalization as a model of administration and bureaucratic operation. The first concerns the use of science in policy. Why is science so closely reflected in the design of an agency when the use of science in policy and administration is the subject of so much controversy? Why does the administration of the environment and public health materially rely on something that is so disputed, and even distrusted (Ezrahi 1990; Hilgartner 2000; Jas et Boudia 2014)? This puzzle is a version of what Bijker, Bal, and Hendriks (2009) called "the paradox of scientific authority," or the intriguing observation that scientific advice continues to appear "effective and influential in an age in which the status of science and/or scientists seems to be as low as it has ever been" (Bijker et al. 2009, 1). In the case of the EPA, the paradox is that science has been incorporated into the standard way in which the various offices develop standards and rules, and thus make policies, at a time when it has become more difficult to agree on the right forms of knowledge to conduct environmental policies, as well as to trust scientific experts (Nelkin 1975). Science is as much a currency of environmental politics as it is a source of authority (Cozzens and Woodhouse 1995; Weingart 1999). Various people inside the EPA, but also the regulated industry and environmental groups, defend their views and construct their interests through science (Tesh 2000). All try to achieve results that advance data and scientific interpretations, while contesting others. Confusingly enough, they all also do so in the name of good science, jeopardizing the very possibility for anyone to gain the ground of objectivity, truth, and scientific authority. Regulatory knowledge is a field of contention (Demortain 2017; Wagner et al. 2018). So, why does science appear to be an instrumental resource for regulating controversial matters when its interpretation is the battleground for so many disputes? How, specifically, could an agency foster a model for the use of scientific knowledge in policy, when this knowledge is contested?

A second puzzle concerns the science of risk. While it appears to be a well-defined science—resting on the calculation of odds using statistics and probabilities to adjust actions and attribute responsibility (Ewald 1986,

1991)—and one that has become increasingly sophisticated at that, it is not strictly reducible to a science. It involves, by nature, evaluative and normative choices within the very act of calculating, of choosing models of extrapolation and statistical methods. These choices help establishing criteria for definitive decisions, separating what is safe from what is not safe, acceptable and nonacceptable. Observers of risk regulation, in the longer run, confirm that such an instrumental paradigm has been dominant (Fisher 2007). The linear model "involving usually a scientific process of risk assessment and then political processes of risk management and risk communication ... has come to dominate much of the policy and legal discourse" (Fisher 2013, 125). However, folded within the logic of calculating uncertain hazards, one finds a multiplicity of other epistemologies and styles of statistical thinking (Hacking 1990; Jasanoff 1993a; Bernstein 1996), or what O'Malley (2004, 21) calls "configurations of uncertainty."[6] Analyzing risk also may mean ranking them, rather than calculating them individually with utmost precision—a commensurative and prioritizing approach. The decisionistic approach to risk also may be supplanted by a predictive approach, believing in the possibility to eliminate uncertainty and illuminate choices thanks to full, exact knowledge. Finally, a deliberative logic, finding uncertainty in the preferences of decision-makers and the public and organizing the confrontation of preferences in order to make a collective choice emerge, is not completely foreign to risk science. What is the relation between these various conceptions of governing uncertain hazards, and what makes risk such a malleable notion? How do these models of administration engender one another or coexist in the agency?

The third puzzle concerns the invention and institutionalization of models for administrative governance. The risk paradigm is the basis of a now widely accepted administrative gospel, or what Jasanoff (1999) calls the "songlines of risk." It has been institutionalized worldwide (Winickoff and Bushey 2010; Fisher 2013), and many of the international texts describing the paradigm refer their readers to the NRC report of 1983 (Jardine et al. 2003), as well as to the experience of the EPA (WHO 2009; Demortain 2011; Demortain 2012). We know a great deal about how and why organizations facing complex and uncertain situations, such as risk bureaucracies, copy each other (Di Maggio and Powell 1983; Brunsson and Jacobsson 2000), but much less about how they innovate. What makes the EPA the source of these innovative frameworks that make the use of science in policy so credible and routine, when its autonomy, authority, and legitimacy are challenged so often? How did it innovate through this ambiguous notion of risk, and succeed in changing institutional forms that change rarely and slowly (Sewell 1992)?

What is more, the EPA has formalized and applied this knowledge of risk much more than any other federal agency in the United States dealing with health, the environment, and safety, such as the US Food and Drug Administration (FDA; Lehman et al. 1955) or the Occupational Safety and Health Administration (OSHA). These other agencies have also invested in formal risk assessment methodologies and risk-based decision-making. In fact, the FDA did so before the EPA. But at no point have these agencies used a model of decision-making through risk assessment as the embodiment of what they are doing or as a standard of credibility and legitimacy of their actions. What makes the EPA, an agency with presumably less autonomy and power than others (especially the FDA) (Carpenter 2001, 2010), institutionally innovative?

Strategic and Cultural Perspectives on Rational Administration

These puzzles connect to a broader question concerning the way in which rationality in administration takes form and the roles of analytical tools and knowledge in the constitution, both material and symbolic, of an agency and of its capacity to administer intractable public problems. They defy two broad ways of analyzing how and why bureaucracies embrace rational analysis and decision-making tools—one that can broadly be called *strategic*, and the other *cultural*.[7]

One way to approach an administrative organization like the EPA is as an expression of the historical force and progression of formal, instrumental rationality, in line with Max Weber's conception of bureaucracy. Weber portrayed bureaucracy as the incarnation of the rational spirit, with *rationality* meaning the instrumental, formal means-end rationality that, in his view, was reaching into more and more areas of social life. Where he speaks about expertise and technical specialization of administrations, it is a signal that he considers that science and bureaucracy partake in the same kind of rationality, based on a "means-ends decision-making calculus" (Reed 2005, 119), and the same process of expansion and accentuation of this rationality.

The theory of organizational decision-making showed that formal, means-end rationality is not the way in which administration was rational (Simon 1969). Decision-makers in organizations do not spontaneously compute all information accessible to adopt the means that are most adapted to an end. Multiple forms of limits of administration impose a bounded model of rationality (Hood 1976; Forester 1984; Lodge and Wegrich 2016). The rational model came under scrutiny with Charles Lindblom's work on incrementalism (Lindblom 1959). Listing all values related to a policy in order of importance, ranking policy outcomes in terms of how efficiently they satisfy each value, and choosing among policy alternatives and instruments to achieve these outcomes comprise a comprehensive-rational model that is an idealization of the scientific method. It enshrines a "rationality project" (Stone 2012/1988, 9): namely, the project of making policy "with rational, analytical, and scientific methods," based on a model of reasoning in which "decisions are or should be made in a series of well-defined steps" (ibid., 12). But in the actual, material "administrative rationality" (Pfiffner 1960), facts and values are not easily separated, and pluralism in the organization makes it difficult to reach common goals, and measures of efficiency. Decision-making tends to be incremental, circular, multidimensional, and in conformity with the power structure of the organization (Nicolaidis 1960).

Subsequently, a vast literature on strategic decision-making confirmed Lindblom's initial findings, according to which "formal analysis has only a partial, incremental role to play in decision making" (Langley 1989, 623). Assessing the variety of ways in which formal analytical tools were used in the strategic management of organizations—the processes and practices unfolding around them, the kind of leadership and governance structures with which they were associated (Langley 1989, 1991), and their outcomes across a large number of organizations and cases of decisions (e.g., Hickson 1987)—it highlighted that formal analysis has only indirect effects on the formation of decisions (March 1982; Brunsson 1985). In practice, they serve to communicate already-formed decisions in order to justify them a posteriori (Bower 1970; Meyer 1984; Langley 1989), extend the control of the dominant coalition in the organization (Mintzberg 1980; Newman and Rosenberg 1985), or deflect attention from issues that leaders of the organization fail to answer (Meltsner 1976).

In public administration, the application of rational analytical tools "reflects the interests, strategies, and compromises of those who exercise political power" (Moe 1989, 267), and the politics of bureaucratic design more generally (McCubbins et al. 1987, 1989; Macey 1992). It is defined by the strategies of principals to control regulatory agencies and limit their powers (Howell and Lewis 2002; Wood and Bohte 2004). One of the rules or expectations that principals apply, in the case of regulatory agencies, is that the agencies will strictly work to establish facts and refrain from deciding policy values and making law (Yellin 1983). Risk analysis and various kinds of analytical exercises thus get used in regulatory agencies because of the preferences applied by the coalition of actors that conceive and control the agency (Shapiro 2011). Agencies retain some power in this relationship with principals, and of course, they may use rational analytical tools in such a way as to disguise substantive policy choices. But whatever the actual pattern of use of analysis, it appears rational and instrumental in terms of the behaviors and strategies of actors in this interinstitutional game (Allison 1971).

There are three limits to this strategic vision of analytical knowledge in bureaucracy, as a "technology of rationality" (March 2006, 201). The first is that this perspective grants little autonomy to agencies and pays little attention to the content and effects of their expertise. It tends to "minimize the role of bureaucratic action" and disregard the real "capacities to analyze, to create new programs, to solve problems, to plan, to administer programs with efficiency" (Carpenter 2001, 14–15). Second, it simplifies the environment in which the agencies operate, erasing from the picture a whole set of other audiences among which agencies construct their role and define the knowledge they need to use and the outcomes they try to achieve (Arrelano-Gault et al. 2013). Third and finally, this literature keeps instrumental rationality as its horizon, and in a sense, takes it for granted (Cabantous et al. 2010). It leaves aside the hypothesis, deriving from Weber, that formal, instrumental rationality may be in tension with other material rationalities, or systems of meanings and values in an organization (Diesing 1962; Swidler 1973; Clegg 1975, 1989; Kalberg 1980; Brubaker 1984; Townley 2008; Baunsgaard and Clegg 2012).[8]

New institutionalism takes a different approach to the question of rationality, which it does not approach strategically, but rather culturally. Dobbin (1994, 118) argues that the work of Weber on bureaucracy and instrumental rationality was structured by a "tension between the modern tendency to see rationality as acultural and the sociological tendency to see all practices and understandings as part of culture" (see also Hindess 1987). New institutionalists built on such a cultural vision in order to articulate a new vision of rationalization, in which "supposedly universal precepts of organizational efficacy are simply abstractions from social practices that emerged for complex historical reasons" (Dobbin 1994, 122). Rationality is a product of what is normatively considered as the rational form within a given organizational field. The main logic at work in the adoption of rational models is not efficiency and performance, but legitimacy, which is understood as conformation to socially defined norms of what it is to be rational: "Highly structured organizational fields provide a context in which individual efforts to deal rationally with uncertainty and constraint often lead, in the aggregate, to homogeneity in structure, culture, and output" (Di Maggio and Powell 1983, 147). The context of these organizational fields is where it becomes rational to embrace particular organizational forms and practices, true "rational myths" for organizations (Meyer and Rowan 1977, 360). The adoption of such templates leaves a generally quite large degree of decoupling. It does not mean that practices in the organization are regulated down to the minute level in the way the template dictates. But the adoption of a model is a ritual of legitimation—a way of complying with the standards of legitimacy of what an organization should appear to have (Brunsson 2000).

These elements of rational organizational life are induced in organizations (Scott 1987) through a process "rooted in conformity" and "in the taken-for-granted aspects of everyday life" (Zucker 1983, 5). The adoption of seemingly rational models in organizations is the result less of internal adaptation to locally understood problems and locally devised strategies than of the engagement with professional intermediaries—consultant, standard-setters, and prescribers of organizational methods of all kinds—and state organizations, which codify, carry, and prescribe these norms across the organizational field (Greenwood et al. 2002; Scott 2008). They induce norms and ideas about appropriate practice, and promulgate rational myths as they do (Scott 1987).

This process of collective institutionalization of models across populations of organizations verifies in the case of risk management, where a multiplicity of organizations, faced with increasingly hostile publics and demands of accountability, turn to risk as a logic of managing the threats to their own legitimacy and existence (Rothstein 2006). Measuring and analyzing risk is an effective technology for organizations facing uncertainty and risks to their reputation (Power 2004; Power et al. 2009). The organizations that formally embrace risk assessment and risk management are precisely those that manage social problems incorporating high risk of failure and legitimacy attacks (Saint-Martin and Allison 2011; Rothstein and Downer 2012; Huber and Rothstein 2013; Mazmanian and Beckman 2018). The limit of rationality there is less that actors are not as rational as they pretend to be, but rather that they embrace formal conventions and schemes for managing risk under the normative pressure of professional networks (Hayne and Free 2014), without actually adapting their practices. Risk management formulas tend to become void, legalistic rituals to demonstrate attention to risk and rational responses to threats, with limited impact on what organizations actually pay attention to and how they process these threats (Power 2007). Clarke's depiction of disaster management methods as "fantasy documents" is most explicit about this (Clarke 1999). Worst of all, it may direct their attention away from harder-to-detect problems and challenges and toward more conceivable, codified, and measured problems (Power 2007; Rothstein et al. 2006).

The advantages of such a sociological, new institutionalist approach is that it provides a criterion to understand when and where rationalization operates in practice—the kind of diffusion of norms and ways of doing things that is described under the notion of institutional work. This work can apply to any rationality, or what the field calls *institutional logics of action* (Thornton et al. 2012). So it helps to treat symmetrically the different coalitions of actors that engage, with their own register of rationality, in the enterprise of legitimizing an organization.

There are limits, though, to the way that it approaches the organizational production of rationalities. First, this strain of work does not provide any external benchmark through which one may understand that one logic of action takes precedence over another. It is, in fact, not very interested in confrontation and conflict among rationalities and their proponents. New institutionalists overlook an important condition of legitimacy of organizational forms and rules: the fact of corresponding to the substances of the situations from which it is abstracted, or to actual values prevailing in the society (Selznick 1996; Stinchcombe 1997; Scott 2002).

Second, in new institutional scholarship, organizations tend to be described as recipients of models circulating in organizational fields and professional networks, hardly as sites of articulation or even invention of such rational forms. The various logics of conformation and imitation are particularly effective to analyze the adoption of rational forms, less their initiation. A federal organization like the EPA would more likely be a prescriber and authorizer of organizational forms than a recipient of norms stemming from its environment in this framework—sending us back to the original question of the origin of its credibility and authority in articulating modes of reasoning and deciding.

Design, Controversy, and Legitimacy

Rational decision-making seems to make sense, thus, as the result of either of two contrasting processes. We end up with a choice between considering it as the reflection of an organization's strategic behavior in interinstitutional power games, or as ritualistic and symbolic of its belonging to an organizational field (Meyer 1984); or between a process in which "organizational managers look out," or the "society looks in" (Suchman 1995, 577). In both cases, what counts as rational knowledge seems to be given. Not much seems to be happening in terms of production and validation of rational forms of knowledge; and conversely, rational knowledge does not seem to have much performative effects on bureaucracies, their legitimacy and autonomy to act and govern. To better analyze the active work on knowledge forms and the shaping of legitimate institutions, in its political context, we need to abandon the alternative metaphors of strategic rationalization and mindless imitation, to assume instead that knowledge, and science, partakes in the design of bureaucracies—understood as the purposeful forging of an organization for the legitimate administering of public problems. Design, in short, is rationality forged in response to a state of controversy.

Research on design in political science developed in the United States in the 1980s (Howlett 2009; Howlett and Lejano 2013) to investigate the rational composition of policy. It investigates the crafts and methods pertaining to "the process of inventing, developing and fine-tuning a course of action" for public policies (Dryzek 1983, 346). It builds on Herbert Simon's notion that the construction of alternative courses of action is a fundamental pattern of social action (Simon 1969). It considers how the behaviors of actors involved in politics and policymaking is controlled or steered to achieve predefined choices. It also assumes that there is a craft to it, which can be taught and equipped (Alexander 1982; Linder and Peters 1984, 1987; Weimer 1993; Considine et al. 2014). The field has long been animated by a dispute as to whether design could indeed be guided by an instrument's so-called blueprints or instead should be considered a more contextual and pragmatic activity (Dryzek 1983; Junginger 2014). But overall, its roots are in the rational and instrumental tradition of policy studies (Howlett and Lejano 2013; see also Lascoumes and Le Galès 2007). This rational orientation can be seen in the way in which the question of bureaucratic design is treated in the literature: as a question of defining administrative procedures and rules to achieve substantive goals (Weimer 1995). Design is, in this version, an intention-based theory of institutional change (Goodin 1996).

Bureaucratic design has a both a material and symbolic dimension. Materially speaking, the design of the bureaucratic institution involves its programmation, so that the organization produces the kind of outcomes that are legitimately expected from it. This definition of design brings us back to the concerns of some of the earliest students of organizations. Most of the early theorists of organizations were interested in studying formal organizations (Blau and Scott 1962), governmental or industrial ones (Arellano-Gault et al. 2013), and looked for the implications of the formal rules, processes and roles that visibly defined them and helped them coordinate people (Barnard 1968), on the internal life of organizations and on their capacity to make collective decisions and act purposely. Early theorists approached formal organizations, particularly public bureaucracies, as a device that is programmed and structured to act (March and Simon 1958; Bittner 1965). Their objective was to understand what gave them this capacity to act in a coherent and unitary manner, looking first and foremost into the "operative codes" and "universalistic formal rules" that make the organization (Parsons 1956, 63). Of course, scholars of organizations have subsequently discovered that formal rules do not define and regulate the organization alone. Informal rules, organizational cultures, and transorganizational coalitions are as important. But design is precisely about controlling the roles, cultures, and forms of political coordination between the diversity of bureaucratic actors—assembling the organization—so that it produces expected outcomes (Dowling and Pfeffer 1975; Stinchcombe 2001).

Design generates the material elements of administrative organization through which a public image of the agency and of its action is formed. The models, frameworks and other norms of action applying across the various parts of the organization, formally or informally, enhance its credibility in various ways. They provide a vocabulary to represent the work of the agency as one that is able to determine the necessary rules, standards or *decisions* for society (Laroche 1995). They legitimize the organization, because they project a rational image of practice, mimicking the orderliness of science (Power 2016). By showing which generic, abstract actor is included in the governance of an issue, these designs structure the political image of the agency, as a more or less democratic, participatory body (Moffitt 2014). Designs work as a *bureaucratic screen* between internal practices and the agency's audiences, offering them a rationalized representation of what's happening inside and of how the agency is administering things, to legitimize it. They are the instruments to expose the inner workings of the organization and subject them to the evaluation of audiences (Power 2007). They help focus, and control, the attribution of dispositions to the agency and the definition of its overall character (Selznick 1949; Dowling and Pfeffer 1975; Suchman 1995; Carpenter 2010). They form a narrative of rationalization that legitimizes new actions and performs the future of the organization (Brown 1978).

The link between design and legitimacy is better understood by factoring in the structural condition of controversy. As Schön and Rein (1994) have shown, design is a response to uncertainty and the controversy it generates. These scholars were inspired by an emerging set of questions in the field of urban planning, about so-called wicked problems: problematic situations that lack structure and boundaries and are not amenable to standard treatment (Rittel and Webber 1973). These situations are characterized by a dissolution of the fact/value distinction, and irreducible disagreement about policy means and/or ends. Taking inspiration from the work of architectural design and approaching design as a process and form of knowledge (Cross 1982; Weick 1993), Schön and Rein identified what they call "design rationality": They see "a policy designer who constructs, in some relatively protected forum, a representation of a policy or program that will be sent out, upon its completion, into an actual policy environment. As the representation of the policy object takes shape, the policy designer's seeing/moving/seeing reveals new meanings, goals, and criteria, some of which are found to be mutually incompatible, requiring the framing of new problems, opportunities, or dilemmas" (Schön and Rein 1994, 86). Between comprehensive rationality and chaos (Cohen et al. 1972), design resembles a pragmatic process of inquiry into the right techniques to solve a problem in a contentious policy environment (Evans 2000; Sanderson 2009; Dalsgaard 2014). From their perspective, design is a dialectical process of experimenting with a form and adjusting its shape depending on the results of the trials,[9] to bring structure to controversial problems. Controversies stimulate the construction of new bureaucratic forms. They stimulate the reflexivity of those who aim to administer, such as bureaucrats and their scientific advisers, leading to changes of pre-formed institutional practices and contents (Riles 2004).

Risk issues are wicked, unstructured problems (Hisschemöller and Hoppe 1995), in which the limits of knowledge compound with the diversity of interests and viewpoints in the society. These situations of uncertainty invite disagreement about hazards and their causes, but also about the legitimacy of institutions, administrative instruments, and official expertise (Martin and Richards 1995). In a risk policy controversy, not much time elapses between the moment when the purely scientific question emerges and when institutional factors underpinning the issue are pointed to (Nelkin 1984; Tierney 2014). In these contexts of irreducible disagreement, what administrations know, where they collect information and opinions from, and who is included in their actual mode of governance get incriminated, whether they augment or minimize risks, is always disputed. Administrative governance, the perceived credibility of the administration to deal with problems, becomes particularly hard to establish and defend (Douglas and Wildavsky 1983; Stone 1989; Bovens and Hart 1996). Bureaucratic design ensues.

The Sciences of Bureaucracy

In this book, I approach bureaucratic design as an iterative process of formal, purposive assembling of an organization to counter controversies and enhance its credibility as a producer of legitimate outcomes. Science plays an important role in the institution of legitimate bureaucratic forms, precisely because its practitioners perform the structuring of problems, and assembling elements of knowledge to obtain an outcome. Design relies heavily on the competences, rationales, and propositions of those scientists who are closely involved in policy, such as the various risk experts whose work will be at the center of this book. They are an integral component of the process by which conflicting actors and viewpoints, both among the audiences of the bureaucracy and internally, are described and framed in an integrated and legible decision-making process.

Here, I capitalize on a whole strain of research in science and technology studies (Hackett et al. 2008), as well as its demonstration of what one may call the organizational ingeniosity of science. From Shapin and Schaffer's seminal work on experiment and demonstration, in which the scientific method is defined as a set of "patterns of doing things and of organizing men to practical ends" (Shapin and Schaffer 1985, 15), to Collins's study of the alignment in informal networks of experimental replication (Collins 1974), through Star's ethnographic analysis of the construction of shared infrastructures of information (Star 1992, 1995; Star and Ruhleder 1996; Bowker and Star 1999) and John Law's research on scientific "modes of ordering" (Law 1994, 20), science and technology studies have generated a great deal of resource to understand how science works to align people with incommensurable views on shared perceptions and common choices. Looking behind science, and the task of demonstrating and proving, one finds the creation of common standards of information and of generic criteria of facticity to bring people together in joint spaces of interpretation.

This is in fact what students of risk assessment and other decision-making sciences have shown: The conception and application of standard procedures for cost-benefit analysis helps depersonalize decision and create an impression of objective, regular, and replicable knowledge (Porter 1995), in keeping with expectations of efficiency, impersonality, and rule-based functioning at the heart of the American bureaucratic paradigm (Barzelay and Armajani 1992). These forms of mechanical objectivity respond to the questioning of the choices and criteria of professional communities that oversee administrative decisions. Analyzing risk, costs, and benefits in an orderly sequence and organizing the respective role of expert judgment and administrative pronouncements form a "dominant script," thanks to which regulators have achieved a form of objectivity (Kysar 2010). These policy sciences generally aim to settle debates and make issues decidable (Fischer 2000). Their practitioners maintain a tight relation between the sciences' disciplinary ambition to know, describe, and model the world and the organization of administrative knowledge and techniques.

Design is comprised of two main practices. The first practice is the creation of what one may call a knowledge representation (Star 1995), defining common objects of attention (risk) and the forms of knowledge that are necessary to know this object, as well as the remaining uncertainties. Design involves the construction of a scheme, comprised of various mutually exclusive but plural knowledge rubrics. If assembled, these elements of knowledge can produce an objective image. These rubrics are "'black-boxes' whose authority might be invoked automatically, avoiding further open and indeterminate negotiation," and "deleting the ambiguity" inherent in knowledge (Wynne 1992, 748). This first dimension is cognitive and results in the formalization of bureaucratic knowledge.

Second, from this knowledge representation derives organizational roles and identities. Assembling knowledges is assembling the people that embody or carry this knowledge. This second design practice involves the definition of linkages and boundaries between carriers of this knowledge, articulating their work through rules and procedures (Fujimura 1987; Strauss 1988) to align them on common elements of information, and hopefully decision. This second aspect is more regulative and materializes by the emergence of *bureaucratic technologies*: the processes through which a bureaucracy succeeds in producing legitimate outcomes, the accepted "if-A-then-B" kind of algorithms based on which it can create an internal discipline (Landau 1992).

Such design practices are prevalent in administrations that host and use analytical disciplines, from risk analysis to systems, policy, or cost-benefit analysis (see chapter 1). These designerly sciences (Cross 1982) are at the source of the bureaucratic assemblages that will be discussed in multiple chapters in this book—the various frameworks, models, and integrated decision-making schemes that agencies such as the EPA embrace, as they seek to articulate the criteria of the various conflicting actors engaged in risk controversies. They enshrine what one could call the science of a bureaucracy—the expertise of an organization that learns how to reinvent its mode of government amid constant conflicts about the environment, risk, and its legitimacy to administer them.

Assembling a credible bureaucracy does not depend on the willingness, capacity, and rationality of a unique and detached designer. Rather, it involves competing networks of actors that form in the course of controversy to shape the knowledge, technologies, and image of the organization. They typically take shape during moments of intense controversy surrounding the agency—moments when its institutional life is threatened and where its operations and forms can be modified. They define what is at stake in the controversy, analyze the positions of people comprising the controversy, and give the ways of articulating them, as well as the organizational forms and frames to adopt for that purpose.

Schön and Rein identified a unique "design rationality," but there may be different responses to different sorts of controversy. I broaden their approach to consider that design is comprised of a variety of rationales depending on the way uncertainty is interpreted. This extension is necessary to take into account the fact that different traditions are at play in the modeling of institutions (Linder and Peters 1995). Notwithstanding the supposed uniformity of the organizing concept of risk (Jasanoff 1999; Power 2004, 2007, 2014; Rothstein et al. 2006; Rothstein and Downer 2012), it seems appropriate to consider that the generic issue of risk has generated a variety of bureaucratic designs that have taken form over time and competed with one another. These plural designs and distinctive relations to uncertainty, express the variety of material rationalities that Max Weber talked about. The EPA is an agency that harbors several of these.

A scientistic, *predictive* design finds the origin of uncertainty in a lack of knowledge, to be reduced through the search for more accurate and precise information. In organizational terms, it translates into algorithmic, sequential modes of assemblage, in which the successive consideration of a defined set of information is logically conducive to a policy. It involves a minimal set of actors, and an almost machinelike assemblage in which science leads to or dictates policy, illustrated by linear science-to-policy sequences. But controversies, as indicated in this chapter, generally involve doubts or disputes concerning the values and policy goals that should be pursued. Facing such structural uncertainty, a *decisionistic* rationale may appear, embodied by a more modular assemblage involving calculation of risk on the one hand *and* imposition of a choice criterion on the other hand. They mutually support each other, and no sequential order is clearly established.

Two other designs may be employed in the face of uncertainty. *Commensuration* is a form of assemblage that is motivated by the instability of preferences about safety, both outside and inside the agency. Uncertainty is not a problem of knowledge but of unclear characterization of one's utility function and goals, and it translates into disputes about the prioritization of risks in comparative, commensurative frameworks. Where controversy is rooted in the impossibility to agree on values, nonalgorithmic assemblages emerge (Majone 1992; Thacher and Rein 2004; Shapiro 2011). Instead, a *deliberative* design emerges from these controversies that are due to irreconcilable visions of the world and conceptions of what counts as risk. From that perspective, the ideal bureaucratic design is one that orchestrates the collective articulation of these views.

Whichever design predominates, it has a deep, legitimizing function for embattled administrations. It describes people defending heterogeneous criteria of what is a risk and what matters, incorporating them into an abstract order. The frameworks composed by the specialists of analysis and decision-making embody a constitution of sorts, showing how the organization articulates the views of those involved, both internally—among its scientists, lawyers, and political officials—and beyond itself, with varied audiences of environmental policy and regulation—scientists, NGOs, communities, industry groups, other agencies, executive or legislative principals. Frameworks thus displace controversy.

Once instituted, design may become an object of controversy, as audiences of the agency take this screen as the reality, and problematize its effects. The process of design is iterative, as this book will show through giving the history of the risk assessment–risk management framework—how it took form, how it was reflected in the EPA, and how it subsequently became a component of the controversy about the agency and its capacity to handle uncertain issues. One assemblage leads to another, as controversy forces the agency to reinvent its knowledge, technologies, and the overall image of how it acts.

Organization of the Book

A central proposition in this book is that the EPA is an agency that is structurally embedded in controversy. This configuration involves a constant evaluation of the credibility of the knowledge and regulatory measures adopted by the agency to address what appears as uncertain, risk issues on which its various audiences are frequently, if not systematically in conflict. The EPA's proposed standards are contested and debated time and time again before producing any sort of effect. They are subjected to variegated and often opposing demands and moralities. Controversy materializes by the multiple trials, formal and informal, to which the agency is subjected. These trials include court judgments, hearings in Congress, reviews of decisions and programs by the White House and its various offices and political parties and associated think tanks, and public campaigns by public or private interest groups. They express a fundamental state of dispute about the environment and its problems and widely diverging sets of knowledge, criteria, and values to apply in environmental matters. This state of dispute reverberates inside the agency, because the diversity of its various program offices and its professional groups inside each of these offices, make it difficult to construct and impose integrated visions of risks.

At the same time, the EPA is remarkably effective at organizing itself, inventing modes of assemblage of the actors that take part in the resolution of environmental problems. The chapters of this book recount how a variety of designs were activated, depending on who coalesced in and around the EPA bureaucracy, to bring together people with otherwise conflicting views of risk. The book offers detailed histories of each of these designs and design networks, corresponding to successive moments in the political life of the EPA. It offers a political history of rationalization, showing how rational decision-making tools and their proponents actively contribute to shaping the legitimacy of an organization to administer the environment and hazards, in a political configuration of controversy.

Chapter 1 goes back to the history of risk analysis to show that it is an academic and professional field that, at its heart, is a form of design science that responds to political demands emerging from embattled institutions to help govern controversial situations.

Chapters 2 and 3 recount the emergence of two early bureaucratic technologies in use at the EPA—or at least in parts of the agency—and the controversy in response to which they emerged. The first is risk assessment guidelines and the use of what came to be known as *default assumptions,* which help to compensate for the lack of data and to define thresholds above which an intervention is thought of as legitimate and force a decision. It is the first design to have taken form at the EPA, at the end of the 1970s, amid the growing scientific and regulatory controversy surrounding regulatory measures for cancer-causing chemicals. This design has benefited from the work of toxicologists, biologists, and statisticians that embraced the goals and missions of the EPA, and it was the first sketch of a mode of articulation between scientists and policymakers, or between calculations and policy choices. Chapter 3 traces the development of a technique of risk-ranking as part of the attempt to control diverging estimations of risk made by separate offices of the agency. Differences between these estimations produced an inconsistency that typically caused the EPA's work to be challenged. This time, economists and policy analysts led the reflection on the necessary designs, and worked to generalize it in the agency, without astounding success.

Chapters 4–7 focus on the risk assessment–risk management framework, its conception, and institutionalization in the EPA. The framework is an integrated design for risk-based decision-making using elements of both the knowledge deposited in risk assessment guidelines, and elements of economists' risk-ranking approach. This integrated design emerged in the agency first, but it took its final form in RAFG because the NRC became the center of the configuration in which the authority of regulatory agencies, and their right to use science for decision-making was debated. The set of concepts presented in RAFG in 1983, I argue in chapter 4, inaugurated a particular period in the agency, precisely because it provided for ways to ideally articulate the two preceding bureaucratic technologies—quantitative risk assessment and risk-ranking—in an integrated scheme that combined multiple approaches to risk: scientific calculation of hazards, definition of policy criteria to address remaining uncertainties, and comparison and prioritizing of these issues. Chapter 5 shows how instrumental this emergent design was for the EPA administrator to construct a legitimate image of the work of his or her administration. The way in which the administrator extended the framework to include a more deliberative design of risk communication shows just that. Quantitative risk assessment, risk-ranking, risk communication: each of these designs was institutionalized in the agency, changed the organization, and framed subsequent conflicts about the missions, powers, and procedures of the agency. This is what chapters 6 and 7 discuss.

Chapters 8–10, finally, recount how the risk framework unraveled, as it no longer offered protection against new controversies. Chapter 8 charts the relative marginalization of the risk-ranking design under the first administrator appointed by a Democratic president, for whom it was of little use in a configuration in which the main conflict engaging the agency concerned environmental justice and ecological restoration, not risks. Chapter 9 explains how the early risk-assessment mechanism devised in the 1970s lost its emphasis in the context of renewed controversies over the EPA's ability to produce consensual estimations of risk. The chemical industry and its allies systematically pleaded for methods to reduce uncertainty and predict risk. Finally, chapter 10 shows how the agency has become the site of different designs—more holistic and pragmatic at once—in response to the controversies manufactured by opponents of the decisionistic design institutionalized in the agency over the years. The promotion of a formalism in terms of *problem formulation* attests that the EPA nevertheless retains the capacity to give shape to its own mode of administrative action, in a context in which it fails to design what science to use and how to forge protective standards, as today's situation precisely illustrates.

1 Risk Sciences: Expertise for Decision-Making and Dispute

Risk assessment emerged in the 1970s as a regular component of the dominant paradigm of systems analysis, alongside "cost-benefit analysis, technology assessment, social forecasting and the like" (Hoos 1979, 192).[1] From the moment it first took shape, risk assessment was considered to be a symptom of the emergence of "new forms of technology management, the most visible of which are detailed analyses of the anticipated impact of proposed developments" (Fischhoff 1977). Economic cost-benefit analysis, general systems analysis, operations research, decision-theory way of thought, and risk assessment all are "attempts at policy science" (Wynne 1975, 118). They comprise a "family of techniques ... conceived as ways of improving decision-making by broadening the role of logic and empirical inquiry" (Tribe 1972, 75). Rip (1986) later labeled this set of sciences "strategic" sciences, to convey the fact that they shared a similar interest in aiding decision-making.[2]

By shaping and embracing the quantitative assessment of health risks or the comparative economic analysis of their reduction, the EPA has placed itself in the ambit of these sciences, and of this particular way of understanding the administration of the environment and health, as a way of making rational decisions. Sociologists and philosophers, very often critical of these policy sciences, tend to argue that they are representative of an expanding technoscientific or technocratic ideology. This narrative, however, obscures the contextual and historical constitution of these sciences and of their techniques. They were born in the context of public controversies surrounding technologies and their hazards, as well as policies for managing them. Risk research, it appears, is knowledge formed to respond to public controversies about environmental and health hazards, with a view toward solving them.

The Sociological Critique of Policy Sciences

Risk assessment has emerged as the discipline of quantifying the negative events in technical systems. As such, it is intimately linked to systems analysis. Originally, systems analysis was the name that was given to the civic, administrative, and corporate transposition of the particular discipline known as "operations research" in the United States or "operational research" in the United Kingdom (Thomas 2015). *Operational research* was the name that various departments of military and defense ministries in these two countries gave to services in charge of the analysis and optimization of military operations, equipment, and weapons. Operations research sought to apply a theory to an almost infinite set of contexts of action. It was the stuff of physicists and mathematicians, recruited to work closely with military personnel, to review and analyze operations in the field and to find ways of optimizing them by applying a "scientific method."[3] It aimed to provide the means for choosing among various solutions in a situation with multiple options and high levels of uncertainty as to the capacity to attain the defined objectives.

Systems analysis evolved after World War II as an attempt to generalize the philosophy of operations research (i.e., optimization) beyond military issues and organizations. The RAND Corporation, an organization for research and development founded by the US Air Force, became the hub of that intellectual enterprise (Jardini 2000). The concept was theorized within its walls and applied to more and more cases, particularly under the intellectual drive of Charles J. Hitch, a British economist. Hitch theorized the extension of operations research into an approach for guiding choices and rational decisions, becoming a "generic form of policy analysis" (Thomas 2015, 268), a more widely applicable heuristics aimed at articulating and exploring ranges of scenarios and decision options, rather than at identifying the only optimal decision.[4]

Systems analysis started to be applied within government and gained public visibility in the 1960s when Robert McNamara became secretary of the US Defense Department, and Hitch his general comptroller. Together with other "whiz kids" recruited from RAND and the Ford Corporation (where McNamara worked before joining President John F. Kennedy's cabinet), they applied the Planning Programming Budgeting System (PPBS) method that Hitch had developed at the RAND. In 1965, President Lyndon

B. Johnson asked for the PPBS method to be generalized across the federal government. This meant that systems analysis, and specifically the kind of cost-effectiveness and cost-benefit analysis branded by RAND professionals to improve economic efficiency, started to be applied to a variety of new federal programs composing Johnson's "Great Society" vision, from poverty to mass education to urban renewal and health care (Enthoven 1980).

Systems analysis and risk analysis share several important programmatic dimensions. The first concerns the possibility and desirability of rationalizing human actions: from game theory and the concept of utility, what these sciences maintained was the belief that human actions could indeed be directed by optimal decisions, or decisions that would achieve the highest levels of utility for all players in a system. That notion was also apparent in cost-benefit and risk-benefit analyses, which sought to balance risks and costs, or risks and benefits, as if they could be measured on the same plane and their relation or ratio optimized. A second, cross-cutting claim was that mathematical and statistical analysis could indeed be used, as in the prisoner's dilemma, to compute these utility functions. From systems analysis to technology assessment, the motto was: "Data, quantification, calculation."[5] Data gaps or uncertainties were not irredeemable: all of these sciences built on probabilistic thinking and believed in the power of probabilistic analysis, whether objective or subjective, to identity optimal outcomes (Fortun and Schweber 1993). Third, policy science was marked by a belief in the power of a generically conceived scientific method (Mirowski 1999, 2002; Thomas 2015). Systems analysis, technology assessment, and, later, risk assessment all built on a similar belief in the power of fundamental thought processes and modes of reasoning to solve multiple kinds of problems, as well as a distinctive ethos of abstraction, whereby problems are captured through standard, quantitative parameters rather than studied empirically.

Rationalization/optimization, quantification of probabilities, and scientific method were the pride of practitioners of operations research, and hence of systems analysis and policy sciences as well. All three traits were shared by the practitioners of the emerging practice of risk assessment. Furthermore, all of these sister sciences were the object of the same scathing critique by sociologists, such as Ida Hoos.[6] Hoos took up all three traits of policy science, but then derided each of them. First, where analysts saw optimization, she discerned marketing and a lack of accountability. She claimed that systems analysis was an oversold, booming business—an artificial market energized by professionals (not only engineers and economists, but also so-called soft scientists such as environmental designers, urban planners, political scientists, and sociologists) in search of professional prestige and contracts. Likewise, engineers and economists involved in risk assessment were contract-seeking advisers who drew from massive public budgets the funds to perform studies that never got reviewed and had no substantial impact on decisions (Hoos 1983). Systems analysts were everywhere in government, where they claimed to render decisions and policies optimal. Yet, in practice, any actual systems studies came down to an endless, goalless collection and processing of data, which produced obscure results, if any at all.

Second, in Hoos's work, quantification becomes "quantomania": "What cannot be counted simply doesn't count, and so we systematically ignore large and important areas of concern" (Hoos 1979, 193). Quantomania was equated with the drunkard's search for his keys under the lamppost because there is more light there, even if it is nowhere near where he dropped them. Uncertainties and data gaps were pervasive, yet completely ignored by risk analysts, who replaced them by strings of intuitive guesses and assumptions (and all the while accusing citizens and laypeople of irrationality). Actual risks and disasters were firmly denied by engineers, who stuck to their fictitious calculations to show that risks were very low and concrete consequences minor. Risk-benefit analyses and risk comparisons served essentially to force acceptance of ongoing technological developments (notably nuclear energy) on the public.

Third, Hoos reduced the love of the scientific method to an inability to embrace the complexity of social problems. These scientistic exercises, she claimed, shared a recognizable methodological tropism, or "subtle shift from problem or substance focus to methodology or technique focus" (ibid., 193; see also Tribe [1972] on "process reduction" in policy sciences, and Kramer [1975]). Engineers', economists', and natural scientists' faith that science makes complex problems tractable and amenable to decisions was deceiving. It actually stifles social debate on the criteria and values to apply in the selection of what mattered to the population. Hoos held that "society" was not a system like a weapons system (there was no objective, valid definition of its limits; one did not "solve" the problems in that space definitively, and there were no right or wrong solutions concerning health or urban renewal). Systems analysis perpetuated the illusion that grave problems facing society and social groups could be measured, managed, and even solved in an optimal manner, with no involvement by the people concerned with those very problems.

Policy Science and Political Domination

Like Ida Hoos, whom he cited, Wynne came to risk assessment after considering the institutionalization of technology assessment.[7] He questioned "the political model" underlying the conceptualization of the latter (Wynne 1975, 125), claiming that with technology assessment, politics took on a new, scientific, and decisionistic form that distracted the public from fundamentally political questions. Wynne argued that there were clear "commitments and premises which frame the analysis and action of the [technology assessment]-policy scientific movement: first, man is an exclusively materialistic, uniformly rational economic calculator; his society—the only conceivable society—is a utopian version of leisure-class American. The natural environment is to be replaced by an artificial one created by corporate commerce. Mammoth and sophisticated technology administers man's every need; political conflict is emasculated to 'technical' questions over the acquisition of scarce resources; other societies don't exist except in so far as they strive to attain this utopian—or is it dystopian—ideal" (ibid., 115). This paradigm discouraged public participation and assumed that quantitative analysis of impacts and risks was a medium for managing (and limiting) political confrontations and producing supposedly harmonious decisions. The expansion of this "common political-technical language" (ibid., 126) worked, in practical terms, in favor of corporate interests. Technology assessment, as a language of objectivity, cohered with a politics of consensus and an economic model rooted in corporate monopolies and an unbridled quest for capitalistic growth.

Inspired by Jürgen Habermas (1971), Wynne explained that the scientistic language of technology assessment was one medium of legitimation. Under the rhetoric of consensus, and because of its unicity and lack of competing paradigms, it effectuated a dire, almost totalitarian, social control: "A second and perhaps more appropriate model in our case is the one which he [Habermas] terms 'decisionistic'. In this model, the political process still has a place, but its terms are entirely circumscribed by technical predicates, definitions and means.... It is also a more pervasive form of scientism just because it obscures the resultant distortion of the nature of politics and authority under the fallacious impression that politics is still alive and well" (ibid., 123).

Wynne approached risk assessment later, as he set out to study the Windscale inquiry,[8] using the emerging sociological methods of laboratory studies to study "scientific knowledge in important public arenas" (Wynne 2013, xvii). Through numerous publications (Wynne 1982, 1996, 1995, 2002, 2005, 2013), Wynne developed a "sociological analysis of scientific rationality" (Wynne 1982, 127), through five recurring, often critical themes. The first is that risk is a particular, reductive way of framing concerns that exist in the public, and which relate not strictly to the measure of the frequency and severity of predefined events, but also to the value of technological developments for variegated social projects and worldviews. Risk assessment involves a shift: The issue of the presence of technology in our societies is no longer raised as a question of collective choices to be made, but purely as a problem of social acceptance and management of the negative impacts of technology. The political choices that have supported technologies are rendered invisible, and the debates on the common worlds brought about by them are concealed (Wynne 2002; see also Rayner 2003, 2007). There are hidden commitments—a favorite term of Wynne—behind risk assessment. However, by framing the technological problem in strict scientific and metrological terms, it effectively shuts off all possibilities for open deliberation around the ends and values of technology.

Second, risk sciences operate with a standardized model of the citizen as interested only in the utilitarian measure of the consequences of technologies, as only concerned with safety; this model presents the citizen as being incapable of dealing with uncertainty and constructing independent social meanings through autonomous relationships, outside institutions (Wynne 2005,72). The third theme is the claim that reliance on these analytical exercises and on its practitioners severely restricts institutions' ability to learn and compounds a general lack of reflexivity. Risk assessment, being insensitive to the diversity of meanings and commitments toward technologies or the environment, actually fuels the frustrations of the public and communities with regard to institutional policy mechanisms. Fundamentally, it partakes of the declining trust in institutions. The fourth theme is the construal of the public as a figure of error and ignorance. Finally, the fifth theme is that risk analysts or risk assessors are the cadres of the rationalization of society and of its technoscientization. Risk assessment is an instrumental science, pushed by an elite, that is part and parcel of the policy of technological development.

Wynne's arguments provide the beginning of a theorized answer to the problem of why and how risk assessment acquired some sort of currency in the administration and governance of technology-related issues: It is because of its capacity to depoliticize public opinion and to conceal power and domination issues, naturalizing the monopoly of technical experts and bureaucracies in tackling issues at the expense of the wider public. He saw it as a reinvention of the language, whereby technocracy naturalized ongoing technological developments, colonized the life-world, and removed itself from public consciousness and critique. For Wynne, risk assessment and its professional cadres are not assistants to making decisions about technologies; rather, they are part and parcel of a ready-made decision to develop technologies and diffuse them. They are the technology themselves.

Risk: Professional Domination

In 1981, the risk profession became visible with the creation of the Society for Risk Analysis (SRA) in the United States. *Risk assessment,* by then, had become an official label for a new group of professionals close to government and major petrochemical corporations and energy utilities (Hoos 1983). In 1984, James Short Jr., then-president of the American Sociological Association, focused his annual address on the fact that "an entire industry has grown up around risk analysis, complete with professional trappings, governmental agencies, and personnel drawn from a variety of disciplines and professions" (Short 1984). As an organizational sociologist interested in accidents and disasters, on the path toward the development of a theory of "normal accidents" that confronted engineers' faith in the reliability of large engineered systems, Charles Perrow had more reasons than any other scholar to cross the path of risk analysts. His curt and incisive paper (Perrow 1982)[9] was motivated by his attending a symposium in 1981, the proceedings of which were published as *Societal Risk Assessment: How Safe Is Safe Enough?* (Schwing and Albers Jr. 1980). In it, he reviewed the various papers delivered at the conference by leading figures in the field, from Howard

Raiffa to Lester Lave. Perrow was quick to detect the particular worldview (and political inclinations)[10] underpinning the conversation among the new risk professionals. According to Perrow, the speakers had in common a misrepresentation of the public as a mass of ill-informed, irrational individuals who were so frequently wrong about the real safety issues that they should not be allowed to make a contribution to the issue. Perrow also pointed out the biased agenda of risk professionals: "If the public is misrepresented, the issues are more so. The cases that are most frequently discussed are nuclear power, toxic chemicals, genetic engineering, and various forms of transportation. In each of these cases, the risk-acceptance position is the one that favors private business and industry, and the risk-aversive position would harm it" (Perrow 1982, 299).

Risk and Culture (Douglas and Wildawsky 1983), written at about the same time, proposed an equally incisive critique of risk analysis. This book inaugurated an original theory to understand why and how certain collective hazards or harms come to matter more in a social group, given the diversity of such hazards—a question that risk assessors typically overlook, focusing as they do on a limited set of risks, and one that risk perception specialists often misrepresent. These authors argued that what people choose to worry about depends on their worldviews and their belonging to a given social group. From this perspective, anyone is entitled to worry about whatever she or he chooses to worry about; it is just a matter of cultural selection of hazards out of a broad range of hazards and uncertainties that exist in the wider world.

The book is a clear attack on the view, common among engineers, toxicologists, economists, and statisticians, that people worry about the wrong risks. It confronts these experts, and their claim to know risks better, by making the simple point that, as there are so many dangers threatening us individually or collectively, we cannot know them all, or know them fully; hence, uncertainty matters. In quantifying the costs and benefits associated with particular damages (ibid., 69), risk assessors overlook uncertainty, assuming that "what is true for the past will remain true for the future" (i.e., data about past frequency of events is extrapolated as a measure of what will occur). They further claim that the costs and benefits of technologies can be measured in the marketplace, and that people have sufficient information to make intelligent choices. However, none of these assumptions can be verified: relations to risk in a social context are moral and political and depend on changing, variable values. There are, moreover, massive uncertainties in the calculation of probabilities.

So risk assessors' claim of measuring risks objectively is misleading. In the harsh words of Douglas and Wildavsky, "it is a travesty of rational thought," a kind of "naked objectivity" that applies standard methods without consideration and relevance for the social contexts of risk, and one that comes down to "educated guesses." Disengaged from value debates, working at a distance from the public through numbers only, assuming no role in public decision-making (only in giving "advice"), risk assessors stop where responsibility starts (ibid., 80–81).

Mary Douglas remained interested in the emerging profession of risk analysis, its worldview, and its social role,[11] and continued to write about the profession from her own, anthropological perspective in order to show that risk assessment was the locus of a particular political morality. Industrial hazards, the new taboos of industrial societies, were appropriated by a new class of professional public figures: the specialists of risk analysis and risk perception. The kinds of things that were being called risks in the 1970s and 1980s, typically, were pollutions. Douglas saw pollutions as objects that threatened the cohesion and solidarity of a group; they were things that got designated as bad and foreign to us, and were made invisible to protect the integrity of our communities. The quasi-science of risk assessment was essentially a central institution of judgment and blame attribution in modern, stratified societies.

Risk analysts are precisely in the business of finding causes—or faults—for all the hazards that affect us as individuals. They fulfill the social function of blame, and risk assessment is a blaming technology: "its forensic uses fit the tool to the task of building a culture that supports a modern industrial society" (Douglas 1992, 15). It is the sign of an adversarial, blame-your-neighbor type of society. This institutional role is reflected in the particular epistemology and values that define risk professionals: a defense of objectivity and ideological neutrality, the belief in the rationality of the individual, a preference for methodological individualism, and a depoliticized perspective on problems and situations (Douglas 1990). This forensic role combines with an ideological discourse of social domination over the rest of social beings, which are considered irrational (Douglas 1992, 13).

Secondary Scientization and the Reinvention of Expertise in the Risk Society

In the ideological analysis advanced by the critics of risk assessment, risk assessment (and policy sciences more generally) are the reincarnation of a perpetual technocratic project of rationalization, which is seemingly unaffected by the context of conflicts and controversies surrounding science and technologies. It is curiously decontextualized, and variations and historical changes in the actual practice of risk assessment are mostly overlooked.[12]

Risk assessment appeared in a context in which threats to our safety, environment, and health had become primary issues of concern—which was precisely the vantage point from which critics of policy science and risk assessment could claim that risk assessors ignored publics, knowledges, and meanings. Beck's argument in *Risikogesellshaft* (Beck 1986), his opus on the risk society, is that hazards are inherent to the nature of contemporary societies and manufactured by them. The scale of manufactured risks is bigger than any of the safety issues societies have dealt with thus far. They are often irreversible and extend across time, space, and social groups. They are fundamentally difficult to analyze and apprehend collectively. In the English translation of his book, Beck (1992) makes no explicit mention of "risk assessment" or "risk analysis," and names no professional risk group.[13] But Beck does put forward an original argument about the kinds of techniques and expertise that were being reframed at the time as elements of risk research.

The rise of manufactured risks magnifies one of the greatest tensions in the politics and government of contemporary societies: their relationship to specialized, scientific knowledge as a means of dealing with collective issues. One of Beck's main theses is that science has lost part of its credibility and authority to describe the risks that we face and to prescribe what to do about them. In his lapidary terms, "there is no expert on risk" (Beck 1992, 29). The experience of disasters, risks, and uncertainties is dispersed into societies, and thus everyone is an expert on risk and may participate in the social definition of threats. Scientists' typically rational, calculative understanding of these risks is only one perspective on risks. The authority of science is challenged by the difficulty to calculate risks and establish their causes: "ultimately no one can know about risks, so long as to know means to have consciously experienced" (ibid., 72). Scientists have lost their monopoly over the production of validated knowledge. Their social authority is breached and makes space for explicit distrust. The autonomy and professionalization of science are severely restricted, as citizens question the knowledge and proofs that come forward from this space (ibid., 157).

Beck's thesis about the loss of authority of science is actually intertwined with a second thesis called *secondary scientization*. He discerns a historical shift toward a new mode of scientization in the postwar period. The first phase of scientization, during the Industrial Revolution, was an era when disagreements among scientists and engineers were managed in confined spaces, and where the truth-bearing products of science were transferred to society without question. In the risk society, internal errors and criticism among scientists are exposed because the public has come to doubt technoscientific products. Yet science does not recede entirely when challenged. Scientization continues because the social controversies over technosciences and associated risks are themselves couched in the language of science. All the parties engaged in the definition of risk employ scientific language, notably what Beck calls the "scientized ecology movement" (ibid., 162). Science expands under the form of a language of doubt and critique that fills the public sphere. Scientization thus increases while the authority of scientists diminishes. In the risk society, science emerges as a medium of controversy. It partakes in social contexts of dispute made of data comparisons, interpretive wars, and mutual deconstruction of methods.

The secondary scientization thesis can potentially transform the critique of policy science. Some of Beck's writing echoes the critical arguments of Hoos, Wynne, and others. He points to the fact that scientists tend to legitimize and force the acceptance of risks. He agrees with them that beneath the cloak of objectivity and scientific demonstrations, science justifies private interests and commercial values. Science has become "self-service shops for financially endowed customers in need of arguments" (Beck 1992, 173). Self-appointed scientific experts on risks advance convenient assessments of low or negligible risks of continuing to develop and diffuse technologies (Beck 1992, 173).

But Beck also notes that risk controversies affect scientists who act as experts on risk, and thus they become a driver of transformation of scientific expertise, its content, and its modes of expression in public. He does not fold risk research and risk assessment strictly into science, but rather sees them as the product of an expansion of science in a society marked by controversy. They are the scientific disciplines' response to the new conditions of uncertainty, interpretive flexibility (Collins 1981), and controversial scientific statements on risk. *Secondary scientization* means that more people and groups wield scientific arguments in generalized disputes about the definition of risks and secondary adverse effects. A game of expertise and counterexpertise sets in, in which science is prompted to move toward new competences and territories as it is embroiled in controversy. In Beck's words:

> New public-oriented scientific experts emerge, the dubious aspects of the foundations of scientific argumentation are exposed with counter-scientific thoroughness, and many sciences are subjected through their applied practices to a "politicization test" of a previously unknown extent. In this way, science not only experiences a rapid diminution of its public credibility, but also opens new fields of activity and application for itself.... Here the self-contradiction that scientific development has got into in its reflexive phase becomes tangible: the publicly transmitted criticism of the previous development becomes the motor of expansion. (Beck 1992, 161)

Risk assessment must be seen as characteristic of the era of secondary scientization, of an altogether different contract between policy and science premised on the impossibility of speaking truth to power or of conveying knowledge directly from science to decision-makers—premised, furthermore, on the fact that science has been harnessed by publics and interests.

Thus, this process is politicized and can serve to translate disputes, not avert them altogether. Risk assessment, from this perspective of secondary scientization, is the product of the transformation of scientific rationality under the effect of generalized doubts and interpretive flexibility of scientific claims in controversial contexts. It is less the reincarnation of a technoscientific, instrumental rationality that ignores and suppresses social meanings and relationships to technologies and their risks. Instead, it emerges as a product of these controversies: a reaction of scientific disciplines and of professions that have a stake in the development and diffusion of technologies, to the disputes that their products generate in the public sphere. The science of the assessment of technologies and risks is the product of the encounter between science and social conflict.

Risk Research and the Political Context

Risk research came together as an interdisciplinary field in the 1970s (Rip 1986; Krimsky and Golding 1992; Thompson et al. 2005; Bales 2009; Boudia 2014). The field was structured through the SRA. Risk research lay at the intersection of five separate fields that embraced a probabilistic perspective on the diverse kinds of hazards they were studying (Rip 1986). The first was risk analysis itself, derived from the application of systems analysis to nuclear safety issues. The second was safety and reliability engineering—a far less visible tradition, but equally linked to the search by governments and corporations for optimized, standardized safety in engineered systems such as transportation or industrial production. The third root was research about catastrophic events and disasters, particularly in the discipline of geography. Actuarial statistics constituted the fourth root, and medical research the fifth (i.e., toxicology and epidemiology specifically, aiming to define comparative rates of mortality and of untoward effects). Research on risk perception forms an emergent, sixth branch as well. The various strands of research converged toward the use of the word *risk*, rather than *hazard*, out of an interest in probabilistic calculation, judgment, and mitigation. It also stemmed from their search for new audiences and sources of funding.

The creation of the SRA did not merge these groups into a single new disciplinary entity.[14] The areas of risk research focus on very different sorts of hazards, use different methodologies of calculation, and speak to different audiences. The term *risk*, which successive working groups of the society have tried to define, is quite elusive (Thompson et al. 2005). The first leaders of the SRA ensured that it would become a home for three such groups in particular: safety and reliability engineers (the smallest of the three), health and toxicity people, and social scientists (Bales 2009). For several of these (particularly risk analysts, engineers, and toxicologists), it was a way to move away from a peripheral or subordinate position in their originating discipline and toward a home where their identity would be respected more. Risk research, even after the launch of the SRA, remains a disciplinary archipelago (Hood and Jones 1996; Bales 2009).

One reason for the internal differences in the "interdiscipline" (Klein 1996) is that risk research evolves in response to institutional demands for restoring capacities to decide and choose in the face of social protest (Rip 1986). The very first articulation of assessing risks of nuclear accidents was by an engineer reacting to the problem of deciding on the siting of nuclear facilities (Farmer 1967). Civil use of nuclear technology has always been controversial, ever since the project of adapting this military technology to the production of electricity was formulated (Nelkin 1971).

It soon became the object of national disputes waged in the media, as well as local conflicts surrounding the siting of plants and later of nuclear waste facilities.

At a meeting of the International Atomic Energy Agency in Vienna in 1967, F. R. Farmer suggested converting safety into a measurable index of risk, implying that numerical levels of acceptability and unacceptability could be established precisely. In 1969, Chauncey Starr, an engineer, physicist, and then-dean of the School of Engineering and Applied Science at the University of California, Los Angeles (UCLA), felt the need to engage in the mounting controversies surrounding the development of the civil use of nuclear energy and to demonstrate that public concerns about the safety of that energy were unfounded. Starr's paper was developed as part of his chairing the National Academy of Engineering's Committee on Public Engineering Policy (Boudia 2014).[15] In a paper published in *Science,* he computed a risk-benefit ratio for a set of technologies (car, civil aviation, nuclear reactors), implicitly taking on the context of contested technological developments. Comparative measurement of risks was a manner of demonstrating to the public that focusing on that particular hazard was pointless. It was a way of telling groups and communities mobilized against that technology to shift their focus to other concerns. The paper was one of the most visible and powerful articulations of the notion, repeated ever after by everyone in this field, that involuntary or uncontrolled risks are always more negatively and acutely perceived than voluntary, controlled risks, and that the public directs its attention to these risks, however less probable or damaging they may be.[16]

This link between comparative risk assessment and the handling of public controversy also verifies with legally mandated exercises of cost, risk, or impact assessment. Environmental impact assessment burgeoned after a requirement to perform such analysis was included in the US National Environmental Policy Act of 1969, which provided the impetus for federal agencies to hire or consult thousands of environmental analysts (Clark 1997). Toxicology accelerated when assessments of the safety of food additives, pesticides, and chemicals found in air and water were made mandatory by successive laws that shaped a new social and environmental regulatory landscape. Successive executive orders for the quantification of costs and benefits of regulations (starting with Executive Order 12291, passed in 1981, mandating the performance of cost-benefit analysis for all rules, with an anticipated impact on the economy above $100 million) were the most direct reason for the federal government hiring economists. A study of risk professionals in the United States showed that a third of them were working for federal organizations (Dietz and Rycroft 1988).[17]

The point is that, with "policy sciences" broadly speaking, scientific norms and procedures were being imported "into a political setting of intense conflict" (Taylor 1984, 8), where public decisions had been formidably difficult to shape and implement. The idea that political contexts structure risk research was evidenced in the birth of the SRA. Two of the three disciplinary groups inside the SRA, in fact, were dealing with the most intense public disputes of the time: Safety and reliability engineers were directly involved in the controversy over risks relating to the expanding civil use of nuclear technology; health scientists were involved in the controversy over the risks of chemicals in food and in the environment, and they also were decisive in forging ways of rendering the chemical problem controllable (Sellers 1999; Frickel 2004; Shostak 2013). The third area of risk perception consisted of precisely the field addressing—and translating into tractable terms—the problem of the public and the emergence and expression of its collective preferences, as well as the difficulty of applying decisions where they remained controversial.

This embedding in a political context has implications for the kinds of knowledge produced by risk research (Rip 1986). First, the engagement with controversial contexts means that the field had adopted ways of conceptualizing social controversies and their causes, and of demarcating the scientific contribution to decision-making from the realm of politics. Notions of risk evaluation and risk acceptability featured centrally among the concepts that helped the field take form. In his book *Of Acceptable Risk*, William Lowrance, an engineer by training, distinguished between a factual, scientific measurement or assessment of risk and a normative judgment of safety. He recommended distinguishing these two planes of discussion explicitly, so that value conflicts would not disrupt the agreement on facts (Lowrance 1976).

The fact/value distinction was not always easy to apply, but William D. Rowe's contribution improved the concepts further. In his *Anatomy of Risk* (Rowe 1977)—originally a report to the EPA, dating from 1975, to review the attempts to settle the nuclear dispute by means of probabilistic risk analysis (Rowe 1975)—he did away with the objective/subjective, fact/value dichotomies. He argued that the main distinction to make was not between science and policy, but between technical value judgments, those "made by experts in the absence of hard technical information when faced with the difficulties in obtaining further information" (Rowe 1975, 3) and "societal" or "managerial" judgments of appropriate balance between risks, costs, and benefits. This codification of the politics of risk helped the field assert its legitimate role in decision-making, as shown by the themes that the founders of the SRA selected for the society, notably the "application of assessments to such things as cost-benefit evaluations, administrative actions, political factors, etc." (Thompson et al. 2005, 1336). The center of gravity of the interdiscipline is not simply calculating risk with ever more sophisticated methods and data, but contributing to the intellectual process of forming judgments and public decisions.[18]

The close relation with institutions and with contexts of political uncertainty surrounding acceptance of hazardous economic activities meant that there was a tight connection between the political context in which problems of risk are defined and the actual orientations of risk research. As Rip (1986, 6) put it: "The problem-definitions used by relevant actors in different contexts have shaped the direction and content of risk research directly. In turn, the expertise shaped by such problem-definitions will produce particular kinds of policy advice." Risk analysis of nuclear installations illustrates this point—namely, that risk assessment is the field of inventing technologies to frame controversy and is attached to the context of such controversy.

The probabilistic risk analysis of nuclear reactors in the 1970s stemmed from the problematic implementation of the Price-Anderson Nuclear Industries Indemnity Act of 1957 (or Price-Anderson Act). The Act was adopted to create a pool of funds to insure against the costs of nuclear accidents and to clear a major hurdle for private investment in that energy. It stipulated that private companies were to pay up to around $10 billion of damages, while further actions (e.g., appropriation of federal funds) could be decided by the US Congress in the case of damages exceeding that sum. In that context, the definition of possible effects of a nuclear accident on populations, and the quantification of the probability and severity of these damages, were instrumental in defining insurance policies in monetary terms. In effect, this was a quest for low probabilities (Fuller 1975), which led the Atomic Energy Commission, on the occasion of the renewal of the Price-Anderson

Act in 1967, to dedicate $3 million to new probability research, resulting in the Reactor Safety Study (WASH-1400 or NUREG 75/014) by Norman Rasmussen, a professor at the Massachusetts Institute of Technology (MIT).

Rasmussen's study is remembered as the first large-scale application of probabilistic risk assessment (e.g., Bedford and Cooke 2001), considering a wide variety of possible negative outcomes related to the use of a nuclear reactor, ranging from events affecting the reactor itself to thyroid problems in surrounding populations following accidental radioactive emissions. The power of his study, a fault-tree analysis, rests as much in the identification of the initial 130,000 sequences of events, and the selection of seventy-eight more important sequences, as in the actual computation of probabilities for each of these sequences. This is where the value of the method lies, according to Rasmussen. In his conception of risk assessment, a larger error band than in a strict reliability study can be tolerated because the heart of the exercise is the consideration of the results "to see if they are meaningful" (Rasmussen 1975, 5). He articulated a logical method to identify such meaningful events and assign a number to each of them.

It is not possible to separate the report from the controversy. The authors (and sponsors) consciously engaged with the public dispute about nuclear risks. Rasmussen referred in several places in the report to the public discussions of nuclear risks, and he also set an objective of ending the confusion surrounding those risks. He also included in his study a comparison between nuclear and nonnuclear risks, to stress (as he did throughout the report) that the risks concerning nuclear reactors were low probability (Rasmussen 1975, 103). The Reactor Safety Study was heavily and publicly criticized, instantly after its publication. Rasmussen appeared to have too much sympathy for the development of nuclear energy and minimized the risks in a way that was much too evident. In light of his positive assumptions about nuclear energy, the actual computations presented in the report appeared suspect (Hippel 1977). This applied either to health damages resulting from a nuclear accident, such as genetic defects or water pollution (Primack 1975), or to the failure of reactor components (Weatherwax 1975).

Thus, this study neither resolved the controversy about the risks of nuclear accidents nor laid the foundations for a political compromise on the development of nuclear energy in the United States. But it had an impact in the longer run, as the incarnation of a procedure to organize and enable decision-making on disputed issues. "Shortly thereafter [after Rasmussen's study] a new generation of [Probabilistic Risk Assessments] appeared in which some of the methodological defects of the Reactor Safety Study were avoided. The US NRC released the Fault Tree Handbook … in 1981 and the PRA Procedures Guide … in 1983 which shored up and standardized much of the risk assessment methodology" (Bedford and Cooke 2001, 7).

From a longer-term perspective, the development of uncertainty analysis methods, both qualitative and quantitative, was risk assessors' coded response to the criticism of a technical and political nature (Hayns 1999). Several years after its publication, what was remembered from Rasmussen's study was not the actual estimations, but the heuristics that had been applied. The method was tried again several times after that, in successive reactor safety studies, which means that the report still had a legacy—it provided a way of reasoning about safety issues. Risk assessment, in the aftermath of the report, appeared less as a given numerical figure for the risk of nuclear accidents than as a general method and set of criteria (voluntary/involuntary, natural/artificial, high/low probability, acceptable/not acceptable) to be able to identify, classify, and select safety issues—in short, a socially legitimate way to talk about risk. Finally, by hiding pro-nuclear energy bias within an arcane methodology, Rasmussen paradoxically showed the value of isolating technical considerations from value discussions in a structured decision-making process. The study inspired people like Paul Slovic to study the inherently sociopolitical nature of risks (Slovic 1999). It was also a key lesson for Rowe's thinking on acceptability, societal judgments, and a decision-making structure that could accommodate both the scientific inputs and the consideration of values (Rowe 1975).

Conclusion

Risk assessment can thus be envisioned as a science that participates in the development of rational rules for decision-making in the context of the disputes that those who administer these problems face, as much as one that carries instrumental rationality into public policy. Controversies can be considered to be a primary, causal context for the development of policy sciences, and not simply scientific and industrial interests to avert deliberation and contestation by the public (Cambrosio and Limoges 1991). This has clearly been the case with technology assessment, which evolved a constructive technology assessment branch (Schot and Rip 1997). It has also been the case of systems analysis: the generic systems approach devised by Hitch and his colleagues in the 1950s was a reaction to the political contexts in which analysts were placed, as well as an adjustment to become more effective in these contexts. This has also been the case with risk analysis. Risk research and risk assessment are not inattentive to the possibility of dispute. Risk assessors do not design decisions by themselves; rather, they design opportunities for administrations to make decisions. They do not ignore public controversies. Instead, they develop methods and produce calculations to limit and avert them, in conjunction with public administrations and regulatory agencies. There is a political dimension inherent in risk assessment and in its rise—and methodologically, there is a way to do a political analysis of risk analysis—if we understand it as a science that is constrained by, and responsive to, contexts of controversy.

The ways in which risk assessors conceive of new kinds of data, analytical methods, and structures for decision-making and governance, pinpointing where the public, experts, and decision-makers stand, show that design is just such a response. Risk sciences, and policy sciences more broadly, are design sciences: sciences that construct the formal elements of a mode of decision-making to frame controversies and form decisions. They do so differently in the various subareas of risk research, with a different notion of where the controversy comes from and the kinds of uncertainty that cause them. Probabilistic risk analysis, as developed by nuclear engineers, is certainly different from the calculative-protective vision of medical scientists, who adapted probabilistic approaches to the assessment of cancer risks. This is where the mixed histories of the EPA and of risk decision-making begin. Cancer assessment was gradually formalized and put in place at the EPA in the second half of the 1970s in order to decide the dispute surrounding cancer-causing chemicals—the first of many episodes of the invention of the EPA in response to the public emergence of risk.

2 The Invention of Quantitative Risk Assessment

Rational decision-making at the EPA is influenced by the hope that enduring controversies could be resolved and legitimate decisions advanced. It is shaped through a process that involves scientists and their rationales for reordering knowledge and administration. In 1976, a group of medical scientists and toxicologists, brought together on the initiative of the EPA administrator, engaged in the creation of a procedure for computing and regulating the risks of getting cancer from exposure to chemicals—namely, the cancer assessment guidelines (EPA 1976). This technology for making decisions, still in use today, structures judgment of the hazards posed by chemical substances and helps to arrive at regulatory measures concerning them. The technology involves a series of steps to devise estimations of the level of concern, truncating uncertainty as much as possible to provide a firm ground for decision-makers. This process rests on the assemblage of various people and expertise: biochemists and statisticians, but also policy analysts and administrative officials making the final decision.

The common narrative of the invention of quantitative health risk assessment links it to the rise of risk science and the expansion of engineering methods for probabilistic risk analysis in this sphere of the medical and biological sciences (e.g., Nash 2017). But the history of the EPA's cancer risk assessment guidelines shows that this bureaucratic technology reflected a particular political situation, namely, the agency's difficulty in establishing noncontroversial facts concerning cancer risks before the courts, in the context of the medical disputes raging around carcinogenesis. This circumstance provoked the design of a new, probabilistic organization: one in which a decision emerges from the examination of data and the ascertaining of a level of risk. This technology was conceived by a network including the agency's lead economist and policy analyst, Alvin Alm, and toxicologists from inside and outside the agency. It was to produce long-lasting effects on the internal organization of the agency, as well as its credibility as a regulator of cancer-causing substances, putting together an organization that could govern risks according to degrees of certainty and harm, in a legible manner. Ideally, this was an organization in which cancer risk assessors acted transversally, informing the action of officials in the diverse regulatory programs of the agency.

The Delaney Problem

The environment made it onto President Richard Nixon's agenda under pressure from the public environmental movement that had built up in the 1960s in the United States, culminating in the first incarnation of Earth Day on April 22, 1970. He was also forced to move forward on this issue because of the popularity of Senator Edmund Muskie, head of the Public Works Senate Committee and future Democratic candidate for president, who was championing the environment. The 1969 National Environmental Policy Act was the first piece of general environmental legislation applying across all sectors of government. It required all agencies of the federal government to consider the detailed impacts of actions that were expected to have an effect on the quality of the environment, and it established a Council on Environmental Quality in the White House. Nixon also decided to add to the program of the White House Council on Government Reorganization, chaired by Roy Ash, formerly head of Litton Industries (a company developing military technologies and equipment, and a contractor to the US Department of Defense), a project to institute a new department for all matters related to nature and the environment (Landy et al. 1994).

Both Ash and Nixon were in favor of creating a large, new Department for Environment and Natural Resources. An environmental committee inside the Ash Council was appointed to design this department. The group was headed by Amory Bradford and also included Douglas Costle, a future EPA administrator (1977–1981),[1] and J. Terry Davies, a political scientist and future EPA assistant administrator in the 1980s. It soon developed a set of original ideas, which differed from the plan to create a new department. It argued that such a new department would be too heterogeneous and unwieldy, with imbalances between its environmental and developmental perspectives. Moreover, it would tend to disrupt the distribution of subjects among committees in the US Congress. They pushed instead for a global approach to the environmental problem. A notion of *comprehensive environmental management* implied doing away with the separate treatment of different kinds of pollution in order to sensibly improve the overall state of the environment. The new administration was thus envisaged as an agency rather than a department so that it could approach the environment as a whole and embody public policy in this regard.

That plan was finally accepted by Ash, and by Nixon. When it was formally inaugurated in December 1970, the EPA thus compounded the bureaus that administered statutes relating to the environment in a handful of departments. Tellingly, the act of creating the new agency was a reorganization or redistribution of fifteen existing offices and groups from the previous departments to the new agency. The US Department of Health, Education, and Welfare contributed the groups in charge of air, solid waste, radiological health, water hygiene, and pesticide tolerance. The US Department of the Interior conceded its unit for water quality and pesticide label review. The Atomic Energy Commission and the Federal Radiation Council transferred its service for radiation protection standards. Pesticide registration came from the US Department of Agriculture (USDA). The creation of the EPA was not accompanied by a revamping of all of these statutes and offices. So the legislative tasks, goals, resources, and processes of each bureau, as well as their personnel, remained in place.

These choices had two major implications for the EPA's authority. One was that because the EPA was an agency, not a department, the EPA administrator was not a member of the president's cabinet. This has been a recurrent problem over the life of the agency, and Congress debated elevating the agency to cabinet status on several occasions (in 1992, 2001, and 2012), but it never occurred (though the administrator of the agency has cabinet-status rank today). This meant that the EPA administrator was kept out of a number of strategic policy choices made by the White House, which saw it as a lower-ranking administrative body to be supervised. From the very start, the EPA was a targeted candidate for supervision by the White House Council of Environmental Quality, the Regulatory Council, and the Office of Management and Budget (OMB), and attempts at displacing agency decision-making toward other parts of the executive (Percival 2011).

The other implication was that an environmental protection agency was simply very difficult to steer. Its administrative infrastructure was made of separate bureaus, each implementing regulatory statutes with distinct rules and supervised by different congressional committees, all intent on pushing through "agency-forcing" legislation. The strategy was one of legislative micromanagement, leading to the definition of often-ambitious goals and tight schedules for the agency to meet in order to ensure delivery and minimize the risk of capture of the regulatory body by the regulated industries (Stigler 1971; Rosenbaum 1995). In practice, this meant that the various offices of the EPA had very different objectives, constraints, patterns of staffing, and expertise (McGarity 1991), as well as different regulatory cultures. Some were seen as more inclined toward protective policies, and others as closer to the industries they regulated. It also meant, however, that they were largely autonomous and difficult to steer from the top. They were fiefdoms that could easily frustrate the deployment of the administrator's chosen policies.

The agency's first administrator, William Ruckelshaus, chose an "activist suit strategy" (Marcus 1980, 88) to build up the newly created institution's credibility. His ambition was to quickly amass a number of successes on the legal front, taking on polluters. The strategy was meant to break with the image of its constituent part and to avoid falling captive to businesses' and interest groups' symmetrical criticism, through radical, unambiguous actions. This was announced forcefully by Ruckelshaus in December 1970, when several cities were threatened with prosecution if they failed to stop the discharge of pollutants into rivers within 180 days, and when several corporations were charged with pollution control violations (Marcus 1980, 88–90). Ruckelshaus also decided to opt for a case-by-case litigation strategy, through which he tried to cancel registrations of pesticides based on the argument of cancer effects. This litigation strategy, operated directly from the Office of the General Counsel, had in part been chosen against the Office of Pesticides. The latter, which had been transferred from the USDA, was oriented toward product registration, not driven by public health or environmental protection concerns, and was reputed to be relatively close to the industries it regulated (Landy et al. 1994; Powell 1999). It was one of the offices that seemed less easy to merge within a new environmental agency. To advance his strategy of establishing the EPA's reputation in high-profile litigation, Ruckelshaus decided to circumvent the Office of Pesticides and to have his general counsel run so-called cancellation hearings against a number of pesticides, including dichlorodiphenyltrichloroethane (DDT), aldrin/dieldrin, mirex, and heptachlor.

Cancer Policy: From Fixed Principles and Facts to Flexible Judgment Guidelines

Cancer had emerged as a major national public issue long before the EPA was set up. The question of the widespread effects of various chemicals on the environment and human health lay at the heart of the nascent modern environmental movement in the 1960s. The cancer issue was also a policy issue. Even before the collective mobilization of the 1960s, lawmakers had moved boldly on it. For instance, an amendment to the Pure Food and Drug Act adopted in 1958, pushed by Representative James Delaney, prescribed that any food additive shown in animal tests to be carcinogenic may not be approved. The Delaney amendment, of course, was a radical regulatory intervention and an unmistakable sign of the rise of a new kind of regulation, soon known as *social regulation*: a form of public intervention on industries and markets that is similar to competition regulation, but motivated this time by health, environmental, or social concerns. The rise of social regulation was deeply linked to the growing influence of public interest groups in the 1960s, reflecting a new, more open pattern of interest group politics (Vogel 1988; Bardach 1989; Harris and Milkis 1989). Cancer became an object of official federal policy with the adoption of the National Cancer Act in 1971 (establishing a fund to research possible ways to cure cancer). The Mrak Commission's report "Pesticides and Their Relationship to Environmental Health" (1969), a report of the US surgeon general in 1969 (also on the evaluation of environmental carcinogens), and the 1971 report of the Council on Environmental Quality confirmed the "war on cancer" agenda. They also contributed to changing the framing of the cancer problem. From a focus on finding the medical "silver bullet" for eradicating the disease, it became one of controlling toxics as the silent, environmental cause of cancer and of acting preventively instead of curing. The EPA was thus born in this atmosphere of environmentalism under political pressure from members of Congress, environmental groups, and public demands for regulatory intervention against polluting, health-damaging chemicals.

Establishing a generic cancer policy applicable to all cases and meeting the high levels of protection and safety that the statutes imposed, as Delaney had, proved immensely difficult. The EPA was soon confronted with the challenge of proving in court what was immensely controversial in science and medicine: the causal link between exposure to chemicals and the development of cancer. The dominant assumption in medicine was that exposure to carcinogenic chemicals could bring about cancer effects at any dose. But for some carcinogens, nongenotoxic ones in particular, it appeared that the existence of a threshold was still a possibility. And there was no definitive standard of proof to establish the existence of this threshold in the dose-response relationship. It was a matter of choice between statistical methods to extrapolate between tested doses.

That whole issue aroused strong emotions. It turned into a full-blown medical and toxicological controversy, compounded by legal challenges and doubts about whether administrations had sufficient authority to decide this threshold question, which has a direct impact on decisions on whether to regulate a chemical. Because of this problem, the Delaney amendment was particularly hard to implement for the FDA. It faced strong opposition on the grounds that the economic activity associated with the production and use of these chemicals was often substantial, and it seemed difficult to ignore the benefits associated with a substance, even though there were also risks.

Opponents of the amendment argued, on the contrary, that regulatory measures on products and substances should be based on the doses to which people were actually exposed. In other words, they should be regulated according to the risk, which was a function of the hazard. Harm mattered, but it was assumed to be more appropriate to ban things only when they were proved to be regularly or frequently harmful under one or the other condition. The pharmacology division of the FDA developed a procedure for establishing not absolute, but acceptably safe, levels of chemicals (Lehman et al. 1955; see also Rodricks 2007; Carpenter 2010), sketching out a kind of risk-based procedure for ruling chemicals. The procedure consisted of drawing dose-response curves based on experimental results. The curve normally was supposed to have a threshold corresponding to a dose that was conceptualized as the *No Observed Adverse Effect Level (NOAEL)*. To take into account differences between experimental animals' and humans' reactions to a similar substance, as well as to define a dose that it was deemed acceptable (by toxicologists at least) to expose humans to, the dose was divided by 10, 100, or 1,000. (The convention was to divide by 10 the safe dose found in animals, to determine the safe dose for humans, and/or by a supplementary factor of 10 to determine the safe dose for the most sensitive human, and/or by yet another factor of 10 to protect the most sensitive human against potentially lethal and irreversible effects). The resulting calculated dose was called the *acceptable daily intake*.

With regard to the EPA, the Criteria Office, one part of the Office of Health and Environmental Assessment (OHEA) of the ORD, was routinely performing this sort of assessment for noncarcinogenic substances. This was, in essence, the basis of what was then starting to be called risk assessment, in lieu of the former, more qualitative exercise of "safety evaluation" (Weill and McCollister 1963, 486). But for substances with a cancer effect, the EPA devised a less flexible policy, abstracted from the arguments used by its legal counsels in the first procedures to cancel the registration of pesticides. The cases resembled each other and raised the same kind of questions and challenges for the EPA (namely, the scientific challenge of proving scientifically the existence of a link between exposure to the pesticides and risks of cancer). The EPA's associate general counsel of the Office of Pesticides asked Umberto Saffioti, a scientist in the National Cancer Institute, to abstract principles that could serve as a standard of legal proof of carcinogenicity. Choosing Saffioti was by no means neutral, because he had chaired a scientific panel under the auspices of the National Cancer Institute a few years earlier, which had concluded that no safe level of exposure to carcinogenic substances could be established.[2] Saffioti's principles were still used in the proceedings of the EPA for cancellation of the registration of the pesticides aldrin and dieldrin. The principles were particularly precautionary and represented an aggressive legal strategy. In formulating them, the EPA's attorneys had clearly taken strong stances in medical toxicological debates to ascertain the following facts that were immediately convertible to legal strategies:

1. A carcinogen is any agent that increases tumor induction in humans or animals.
2. Well-established criteria exist for distinguishing between benign and malignant tumors; however, even the induction of benign tumors is sufficient to characterize a chemical as a carcinogen.

3. The majority of human cancers are caused by avoidable exposures to carcinogens.
4. While chemicals can be carcinogenic agents, only a small percentage are.
5. Carcinogenesis is characterized by its irreversibility and long latency period following initial exposure to a carcinogenic agent.
6. Individual susceptibility to carcinogens varies greatly.
7. The concept of a threshold exposure level for a carcinogenic agent has no practical significance because no valid method of establishing such a level exists.
8. A carcinogenic agent may be identified through analysis of tumor induction results with laboratory animals exposed to the agent, or on a post hoc basis by properly conducted epidemiological studies.
9. Any substance that produces tumors in animals must be considered a carcinogenic hazard to humans if the results were produced in a valid carcinogenesis test.

These principles and supposed facts about pesticides and cancer were as scientific as they were policy-based (Jasanoff 1982). They were developed by leaving aside many gray and controversial areas regarding the determination of carcinogenicity, under the concerted search for effective legal arguments. They were meant to determine mechanically how the EPA would now treat chemicals, as well as securing victory in court. Moreover, because there were hundreds (or even thousands) of them, the combined impact of these decisions on health or the economy could become substantial. They were also an essential instrument for representing the EPA's posture toward the toxics problem as a whole; an image that could supplant variations and inconsistencies across ad hoc responses to individual, controversial chemicals like DDT. In short, they were a bureaucratic technology to bring closure to controversial issues.

They were not, however, a very effective technology because they elicited controversy in turn. First, it was not that easy to defend these principles before the courts. Although credible, these facts could very well be countered by other scientists or medical researchers, whom the industry readily cited in court to counter the EPA. They were, moreover, insufficient to close legal cases. Mirex, for instance, was not formally banned until June 1978, and only after the manufacturer, Allied Chemical, decided to suspend its production because of the high costs of defending its use in court, and after the state of Mississippi, which bought the main unit manufacturing Mirex from Allied Chemical, agreed to rescind the license to manufacture it in exchange for stopping the cancellation hearings. The scientific case constructed by EPA's attorneys did not seem to have been as effective as the successive scandals in the media about the presence of Mirex in the Niagara River (Severo 1976).

Reactions to these principles had appeared within the agency. Scientists in the Office of Pesticides, one of the EPA's program offices that counted a large group of scientists, were frustrated that scientific and medical issues were being arbitrated by lawyers in search of firm legal ground to argue their cases. Other officials in this office simply deemed the principles inappropriate in the context of FIFRA, which mandated weighing the benefits of pesticide use for agricultural production against the risks. Even in legal terms, the principles were not necessarily the right instrument because blanket principles overlooked the wide diversity of types of cancer and modes of cancer onset, and making such blanket judgments could easily be found to surpass the agency's delegated authority under the law it was administering.

At the time, Alvin Alm was managing the Office of Planning and Management,[3] having moved there from the position of chief of staff of the Council of Environmental Quality in the White House in 1973. He was a public administration graduate from Syracuse University, and he already had had ten years of experience in public administration of environmental programs. People who got to know him and worked with him closely describe him as "a 'let's get things done', let's be rational, let's get things organized, a managerial sort of a person"[4] and an "astute analyst" (Fiorino 1995, 56). He was, above all, a staunch proponent of ex ante policy analysis and of the sort of comprehensive synoptic analysis that informed the policy analysis school movement of those days in the United States. At the EPA, Alm assembled an effective team of policy analysts, as described by his administrator, Russell Train: "Al Alm had put together a top-flight economic analysis staff, and the analytic data on costs, both to business and to the public health, that we were able to present were outstanding. In fact, we won many an argument simply because of the quality of our data" (Train 2003, 180). It was Alm who advised the administrator to move from a strategy of asserting definitive cancer facts in courts in aggressive legal strategies. Alm, who was to have a major influence ten years later on the formal adoption of risk assessment methods throughout the agency (see chapter 6), recommended a formal strategy of measuring the risks and benefits of products, as well as the establishment of a scientific group inside the agency to perform these assessments. With these recommendations, Alm was in fact instituting a move from a system of black-and-white decision-making (is a substance carcinogenic or not?) to a system in which the reality to be regulated is captured through a scale of risk, or shades of gray. This implied complex organizational and cognitive changes, such as the explication of decision options, and the complicated design of cross-agency reasoning procedures, to collectively choose among these options. Alm was, at this point in time, sketching out the elements of a probabilistic organization.

The EPA administrator, Russell Train, refused to produce the official notice for the cancer principles (Dickson 1987) and withdrew his support for the Office of the General Counsel's litigation strategy. His reaction was in part guided by the fact that he was spending long hours in hearings before Congress justifying the agency's actions in these legal cases. He followed Alm's recommendation to clarify flexible modes of judgment about toxicity through a *guideline,* a new kind of instrument that aimed less at establishing determinate rules than at clarifying options that rule-makers could use in developing a rule. A guideline, ideally, was to be comprehensive and detailed, to cover the manifold options that decision-makers could resort to under various statutes. In effect, it was a tool to assemble the necessary elements to make the organization more probabilistic: one that could adjust its strategy to the level of risk emerging from the calculations and the level of certainty underpinning these calculations—a flexible organization in which decisions were produced by a chain of actors along which information was transmitted to design the decision, rather than a pyramidal organization implementing a fixed strategy defined at the top. This tool was particularly welcome at the EPA, which had to decide on the carcinogenicity of substances that were regulated under different Acts, some of which required the setting of ample safety margins, and others the balancing of risks and benefits associated with the use of a substance. A guideline, thus, could fit in the interstices of different laws. Adaptable and flexible, it would cover "the whole class of chemical carcinogens" (Rushefsky 1986, 209) and ensure the consistency of the approach, its scientific soundness, and fairness to various chemicals and companies.

By the time the EPA started to formally develop guidance, the U.S. Supreme Court had effectively allowed regulatory agencies to set this kind of informal generic policy (meaning that they were issued through a procedure involving giving notice in the Federal Register and public comment; see McGarity 1979, 752). A guideline, however, also had to demonstrate the EPA's commitment to be as protective as the "zero-risk" Delaney amendment or the "principles" previously used in courts by EPA lawyers could be (US Congress 1983b, 372). The decision mechanism designed in the guideline—one in which a variety of information was considered in order to shape predefined decision options—was the way to ensure this.

The Interim Cancer Assessment Guidelines of 1976

Train opted for the instrument of the guideline, and furthermore, he agreed to establish a Carcinogen Assessment Group (CAG) within the ORD, as Alm had also suggested. The CAG was established as a branch of the OHEA. A sign of the group's policy centrality was the fact that it was based in Washington, D.C., at the agency headquarters (as opposed to the two other branches of the OHEA, based in Cincinnati and in Research Triangle Park in North Carolina). The group was comprised of scientists with a specialization in mutagenicity and reproductive toxicity. It was managed by Elizabeth Anderson, a chemist and toxicologist. The group chair was Roy Albert, a medical doctor and scientist from New York University who also had experience in research on radiation hazards at the Atomic Energy Commission. As part of an agreement with the EPA, he participated in the CAG's activities a couple of times every month, thus giving it scientific medical support and lending it credibility.

The group published its interim guidelines for cancer assessment in May 1976. The structure of reasoning that the guidelines recommended was not immediately apparent from the text. It curtly indicated that the assessment should have a summary that answered two main questions: "(1) How likely is the agent to be a human carcinogen? (2) If the agent is a human carcinogen, what is the estimated impact on human health?" (EPA 1976, 2). Anderson later got into the habit of converting the questions into steps, structuring an integrated, linear process of assessment.[5]

The first question was solved by applying a particular assessment technology abstracted as a *weight of evidence (WOE)*. In this exercise, the assessor decided what weights to grant various kinds of studies relating to the same chemical, yet mostly employing different methodologies and producing different results. A WOE essentially created a hierarchy of knowledge. Studies indicating the presence of tumors (i.e., negative studies) were supposed to be granted more weight than those that did not (i.e., positive studies). If two studies were available to the risk assessor, with one study on rats indicating cancer, and another on mice indicating no cancer, then the assessor was to continue working on a cancer hypothesis and not let the negative study overturn the hypothesis. The judgment about the carcinogenicity of the substance was to be "firm," or backed by "substantial evidence," if malignant tumors were found in several animal species. The evidence was to be considered "suggestive" if the only tumors that were seen were benign and "generally accepted as not progressing towards malignancy" (EPA 1976, 2).

The WOE method addressed rather than ignored the uncertainty or interpretive flexibility inherent in the consideration of experimental studies. Such studies, unfortunately, were rarely entirely positive, and were themselves rare. They failed to establish firm causal relations between exposure to a given substance and cancer incidence in an individual or group. Experiments were not devoid of ambiguity, moreover, for tumors could be interpreted in various ways: "Different substances act at different stages in the process. Some carcinogens may act as initiators, others as promoters. Still others may be complete carcinogens, capable of acting at both the initiation and promotion stages" (Harrison and Hoberg 1994, 24). There were also different kinds of tumors, benign or malign, and the interpretation that benign tumors were a sign of carcinogenicity was contested. Carcinogen identification depended on the capacity to interpret these tumors in animal experiments. WOE did not assume that the studies provided unambiguous results, and it admitted that contradicting signs would be available to the assessor. Neither did it assume that the assessor would be able to decide on the basis of only one study, for evidence was clearly an assemblage. This approach, couched in the scientific language of proof, was in fact a powerful strategy of legal demonstration: a standard of legal proof that the courts soon agreed to follow (McGarity 2003).[6]

The second question involved some quantitative analysis, employing mathematical and statistical techniques. In those days, when EPA staff described the approach, the overall carcinogenicity assessment task was called a "health assessment" or "effect assessment," with sporadic references to the production of "risk assessment documents." Risk assessment, or quantitative risk assessment, covered only the second step of the process, or the quantitative measurement of the impact.[7] With these guidelines, the CAG was extending risk assessment beyond what any other agency was effectively doing.

Much like the WOE, quantitative risk assessment was both computational and judgmental. It involved assessing, first, the number of people exposed to the chemical in the environment. Precise, unambiguous data were missing to determine this. For instance, numbers indicating what percentage of the population was exposed to a given dose of the product were generally absent—there was no system in place that allowed empirical information on this topic to be collected. Therefore, estimates were needed to define an exposure pattern: Who is exposed, and for how long? Are kids and pregnant women more exposed than others?

Second, quantitative risk assessment involved drawing a dose-response curve. Only high doses of the chemicals (not the doses that people are normally exposed to in daily life) were tested in animal experiments. Therefore, the toxicity of the substance was known only at these doses. Mathematical models were used to extrapolate from these points to lower doses, or to the doses that one could realistically expect to find in the environment. The so-called dose-response curve was the line linking the resulting toxicity points. The guideline mentioned just two available mathematical models to perform this extrapolation: the linear model and the probit model.[8] The linear model "assumes that there is a finite, though diminishing, risk at all doses above zero" (Harisson and Hoberg 1994, 23). In other words, when applying the mathematical equation constituting the model, one would obtain dose-response points that, linked to one another, would produce a linear curve, showing no level at which the substance would seem to be safe.

Two mathematical models were mentioned in the guideline, leading Rushefsky to call it very flexible (Rushefsky 1985). In practice, however, CAG members generally used the linear model. As Dickson notes, agencies have generally embraced the no-threshold hypothesis, or at least refused to engage in extensively researching and deciding the issue of whether dose-response relationships could exhibit a threshold in some situations. They have also generally concluded that a nongenotoxic mechanism (wherein a threshold may be found) had not been adequately proven (Dickson 1987). The linear approach was soon institutionalized at the EPA.

An OHEA report of 1980, explaining how it would calculate unit risk estimates for air pollutants as part of the criteria-setting exercise that the

Air Office had to perform under the Clean Air Act, put forward what will henceforth be known as the *multistage model*, developed by the biostatistician Kenny Crump under contract with the agency (EPA 1980a). The exposure calculation and dose-response combined would lead to the expression of a cancer risk in terms of excess lifetime incidence. The implicit EPA guidance was that an excess incidence of one cancer in a population of 1 million people justified intervening. Altogether, the guideline expressed a policy, not just a scientific method, or what the document calls a philosophy: "Evidence has accumulated that indicates that the non-threshold concept can also be applicable to chemical carcinogens ... We, thus, have a comparable conceptual basis for the regulation of chemicals as for ionizing radiation where the philosophy has been to eliminate or reduce exposure to the greatest extent possible consistent with the acceptability of the costs involved" (EPA 1976, 2).

There are various ways of explaining why the EPA opted for a *conservative* (the term that engineers were using to denote a model including wide margins of error) policy of this sort, most of which are related to the need to make choices in the presence of uncertainties. It was not known in detail whether substances with a promoting or initiating effect behaved differently; it was not known for sure which individual substance exhibited a threshold and which would not (Harrison and Hoberg 1994, 25). By applying a linear model, the EPA was sure that it was erring on the safe side. But the other influential reason was the experience available in the area of radiation, where so-called low doses had been a major debate for decades, resulting in the adoption of the ALARA (meaning "as low as reasonably achievable") principle concerning exposure to radiation. In the absence of certainty of where safety precisely lies, one should derive a measure of protection from plausible levels of risk. Roy Albert, who had experience in the area of radiation, consistently defended the application of this reasoning to environmental toxicants.[9] In effect, the guideline was not very detailed (indeed, it was quite short). It largely preserved and valued the judgment of the CAG scientists.[10]

With this guideline, it became official EPA policy to consider, due to the need to protect the population, that there was no safe dose for carcinogenic substances, and that exposure to them should be reduced as much as possible. As a design, it assembled the expertise of toxicologists and biostatisticians, with the choices of other administrative and policy officials of the agency, to stay on the safe side. It was computing as many numbers as possible, aiming to be precise and reliable by reviewing wide sets of data. With these guidelines, the science was becoming more sophisticated and scientists began to challenge diagnoses. It was a long way from the days when regulators and judges faced with making decisions on cancer risk simply asked if the compound in question caused cancer in a lab animal. But the EPA was also acknowledging the existence of uncertainties, and at least provisional limits on the science, and constructing uncertainty-compensation rules out of consideration of its public health mission.

The substantial choices laid down in the guideline were protective, but the procedural design offered protection for the industry too, as well as a measure of the EPA's awareness of what was due process and the limits of its authority. The two steps of risk assessment represented only the first half of a decision-making procedure—described in the preamble of the guideline by the EPA administrator—wherein the decision-maker may balance the risk estimate with a consideration of the costs and benefits of the future regulatory intervention. Again, this had been the suggestion of Alvin Alm, and Russell Train had followed suit to give reassurances to the regulated industries. In his preamble, Train states that the decision-making procedure described in the guideline separated the more robust biological and biostatistical aspects from the rest: "I believe it is important to emphasize the two-step nature of the decision-making process with regard to the regulation of a potential carcinogen. Although different EPA statutory authorities have different requirements, in general two decisions must be made with regard to each potential carcinogen. The decision is whether a particular substance constitutes a cancer risk. The second decision is what regulatory action, if any, should be taken to reduce that risk" (EPA 1976, 3). A footnote in the risk assessment part of the guideline pointed out that "[t]his health risk assessment is part of the risk-benefit analysis [which is performed] after a determination that a health risk exists" in the case of pesticides. It was a trace of the intra-agency conflict on cancer principles, and it showed that the guideline had been devised to resolve the issue. To better articulate the whole risk and benefit assessment strategy, the guideline included a second facet for the economic impact analysis of proposed actions on carcinogenic pesticides.

The EPA documents of the mid-1970s did not feature the graphs that later documents of the 1980s would typically include, but if one had been

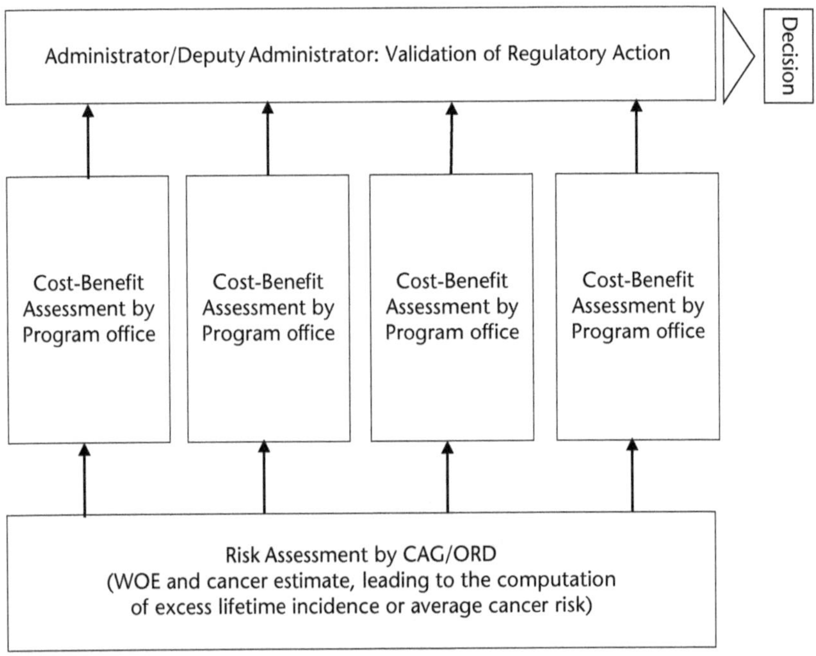

Figure 2.1
A graphic representation of the decision-making system in the EPA (1976) interim guidelines.

produced, it would have looked like figure 2.1. In the bureaucratic assemblage described in the guideline, CAG and the scientific assessment of the risk has a transversal role, feeding regulatory offices with a computation of the risk. Regulatory offices then construct the regulatory decision adapted to the Act that they implement.

The assemblage thus gave the EPA several advantages. It would instill various options in risk assessment and demonstrate its flexibility. At the same time, it also worked as a bureaucratic screen, protecting EPA's internal capacity to overcome uncertainty through expert judgment, as opposed to further research. It also helped to flexibly combine two separate modules for risk assessment and cost-benefit assessment: the cost-benefit part was an integral part of pesticide assessment. It could be left out in other cases. Such was the technology for decisions in a probabilistic organization. Instead of compelling a scientific fact (any chemical producing a tumor is hazardous) to impose a decision (they should be banned), it used an algorithm to produce several informed judgments, in sequence, from which a legitimate decision seemed to flow. The algorithm was a legitimizing design, in that the person who had authority (the administrator) built on the judgments of medical scientists and policy analysts.

Guidelines of Controversy

The cancer assessment guideline was adopted as part of an informal rulemaking procedure involving official notice and consideration of public comments. They were not reviewed by external scientists, however. They were still defended as being based on a broad scientific consensus that the linear, no-threshold hypothesis was the most plausible one, as well as being a reasonable one to apply in the context of health protection.

But this was not sufficient to turn it into a standard practice across the agency. Its application was impeded by interoffice disputes, rooted in differences in the degree to which they embraced conservative assumptions. The Air Office was the first program office to implement the guideline. As mentioned previously, an OHEA report drawn up in 1980 explained how it would calculate unit risk estimates for air pollutants, as part of the criteria-setting exercise that the Air Office had to perform under the Clean Air Act (EPA 1980a). In that document, the OHEA outlined the multistage model. As a result, a substance with either "best evidence" or "substantial evidence" would be listed as a hazardous air pollutant, on the grounds that there was a "high probability" that it was a human carcinogen. Quantitative risk assessment was thus a preliminary indication of significant risk (EPA 1979b, 25). The office proposed "National Emission Standards for Identifying, Assessing, and Regulating Airborne Substances Posing a Risk of Cancer" (Federal Register 1979). The "assessment group" formula underpinning the CAG was replicated in the form of a series of teams, each of which would carry out the corresponding analyses for regulatory offices: the Exposure Assessment Group, Mutagenicity Assessment Group, and Teratogenicity Assessment Group, which engaged in the experimental development of more guidelines. By 1980, three guidelines had been developed and were in use (EPA 1976; Federal Register 1979, 1980b, with a fourth in the making [EPA 1983a]). They routinized the assessments performed by scientists in regulatory offices or in ORD, though without harmonizing them strictly.

The practical work remained quite literally case by case, with the guidelines providing directions for treating more ambiguous cases.

The Air Office was the main client of the OHEA, but the relationship with other program offices was less cooperative. Certain offices were doing their assessments independently, without drawing from the science available in the ORD or using its guidelines. The Office of Pesticides was particularly autonomous for scientific evaluation purposes: by law, agrochemical companies were required to test their products and provide data and studies to the office. The office, therefore, had plenty of knowledge for reviewing and did not need to have the ORD labs perform such tests. It also had a substantial scientific staff inherited from the USDA, including toxicologists and health scientists, allowing it to analyze the industry data and perform many screening assessments, day after day. It produced more assessments than any other program office of the agency. The nature of the Act, which allowed the EPA to take into consideration the costs and benefits of registering a product as a pesticide, as well as the regulatory culture ingrained in many of the staff (including those that transferred from the USDA), meant that their safety assessments were regularly less protective than those of the more health-oriented people in CAG.

The modus vivendi, for many years, was not to question the fact that OHEA was not involved in the health and environmental assessment of pesticides. The Air Office was in the opposite situation, relying quite strongly on the scientific forces of the ORD.[11] The Water Office, finally, was intermediate. Interoffice problems were particularly acute for more controversial chemicals, particularly cancer-causing chemicals found in air, water, or contaminated sites across the country. Each office was led to make its own assessment of the safety of chemicals, specifically when caught in the crossfire of the many public controversies on the subject. In those years, gross, highly symbolic pollutants were addressed, leading to a short list of "chemicals of the month" regularly making the headlines, and which the agency was forced to address. The inconsistencies were particularly problematic where the EPA was actually setting risk levels for states to act on, as under the water legislation, or the EPA's own regional offices, as in the Superfund program.

The guideline drew up a cross-office process, linking scientists in the ORD to regulatory bureaucrats, lawyers, or policy analysts in program offices. In practice, changing the organization to reflect this natural liaison between various sorts of knowledge and competences was hard to obtain. A reform of the ORD was attempted to address this issue. In the 1970s, the ORD budget was in the region of $200 to $300 million, well within the range of the research budget of departments with large scientific programs, such as the US Department of Energy or US Department of Defense. In those years, the ORD was the largest department of the EPA,[12] but it was not integrated with the rest of the agency. The EPA inherited laboratories from various existing departments; it never actually designed the ORD, its research orientation, or the actual laboratories. The problem of the relevance of the research carried out in the ORD for the work of program offices emerged almost immediately during the early life of the agency.

A report by the NRC (1977) pinpointed the problem. Despite its title ("Decision-making in the EPA"), the report focused to a large extent on the question of the science that the EPA used for its regulatory work, and put its finger on problems that were there to stay: namely, the need to have a system in place to review the data and scientific studies that regulatory offices used, to ensure that all of the science was used; and that leaders of the agency must have direct access to the science.[13] In August of the same year, the SAB of the agency published its own report on "the research, development, monitoring, and technical support system of the US EPA" (SAB 1977). A "dangerous ally" (Jasanoff 1990, 85) of the agency, the board did not spare it from criticism, noting the continuing problems that plagued the relationship that various parts of the agency had with science and research. It highlighted a lack of communication between natural scientists and lawyers and recommended that the former present their data and translate their information into adapted formats—meaning that the technical information should be organized in the form of a cost-effectiveness comparison of various control options. It found a lack of logical connection between the research performed by laboratories and the needs of the regulatory offices; questionable value of some of the research performed in labs; duplication of research among the various laboratories; laboratories functioning like autonomous empires; and a lack of assessment of the uncertainties, costs, and benefits (SAB 1977). In other words, the ORD laboratories did not appear to be functioning as laboratories of a regulatory agency, and their activities seemed to be steered to some extent by the leaders of the various program offices or of the whole agency. A reorganization of the laboratories was eventually decided, leading to the creation of five "megalaboratories." The OHEA was promoted to that status, in which the CAG took on a prominent position.[14] With this decision-making guideline, cancer-causing chemicals were in the process of becoming a generic object for the agency, and CAG was virtually becoming a transversal office, providing science to the various other offices of the agency. Science-based decision-making was becoming a new, virtual organigram for an agency otherwise made of regulatory siloes.

The guidance developed by the EPA found confirmation in a multiagency document that was soon compiled by an interagency group, the so-called Interagency Regulatory Liaison Group (IRLG). The origins of that group go back to May 1977, when the EPA, FDA, and Consumer Product Safety Commission (CPSC) held a joint press conference on chlorofluorocarbons that helped the leaders of each agency to realize the benefits of such occasional coordination and common action, and also led to the idea of having more regular meetings. The IRLG was formally created in September 1979 as a "systematic, but short-lived, attempt at comprehensively coordinating regulatory policy" (Gore 1993, 19), with eight working groups, including one on risk assessment. It developed common cancer guidelines based on that of the EPA and the comparable approach developed by OSHA.[15] The risk assessment work group, which included Elizabeth Anderson of the CAG, decided to focus on cancer and produce a state-of-the-art methodology.

The first draft of the document threatened the generic cancer policy of OSHA, which had not embraced quantitative risk assessment as the EPA had done (Landy et al. 1994). Saffioti, who was instrumental in the development of cancer policy in that agency, regarded quantitative risk assessment as spuriously precise and certain. He only agreed to the inclusion of a short discussion of quantitative risk assessment in the document, highlighting uncertainties much more than the former draft had done. Roy Albert, who had joined the group for the revision of the first draft, did not facilitate the rewording. Although they were proponents of quantitative risk assessment, he (and Elizabeth Anderson) considered the EPA and CAG to be ahead of any other agency and were wary of the difficulties that the interagency policy could create for the CAG's work.

The resulting document was entitled "Scientific Bases for Identification of Potential Carcinogens and Estimation of Risks" (IRLG 1979). It reflected both the composition of the group, which read "as a list of prominent cancer researchers and advocates of risk-averse cancer policy" (Rushefsky 1986, 94), and the tensions spawned by each one's defense of the agency's guidance. The former led to a broadly conservative, cautious guideline, which erred in all choices on the side of caution in order to avoid false negatives; the latter resulted in a lack of discussion about the range of available options (e.g., on the weight to grant to different kinds of negative or positive studies, or the choice of sample size and statistical confidence levels in animal tests). With such lack of detail and flexibility in the options, the guideline hardly reflected the uncertainty in the exercise of carcinogenic identification, and thus the policy nature of many of the determinations involved. The reason for this was clear: the group had explicitly been instructed not to address policy. So the inherent policy dimension of the technical choices laid out in the text could not be assumed explicitly. It led the group to forge an ad hoc distinction between science and policy. The draft work plan of the IRLG, back in 1977, had already stipulated, "Whether a particular type of risk may have to be accepted in certain circumstances is a policy decision, but the issue should not become entangled with the scientific problem of risk measurement" (IRLG 1977). At the end of the day, the guideline was seen as disguising policy choices in scientific terms (Landy et al. 1994).

The White House Regulatory Council still endorsed the report. It argued that agencies should deal with cancer risks quantitatively and use the IRLG report as a guideline. But the guideline, like OSHA's generic principles or EPA's interim guideline of 1976, proved to be "exceptionally controversial" (Rushefsky 1986, 89). At the very least, the chemical and oil industries had decided to engage in a forceful debate about conservative policies. A number of large chemical and petrochemical businesses banded together in 1977 to form the American Industrial Health Council (AIHC), a vehicle used to counter the development of guidelines by the social regulation agencies. The main threat, for the industry, was that guidelines would in effect distort the regulatory frameworks, taking a precautionary stance toward all chemicals, beyond the letter of the Acts. The 1977 Clean Air Act, for instance, addressed a few extraordinary air pollutants. As guidelines represented implicit extensions of regulatory policy to potentially unlimited numbers of chemicals, they had to be stopped.

The AIHC pointed to the confusion in the IRLG report between methodological and regulatory considerations to debunk its conservative preference. It accused the working group of not admitting its policy biases

1. Identification of the hazard
   Epidemiology
   Toxicology
   In vivo tests
2. Characterization of risk
   Potency
   Exposures
   Susceptibility
3. Reduction of risk
   Information
   Regulation
   Substitution

Figure 2.2
OSTP's 1980 linear decision-making process for carcinogens (adapted from Faustman and Omenn 2010, 1083).

and of hiding them in arcane technical language (Landy et al. 1994). The industry pushed instead for the application of guidelines developed by another group under the aegis of the White House Office of Science and Technology Policy (OSTP). The group worked under the leadership of Gil Omenn, a medical doctor and scientist, professor of medicine at the University of Washington in 1979, and deputy science advisor to President Jimmy Carter between 1979 and 1981.[16] Omenn had been associated with the work of the IRLG's risk assessment group, but he was seen as defending a position that was contrary to OSHA's approach. The OSHA official in the IRLG spread the word that Omenn was the industry's mouthpiece within the group and that he should be dissociated from its work. He stopped being formally invited to the meetings and thus practically left the group. He was thus free to work separately on OSTP's own guidance document.

The OSTP document, published in the medical literature as Calkins et al. (1980), differed markedly from that of IRLG. First, it addressed regulatory questions, whereas the IRLG had decided from the beginning to leave policy out of its scope. The report presented cancer assessment as a component of a two-stage approach: first research, and then assessment by regulatory agencies of the significance of the risk and benefits behind the findings. In operational terms, the framework contained three steps: *identification* of carcinogenic substances and *characterization* of their risks by scientists, followed by the act of *reduction* (see figure 2.2). While similar in appearance to the two-step approaches of the EPA and the IRLG, this guideline introduced much greater flexibility in its design, with the request for information (presumably from the manufacturer of the substance) being a possible regulatory outcome.

The OSTP guideline was also much more positivistic in the identification of carcinogens. It did not recommend any juggling with positive and negative studies, as in the weight of evidence judgment. It argued that classifying a substance as carcinogenic should happen only if preponderant scientific evidence indicated as much. Furthermore, the text did not recommend any model for extrapolation. More crucially, the OSTP recommended that all authority for cancer risk assessment be consolidated under the national toxicology program—including the research performed by the EPA and OSHA[17]—much as the industry advocated. This guideline, in essence, disassembled the EPA: in a typical scientistic, predictive fashion, it saw no uncertainty, and no need to create a particular organizational space, within which judgments could be formed. Graphically speaking, the linear, step-by-step process exhibits no boundaries. It does not allow delimiting, crafting a legitimate space of intervention of the agency.

Industry Campaign

The chemical and petrochemical industries waged a concerted battle against OSHA and EPA guidelines, coordinated by the AIHC and supported by other groups such as the Chemical Manufacturers Association. Every EPA publication was met by counterpublications. Also, all the public consultation meetings organized by the EPA for air or water pollutants were crowded with industry representatives forcefully denigrating, in particular, the generic and default conservative approach embedded in the agency's 1976 guideline. In 1980, for instance, the EPA organized a public hearing on the proposed standards for air pollutants, established in the framework of the Clean Air Act and based on the work of the CAG. The AIHC sent several people to speak at the hearing, and dozens of representatives of its member-companies filled the auditorium. Industry people took issue, first, with the WOE approach, arguing that the EPA was in essence classifying chemical substances as carcinogenic or not based on a single study. Wayne Jaeschke, head toxicologist at Stauffer Chemicals, repeatedly referred in his talk to the supposed "one mouse criteria" that the EPA was applying in its assessments: "The one mouse criteria that EPA has set up is totally unsound, that this is a matter where there are no rationalized clear-cut principles. That's been made abundantly clear by the litany of testimony here on metabolism and other factors, and it certainly seems to me that this is a matter of scientific judgment ... You must have the best scientific judgment" (EPA 1980c, 22). The CAG's Anderson countered this, showing that the decisionistic system designed in the guideline was not applied without judgment or lack of adaptation to the circumstances: "Any information we have about mechanisms of action, short-term test data, all of the information is put together to see what sense we can make of the entire picture, so there is no search in a haystack for one study that happens to show a positive signal and then just action going straight ahead on that basis" (ibid., 24).

The AIHC had also sent Richard Wilson, a Harvard University physicist, to the hearing. As an early proponent of cost-benefit analysis, Wilson discarded the no-threshold hypothesis, along with the lack of flexibility in the guideline: "I don't say that you should automatically let the risk assessor choose what he wants, but you should not rule him out by some legal process saying, you must take the linear hypothesis. Take the linear hypothesis, if you can accept something, fine, but don't leave out the possibility of someone coming back in this particular case ... it should be allowed, and I think there would be such cases." Anderson, at this point, countered that the current IRLG guidelines and EPA's general approach recognized that "where such data can be generated, it certainly would be used" (ibid., 34–35). Wilson also advanced his own estimates of cancer risks at the population level in the United States for the purpose of deconstructing that of the EPA. There were, in his view, no more than ten cases of cancer per year that one could associate to industrial chemicals in the air, such that EPA's conservative policy was not warranted.

These acrimonious exchanges showed the interest of the guideline as a bureaucratic technology. It provided a reference point for audiences that wished to engage with the agency and criticize its expertise and decisions. But they also left a lot of flexibility inside, for agencies to shape decisions case by case. It worked as a screen, giving something legitimate to see outside ("The EPA uses this or that method of extrapolation"; "Health scientists assess, managers decide"), but concealing some of the details of what was happening inside for each substance. This, of course, created frustration on the part of these actors who wanted to control what the agency was doing and reduce its autonomy, leaving them to complain that the agency was not doing what it was pledging to do or that it was applying a cookbook approach,[18] blindly following linear extrapolation methods, and so on.

The regulatory agencies' efforts to establish guidelines had diverse fates. Despite great initial enthusiasm around the IRLG guidelines (Clark 1979), the agencies failed to formally adopt them (see Landy et al. 1994). From this point onward, the agencies parted company.[19] The White House Regulatory Council continued working on a possible cross-agency platform for risk assessment, resulting in the later publication of its principles for chemical carcinogens (Federal Register 1979). Still, the EPA effort earned it some legal security for its regulatory actions. The key court case in that period was the Supreme Court's 1980 ruling on benzene, following the OSHA's adoption of a workplace exposure standard for the chemical of one part per million. According to the majority opinion of Justice John Paul Stevens, no exact quantification of the risk was expected, but the agency "does have an obligation to find that a significant risk is present before it can characterize a place of employment as 'unsafe'" (448 U.S. 607, 656).[20] Throughout its judgment, the Court took the establishment of a quantitative dose-response curve as the main way to demonstrate this significance. The judgment promoted quantitative risk assessment as a means of executing protective regulatory decisions.[21]

With the benzene decision, the Supreme Court effectively required agencies to use a quantitative or numerical measure of the risk to decide on whether to regulate a substance. A system of classification based on qualitative expert judgments was no longer sufficient. These measures of the risk were needed for agencies to distinguish between significant or insignificant (de minimis) risk. In 1979, the court ruling in the *Monsanto v. Kennedy* case, involving the FDA, essentially authorized the agency not to ban a carcinogenic substance—despite the Delaney amendment that ordered it to do so—if the insignificance of the risk were demonstrated in a quantitative risk assessment. The Supreme Court, in this case, decisively promoted quantitative risk assessment as a tool for finding criteria—if not thresholds in a dose-response curve, or at least qualitatively appreciated levels of significance—and constructing decisions.

The CPSC was next in line. In a 1983 decision, the Fifth Circuit Court of Appeals ruled against the agency's decision to ban a kind of foam insulation using formaldehyde.[22] The court established several important rules applying to risk assessment: Agencies should not base their risk assessments on a single study; risk assessments should yield a most probable estimation, rather than a range of values, using the upper value to define its standard; agencies should not use a single model (like the multistage model of the EPA) for their extrapolations; and finally, they should use real exposure data rather than worst-case assumptions.

On the whole, the EPA's decisions were spared from criticism in court. In an essentially supportive Supreme Court ruling,[23] the agency was granted the right, in areas where risk assessments are needed, to make policy judgments when findings of fact could not be made with absolute certainty. The ruling did not "speak to the content of the policies that may fill the factual void" (McGarity 1979, 781), but it still supported the agency in the belief that the guidelines were a legally acceptable benchmark of analysis and decision for the courts, and that therefore, it could extend their use across regulatory programs. It provided a reference for the courts to examine, and in a context where the courts only reviewed the procedures applied by agencies without scrutinizing their scientific reasoning and interpretations, they found nothing to argue against.

By providing a clear benchmark of scientific analysis to the courts, the EPA facilitated the institutionalization of what has come to be known as the hard-look doctrine: a doctrine in which courts are supposed to take a hard look at whether and how agencies applied the intentions laid out by Congress in the statutes they administer, and considered all aspects of the problem and the contentions of all parties (Glicksman and Schroeder 1991; Foster and Huber 1999; McGarity 2003; Freeman and Vermeule 2007). Counterintuitively, the hard-look doctrine also recognized the fundamental lack of the courts' expertise to substantively review the science used by agencies in their decisions, as well as their lack of legitimacy to prescribe agencies' criteria for their decisions, in cases where Congress's intentions were unclear. The hard-look doctrine justified applying the "arbitrary and capricious test," originating in the Administrative Procedure Act of 1946, whereby agency regulations were generally upheld so long as the agency demonstrated that it had applied a minimal standard of rationality. Risk assessment guidelines provided just such a standard of rationality, and wherever the EPA could convince the courts that it had formally applied its own guidelines, it stood a good chance of not losing its case in court—a good reason for not altering its frameworks.[24]

Conclusion

Risk assessment, in the form of a guideline, emerged as way of making decisions in the EPA in the 1970s in order to defend the goals and legitimacy of the agency in difficult contexts of uncertainty, close supervision, and intense industry opposition. In the words of Terry Yosie, staff director of the EPA's SAB in the 1980s, the guideline "represented one of Washington's most venerable principles: be sure that your potential adversaries (be they industrial firms, environmentalists, Congress, or the OMB) debate your ideas, and your agenda" (Yosie 1989, 3).

The risk assessment methodology, standardized in the guideline, even if not properly and systematically applied in all corners of the agency on all chemical substances assessed, provided a bureaucratic screen, both representing to the outside what the agency was holding as its ideal form of assessment—decisions adjusted to degrees of risk and certainty emerging from consideration of data and calculations, but also guarantees of "objectivity," "consistency" or "feasibility" (Stephan et al. 1983)—demonstrating its commitment to approximation of this ideal, *and* concealing, in the meantime, the individual decisions that did not follow the model. The guideline deflected accusations of poor science and biased regulatory decision-making. It helped tone down the level of judicial scrutiny. It became a focal point in the ongoing discussions with the industry, helping the agency set these debates and negotiations on its own terms. In all these respects, it was an effective bureaucratic tool that acted to durably reflect within the agency. It did not solve all of the EPA's problems though, particularly not the continual lack of consensus among the agency's diverse offices. Other designs were needed for this problem of integration of the organization and of its knowledge that economists and policy analysts took care of articulating.

3 Prioritizing Toxics: The Prehistory of Risk Management

The interest in risk as a legitimate design for administering environmental problems in the EPA did not materialize only among medical scientists, toxicologists, and statisticians with regard to the application of probabilistic analysis to chemically induced biological mechanisms. At about the same time as the health scientists of the ORD were articulating a standard risk assessment method and pushing it at the agency, the EPA administrator and some of the agency's policy analysts were grappling with another kind of controversy: incoherence caused by offices delivering divergent standards on the same chemical and risk. This incoherence stemmed from the differences among offices operating on the same generic regulatory object—chemicals—but with contrasting regulatory regimes and different modes of relationship with regulated industries. It gave substance to another form of uncertainty, as damaging for the EPA: that concerning the diversity of valuations of the risk, and the diverging preferences for tackling one or another hazard first. The diffusion of chemicals across environmental media, and thus across the boundaries of separate regulatory offices and their regimes, revealed this uncertainty, and created occasions for conflict, as the EPA was not designed to deal with chemicals in a coordinated manner. There was a major risk of controversy in this situation: divergent estimations of risk within the agency fueled disputes outside it about the reality and acceptability of chemical hazards and, most important, disputes about the appropriateness of the design of the agency and the legitimacy of its action toward chemical hazards—all of this occurring at a time when the debate on the excess of environmental regulation and the virtue of having put these new risk agencies in place became urgent.

This chapter explores the formal bureaucratic assemblage that emerged in this context. The agency's policy analysts, with support from the EPA administrator, articulated a new, cross-agency discipline of risk-ranking. It did not reorder the agency in the way that quantitative risk assessment was in the process of doing. In particular, it did not create any strong base of knowledge across offices or coordination among them. But five years ahead of the installation of a new discipline of *risk management*, which risk-ranking preceded, the experiment helped establish the value of economic knowledge and of processes of centralizing information for the agency's policies, and showed the advantage of keeping economists and policy analysts close to the leaders of the agency in order to steer the agency and control the image of what it delivered overall.

EPA's Inborn Consistency Problems

The internal coherence of the EPA has been a problem ever since its inception in December 1970—and in fact, even before that. As the previous chapter recounts, after protracted negotiations, the Ash Council settled for creating an agency that assembled all regulatory programs aimed at combating environmental pollution under a single administrative umbrella. Whether a department or an agency, one question remained: Would this new governmental organization have new missions and powers, or would it simply gather those of existing services?

One of the organizational plans that the group contemplated was more disruptive than the others. It consisted in creating, from scratch, a functional agency in which the personnel of the units inherited from the various departments would be redistributed in newly formed, nonprogrammatic offices for dealing with abatement, monitoring, research, and standard-setting, separately. That plan had been devised by Alain Enthoven, a systems analyst and former adviser to the US Department of Defense, where he had been busy introducing systems analysis methods under the leadership of Robert McNamara. Enthoven introduced his functional scheme just two months before the EPA started its operations. Terry Davies pushed for an even more integrated design, with an agency structured around key functions, including disseminating information and performing analysis. Douglas Costle, who led the environmental committee that designed the new administration, was no enemy of this functional plan. In contrast to Enthoven, however, he feared, as an experienced bureaucrat, that too strong a disruption of existing bureaucratic structures would result in chaos and opposition. He thus settled for an intermediary, incremental strategy: launching the agency on the basis of a program-office structure "with substantial continuity with its programmatic past" (Marcus 1980, 104), but subsequent reinforcement of integrative services.

William Ruckelshaus, upon being appointed as the first administrator of the agency, was sold on Costle's plan to make the agency evolve toward cross-program integration. He understood that the agency would be much more governable and project a united image to the public if offices were not separate, undirected fiefdoms. As Daniel Carpenter's work shows, agencies are seldom free of legitimacy, identity, and reputational challenges, notably at their creation (Carpenter 2001, 2010). Being federal bodies, they are necessarily assailed by positive and negative discourse that constructs their image and legitimacy. Image is, moreover, an instrument of management of these organizations. Administering them, therefore, implies the construction of a single image of their action and outcomes. After five months in office, following Costle's advice to proceed in various stages, Ruckelshaus outlined a new organizational chart, adding three functional offices to the existing program-oriented ones: Planning and Management, Enforcement and General Counsel, and Research and Monitoring (see figure 3.1). As the

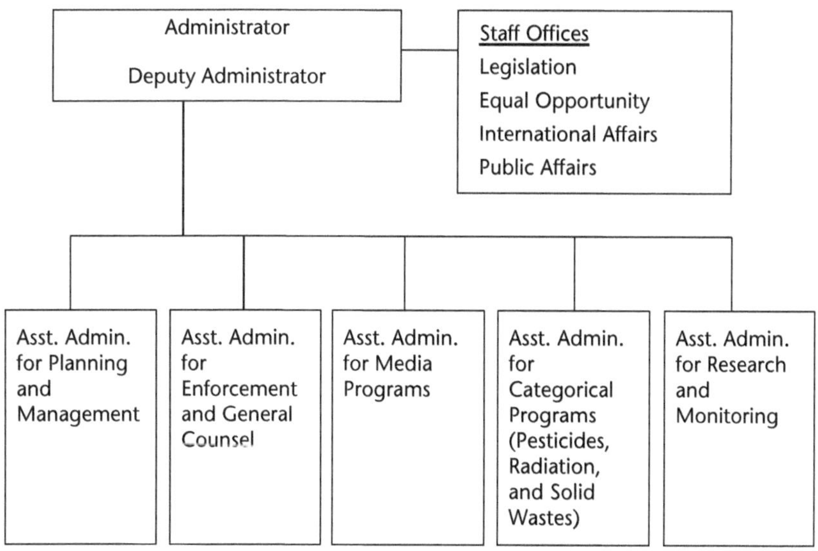

Figure 3.1
Ruckelshaus's second organizational plan (adapted from Marcus 1980, 105).

previous chapter illustrated, the first two were instrumental in constructing the EPA's authority, insofar as they were responsible for conceiving of two essential instruments: the cancer principles, first, and the risk assessment guidelines, second.

But Ruckelshaus did not fully apply stage 2 because he had planned to group all media-related programs under one umbrella office, Media Programs. He ended up, instead, with separate offices for Air and Water. He never implemented stage 3 of Costle's strategy—scrapping program offices to institutionalize a purely functional organization. According to Marcus (1980), combating the offices to create a whole new structure for the EPA was not of high enough importance to Ruckelshaus in the context of his chosen, litigation-oriented strategy for the agency. This strategy obviously corresponded to Ruckelshaus's inclinations and ways of working. As a lawyer, he could easily launch prosecution actions with top legal aides, which would make an impression on the public, industries, and states. It partly dispensed Ruckelshaus from going any further in implementing a reorganization plan for the agency. The organization of the agency along programmatic lines was thus never disrupted. By 1981, the agency's organizational chart mixed functional and programmatic offices, on top of regional ones (see figure 3.2).

During his first mandate, Ruckelshaus nevertheless experienced the negative effects of this organizational scheme for the deployment of his policies. He had felt frustrated early on due to not being able to count on dedicated scientific advisory services for reviewing the standards prepared inside the Air Office. He had only three days to review the proposed standards, with hardly any support. Marcus shows that this experience, coupled with pressure from the White House to factor economic implications into the EPA's decisions, forced Ruckelshaus to think about ways of controlling decision-making by program offices. The result was the "1000.6 approach" (named after the 1971 EPA order that describes it), which instituted a central decision-making steering committee designed to turn the standards prepared by program offices into options for consideration by higher-level decision-makers. With that approach, the program offices' work was only an initial sequence in a broader decision-making process involving ex post reviews of decisions drafted by ORD scientists and lawyers in the Office of Enforcement. In this way, the EPA administrator and associated decision-makers were afforded the possibility of choosing among options based on measured costs and benefits, as well as political and judicial opportunities.

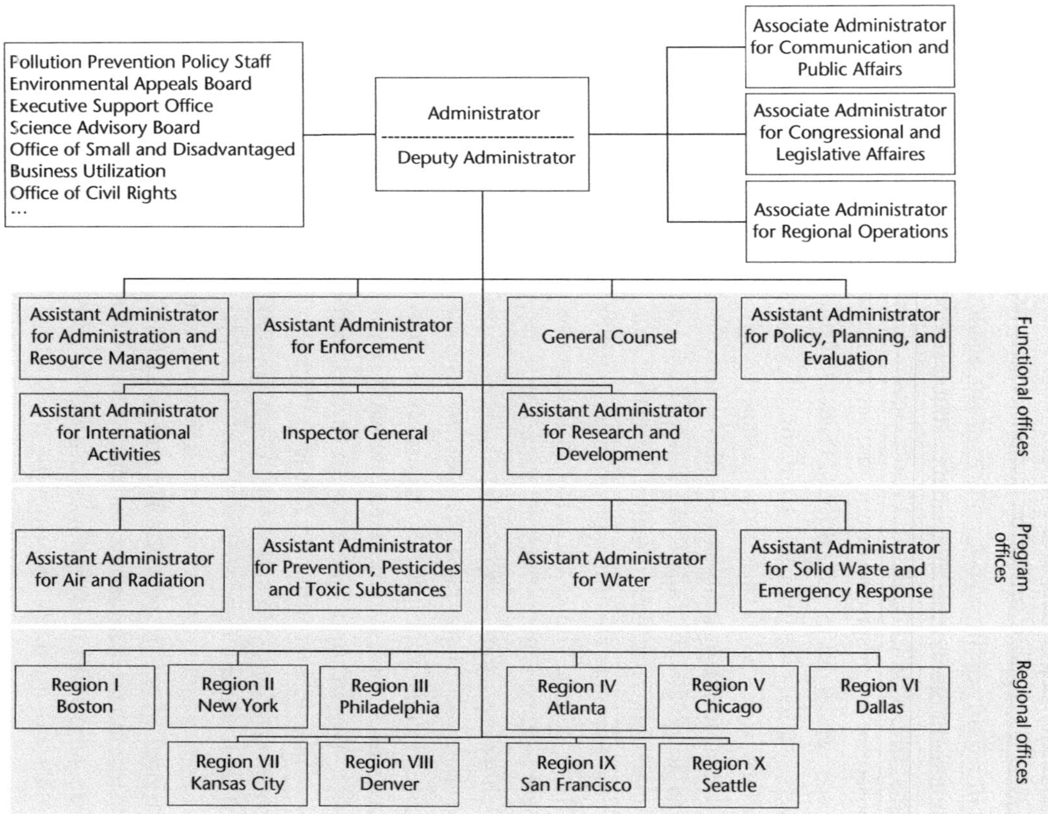

Figure 3.2
US EPA Organization chart circa 1981.

Unfortunately, the picture soon became even more complex, with the enactment of major laws throughout the 1970s containing distinct provisions, standards of safety, and regulatory instruments to address a similar regulatory object, chemicals. Chemicals were in the process of becoming the boundary object (Star and Griesemer 1989) of the EPA as a molecular bureaucracy (Hepler-Smith 2019): an object with a distinct identity, but dealt with differently by the various offices of the agency. Initially, the EPA had to implement preexisting laws, such as FIFRA, passed in 1947 and amended in 1972. Under that Act, the EPA had to review existing and new pesticides proposed for use in agriculture and manage their use in the fields. This mission was carried out by the Office of Pesticides, transferred from the USDA (while pesticide labeling came from the US Department of the Interior). Another was the 1970 Clean Air Act—specifically, the NAAQS program—that sets atmospheric goals for a number of specific pollutants (the so-called criteria air pollutants: carbon monoxide, lead, lead compounds, nitrogen dioxide, ozone, particulate matter 10, particulate matter 2.5, and sulfur dioxide), as well as creating a program to control emissions to reach these quality standards. The EPA was to define threshold levels for the presence of these pollutants in ambient air. Yet another program was the Hazardous Air Pollutant program, in which standards were set for the emissions of 187 listed chemicals. These standards prescribed the use of a control technology that was deemed appropriate to reduce these emissions. Until the EPA was formed, the Act was administered by the US Department of Health, Education, and Welfare, from which the Air Office inherited its staff.

The amendments of the 1972 Federal Water Pollution Control Act (better known as the CWA) aimed for the elimination of pollutant discharges from point-sources into surface waters. They also charged the EPA with drawing up a list of national standards for toxic water pollutants. The main instrument in this respect was effluent guidelines, encompassing a toxic effluent standard (a limitation for these effluents, including an ample margin of safety) and a prescription to use a given technology that could help to reduce the discharge to below the threshold. The amendments to the CWA that were passed in 1977 confirmed this approach of setting criteria for a list of 129 priority pollutants.[1] The exercise was called criteria formulation. Based on these criteria, the EPA would choose the best available technology to bring down these pollution levels. The criteria were set by EPA and the individual states, which could choose a different criterion for each specific body of water. In practice, the EPA gave some discretion to the states, which at times adopted a more precautionary approach. The staff administering the CWA program came from the Department of the Interior and the Department of Health, Education, and Welfare.

However, with the creation of the EPA came multiple other new environmental policies and statutes. In 1976, the US Congress adopted the TSCA. As a result, the EPA had to look at many more chemicals that were about to be commercialized or already were so, independent of the dose and of the effects they produced in particular media. Part of the motivation for passing this Act was to include chemicals that were not already covered by separate legislation for air, water, waste, or pesticides. Sections 6 and 9 of the Act required the EPA to regulate chemical substances under the authority provided by other Acts, where possible. But it also established two huge new programs for reviewing the risks of existing chemicals and screening and licensing new chemicals (Boullier et al. 2019).

In 1977, Congress also reformed the Clean Air Act. The amendments soon set further ambitious precautionary goals for the agency (Graham 1985). First, they established a broad definition of "hazardous air pollutant" as "an air pollutant to which no ambient air quality standard is applicable and which in the judgment of the Administrator causes, or contributes to, air pollution which may reasonably be anticipated to result in an increase in mortality or an increase in serious irreversible, or incapacitating reversible, illness."[2] Second, it provided that the EPA administrator decide within one year (or two years for radioactive emissions) whether radioactive pollutants, arsenic, cadmium, and polycyclic organic matter were hazardous within the meaning of section 112.

In the same year, the RCRA was passed in order to expand the program for managing and regulating municipal and industrial waste, including toxic or hazardous waste. The program involved defining systems for managing the production and reducing the volume of waste, reducing the use of landfills, managing a permit program for producers and transporters of waste, and for managers of landfills, and so on. Finally, with the 1979 creation of CERCLA, also called the Superfund program, the EPA became concerned with the presence and effects of chemicals at local, contaminated sites. The Act provided for very wide-scale activity, covering an estimated 5,000 abandoned dumping grounds across the country that were deemed to be potential public health hazards. The Superfund program

Table 3.1
Main statutes administered by the EPA, listed by office

Office	Act/Regulatory Program
Office for Air and Radiation	Criteria air pollutants (NAAQS) Hazardous air pollutants (NESHAPS)
Office of Pesticides and Toxic Substances	Pesticides (FIFRA) Toxic substances (TSCA)
Office for Solid Waste and Emergency Response	Contaminated sites (Superfund/CERCLA) Hazardous waste (RCRA)
Office of Water	Drinking water (SDWA) Surface water (CWA)

involved emergency action in case of oil spills or other similar pollution, long-term remedial and containment action, and enforcement against parties responsible for hazardous waste and spills. It generated much more activity at the EPA headquarters and in EPA regions than any other program, and it obtained extensive political visibility through the Love Canal and Times Beach scandals.

The various Acts[3] that the various offices of the EPA had to implement by the end of the 1970s (see table 3.1) had little in common. They dealt with significantly different environmental problems, some of which were closer to conventional, well known, and hardly hazardous forms of pollution of natural resources, while others involved the management of the uncertain effects of toxic substances. Moreover, even where these substances were targeted for administrative action, the measures taken by the offices in the various programs would still differ. The programs were developed by coalitions of senators and congresspeople, under pressure from different interest groups and reacting to specific controversies and scandals, with little continuity among them. The result was very different standards of evidence and safety criteria, as well as different levels of authority and different levels of resources, for the EPA. The Clean Air Act required that the public be protected by establishing an adequate margin of safety, without regard for the costs or benefits of the use of the chemicals in question. The agency was not able to define a level of risk that would match measured benefits; rather, it simply had to aim for absolute safety. The CWA was technology-based: The agency had to prove that a given technology would reduce pollution to a level deemed acceptable. The laws for pesticides and toxic substances were based on yet another standard: They were intended to secure safety in a relative sense, avoiding those risks that would be considered unreasonable in relation to a variety of costs and benefits associated with the regulated chemical. Such was the case for FIFRA and TSCA, which required the EPA to choose the least burdensome options of control.

Taking on the Chemical Revolution

The multiplication of these Acts that target chemicals in various forms and media was both an opportunity and a problem for the EPA. It was a political opportunity because it gave shape to a unified political agenda to which the agency could lay claim, and on which it could build its identity and legitimacy. This is the route that Costle chose to go down when he was appointed as EPA administrator, succeeding Russell Train in January 1976. Costle had worked in the OMB in the White House, and then was a member of the environmental committee inside the Ash Council. As mentioned previously, he had been instrumental in choosing to create the agency in the first place, and thus in departing from the original plan to create a cabinet-level Department for Natural Resources and the Environment. He was a proponent of a gradual integration of the agency for effective and comprehensive environmental management.[4] In the context of a multiplication of agency-forcing statutes that compelled the EPA to work on toxics from multiple perspectives, Costle initiated a strategy of greater integration, which had two aspects: working on a political agenda and working on integrative mechanisms inside the agency.

Costle took a stand on the terrain of cancer risk and health, defending a new identity for the EPA as a health-oriented agency. He developed the theme of the *chemical revolution,* insisting on the potential links between the diffusion of toxics in the natural and biological environments and the prevalence of cancer. This meant that the object of the EPA's overall action had to evolve from manifest pollution and identified chemicals to the more invisible and widespread health threat of toxics. Costle hammered down this theme in successive speeches in 1978. In an oral history interview, he retrospectively emphasized that the EPA was as much a health-protection agency as an environmental pollution agency: "In some ways, it was an intellectual coming-of-age for the Agency to find itself suddenly dealing with a different universe of problems. It was never intended that we would drift away from the original environmental quality-of-life issue, but we did become preoccupied with health concerns" (EPA 2013).

Costle embraced toxic chemicals as the EPA's main issue for several reasons. One was that he felt that cancer and health were the right focus to keep the White House's support and to preserve the agency's budget, as well as its perimeter in the face of a new proposal to merge the EPA into the Department of the Interior (Landy et al. 1994). Train had a relatively high level of independence from the White House and cultivated ties with Congress, which gradually earned him substantial support from the key committees and subcommittees.[5] Costle was much closer to the president, Jimmy Carter, and to his agenda. Carter had made the war on cancer an explicit priority for his mandate. Eight months after the enactment of the TSCA in October 1976, Carter delivered an environmental message to Congress that embodied this new understanding of the problem of cancer and inaugurated a different policy based on comprehensive action on the toxic environment through the coordination of the FDA, OSHA, the CPSC, the EPA, and the USDA. By following up on cancer and health, Costle was showing that his agency would be instrumental to the president's priorities.

The other reason for embracing the chemical revolution and health agenda was that they constituted a plausible common agenda across the agency. Addressing toxics was what the multiple new Acts passed throughout the 1970s had in common. When speaking about the chemical revolution, Costle tried to instill the sense that these Acts were about the same thing: managing our chemicalized environment and its burden on public health:

> Substances show up, not just in one place, but across many media—air, water, land—often simultaneously. The public health concern then becomes the total body burden: what exposure are you getting? ... So what is the nature of our job? It is to use common sense to reduce risk by reducing exposure, to take a harder look at new substances before we introduce them into commerce. But let's not kid ourselves that we are smart enough to know how to draw the bright line, or that there is a single scientifically sound way to do that.
>
> —Costle (cited in EPA 2013)

Toxics, compared to what a later report called the "grossest and most familiar forms of air and water pollution" (EPA 1981c, 1–5), or the pollutants already targeted in such Acts as the Clean Air Act or CWA (conventional

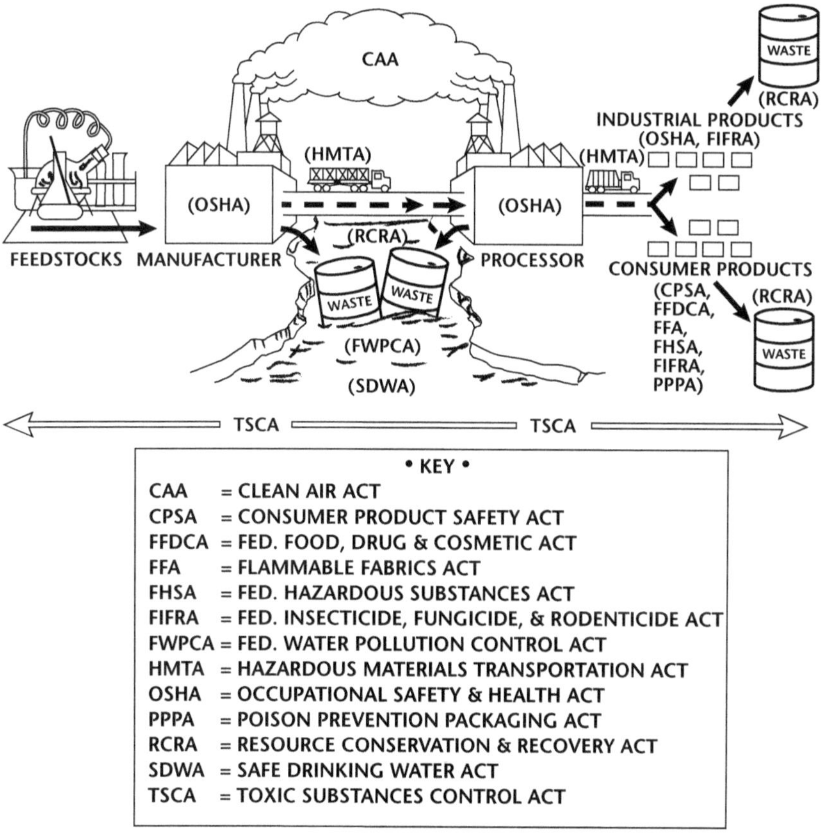

Figure 3.3
"Legislative authorities affecting the life cycle of a chemical" (adapted from EPA 1980b).

pollutants), could not be efficiently managed in only one environmental media. The same toxic could be regulated several times under several statutes because it was present in several media. The possible integration was nicely represented in the *EPA Journal* in 1979, with the representation of the life cycle of a chemical (see figure 3.3).

Balkanization and Credibility

In those days, then, chemicals became a generic risk object of the action and attention of the agency. The multiple statutes adopted over time to deal with chemicals of various kinds and with all environmental media centrally "emplaced" this object in the network of preoccupations that the agency had to deal with (Hilgartner 1992, 48; see also Hutter and Power 2005; Lezaun 2006). Consequently, its reputation and legitimacy became dependent on its action toward this object.

Unfortunately, integrated action on toxics was orthogonal to the balkanized structure of the agency. In 1970, each office within the EPA was different, as all of them originated in departments with contradistinctive agendas and policy cultures. As new regulatory programs were set up, more offices were created, still with different cultures. Each office was headed by a political appointee and had its own weight and importance in the agency. The EPA was probably the most decentralized of all the agencies in the United States—no more than a set of "policy towers" placed side by side (Norton 2005, 20), with large professional cleavages (Landy et al. 1994) and administrative turf wars. Being part of the same agency meant competing for an adequate share of a budget, the administrator's attention, or getting the chance to blame a neighboring office to escape responsibility for a failing standard. The siloed structure of the agency and the weakness of its transversal, functional offices created uncertainty for the management of the EPA's credibility and overall reputation. The autonomy of these offices meant that decisions would be prepared autonomously and not reviewed at the level of scrutiny that they would receive once published. It was difficult for the person signing the decisions, and those called to Capitol Hill to be grilled in hearings about them, to justify a particular decision and to explain discrepancies and shifts in the agency's position on sometimes one and the same chemical.

The ubiquitous dioxin—in reality, a family of organic chemical compounds presumed to be among the most carcinogenic on Earth—was one such case. The EPA had acted early on in its existence to reduce the domestic use of 2-4-5-T, a pesticide in that family. In 1979, the staff of the Air Office suggested adding dioxin to the list of hazardous air pollutants—an action that the agency had not formally decided to take by 1983.[6] In 1981, however, the CAG had produced what should have become an agencywide risk value for dioxin. It estimated that, with a risk as high as $4.25 \times 10{-}5$ mg/kg/day, dioxin was one of the most potent carcinogens known. This rather alarming estimate went beyond anything the Office of Pesticides had reckoned, and it contributed to fueling the controversies that were unfolding at the same time in the media over the ubiquitous dioxin pollution in the country (Anonymous 1978, 1979).

During the 1979 presidential campaign, under the pressure of the ecologist and independent presidential candidate Barry Commoner, the problem of dioxin in municipal waste came onto the national agenda. With the discovery of high levels of dioxin in many sites, from Love Canal to Times Beach, Missouri (the latter was called the "Missouri crisis" in the press), the topic escalated into a nationwide issue (Shabecoff 1983). The subject was highly embarrassing for the EPA, though, because several of its offices were yet to address the topic.

In 1984, the Water Office of the agency issued a criteria document for tetrachlorodibenzo-p-dioxin (TCDD) as part of the CWA, with a different value from the ones advocated by CAG or the Office of Pesticides. As soon as this assessment was published, it triggered a revision, as the document had failed to factor in exposure to dioxin via media other than water, and therefore was useless for the Superfund office. All of this created much confusion around dioxin and made the EPA as much a part of the problem as a solution to it. A difficult hearing in Congress in October 1983 (US Congress 1984) made the EPA's embarrassment fully visible, as it had been with regard to other chemicals to which program offices had reacted, in urgency, and separately, in recent years, from vinyl chloride to the pesticide kepone.

Integration by Analysis: First Attempt

In October 1976, before Carter was elected and Costle began his mandate, the deputy administrator of EPA, John Quarles, had taken the decision to create an Integrated Toxics Strategy Work Group. The group's mission was to think about strategies to improve the agency's action on the problem of chemicals as a whole, as well as to motivate separate program offices to contribute to this goal, in addition to implementing their own statutes. This was the place for the formulation of the strategic problem of "toxics integration." The fate of the EPA, from the mid-1970s with the passing of the TSCA, appeared to be sealed by this double bind: the need to address toxics or chemical safety as a whole, but with insufficient resources to actually integrate what the various offices did. The paradox was defined, internally, as the "toxics integration" problem, a central and "unusually difficult" one for the agency: "[I]t represents the intersection of two problems, the toxic effects problem and the integration problem, each of which is difficult enough alone," and which "exacerbate one another" (EPA 1981c, 1–2).

The more the EPA set out to address toxics, the more it was confronted with its lack of control over its separate offices, their regulatory priorities, and corresponding ways of computing the risks (Beardsley 1987). The paradox and danger for the EPA as a whole were evident: The more a chemical substance spreads, the larger the number of program offices dealing with the substance would be, and the more the EPA would be evaluated overall for its action on the substance. Toxics accentuated both separate program offices' action and the agency's overall credibility.

During a four-month deliberation, the work group on integration identified one way forward: a ranking scheme. The diagnosis was that the EPA had been too much of an administrative victim of Congress's inchoate kind of priority-setting. Each law passed by congresspersons and senators targeted different sets of chemicals, forcing the agency to regulate them separately instead of earmarking global resources and leaving the agency to manage its priorities globally. It was time for the EPA to take control over its own priorities. The working group of the OPRM, with direct support from EPA Administrator Douglas Costle and from his deputy administrator, Barbara Blum, concluded that it should experiment with a ranking scheme (EPA 1981c). This ranking would proceed in two phases. First, an extensive threat list would be established based on unified assessments of the health effects, exposure levels, and ecological damages of many chemicals. Second, an action list would be drawn up in order to guide regulatory action concretely across the agency. In this second phase, elements other than just the science would be taken into consideration: costs, benefits of regulatory intervention, and feasibility also would be the criteria. The notion and design of toxics integration thus embedded the expertise of economists and policy analysts, expertise generally termed regulatory analysis, involving the combined analysis of the benefits of regulatory measures (including the reduction of health risks) and their costs, for a variety of alternatives. At that moment in time, this expertise was being promoted, and an attempt was made to materialize it in the organigram and functioning of the agency, in competition with the biological and biostatistical expertise promoted through the cancer assessment guideline of 1976.

The project was both ambitious and realistic. Computing common assessments of the hazardousness of many substances bordered on the unfeasible. The knowledge about many substances was simply incomplete, and therefore unequal. Establishing a common assessment of a substance with people from across the agency, furthermore, would run into the difficulty of having people agree on the same set of assumptions regarding the linearity of the dose-response relationship, WOE, and the rest. There was, moreover, no evidence that costs, benefits, or feasibility could even be quantified and compared. But the plan was also realistic, and policy analysts adapted their ideals of commensuration to the context as follows: "[I]n view of the uncertainties in defining precisely the factors needed to make such a ranking (effects, exposure, abatement costs, benefits, data credibility, etc.), we recommend a coarse ranking, such as *high, medium* and *low* priorities, rather than a strict ordinal array for both the action and threat lists" (EPA 1977, 23, emphasis in original). Also, the ranking process would be less mechanical (i.e., follow a set scientific methodology) than judgmental and participative.

A major issue was where to locate such an activity in the agency. In a typical decision-science fashion, the economists outlined a number of recommended management options, including having a staff office for toxics integration, with political backing and substantive authority to ensure compliance with its ranking across the agency. The group advised that the new office be chaired by the assistant administrator for toxic substances, or by a college of relevant assistant administrators. Having one or several political appointees as its head, it thus would benefit from greater authority over program offices. The office would assess the hazards of substances independently, drawing from data contributed by individual offices and sending its assessment of priorities back to the offices for them in order to manage their own workload and crises in an informed way, compatible with other offices. It also would create an integrated annual work plan, with repercussions on the activities of all headquarters and regional offices, as well as on the ORD (which would be required, under this plan, to do its own planning only after the Toxics Integration office had drawn up its integrated plan). The EPA assembled according to this plan would look very different from the one designed in the guideline of 1976. It would be organized around the risk-ranking function and integrated thanks to comparative risk information, as in figure 3.4.

Three things happened in practice, none of which was really transformative. First, a Toxic Substance Priority Committee was formed inside the OPRM to monitor ongoing activities at the level of individual chemical substances. In 1979, it published a lengthy status report on EPA chemical

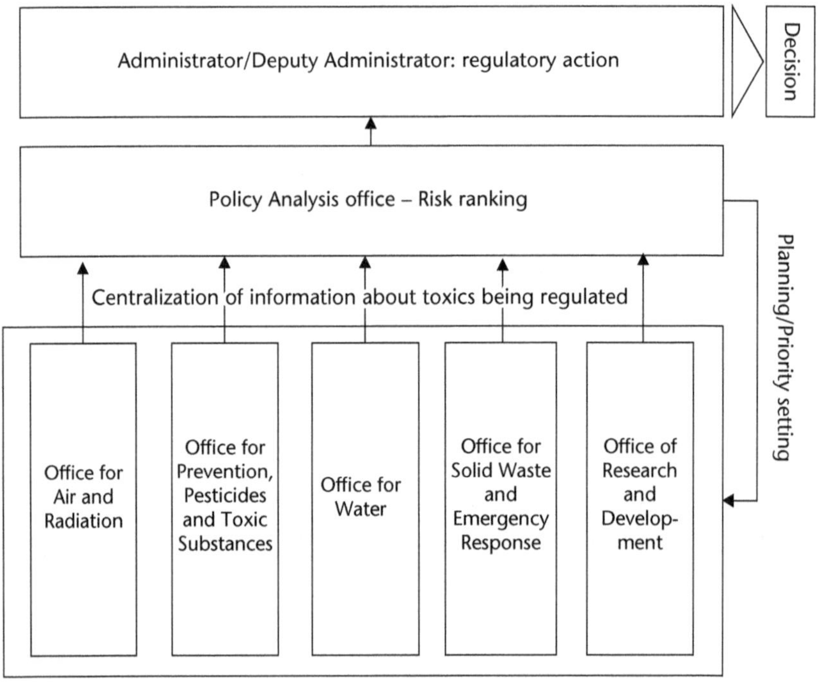

Figure 3.4
Design of an integrated EPA with a central risk-ranking function.

activities, listing the parallel activities in different offices for each substance (EPA 1979a),[7] including substances that mattered greatly to the EPA and to the industry because of their toxicity and the large volumes in which they could be found in the environment. The committee only had the capacity to gather information, and its function was purely advisory. It only barely succeeded in producing coordination among offices assessing the same substance at the same time. A subsequent report on toxics integration in 1981 noted that some chemical substances were assessed for their hazards and risks several times, in parallel, by separate offices. Between 1978 and 1980, for instance, ethylene dichloride was assessed by six offices of the EPA, all working simultaneously on information on its health effects, production and use, sources of release, and exposure. Other chemicals were included in this case, including highly controversial substances and key concerns for the administrators, such as arsenic (Powell 1999). Second, an Office of Toxics Integration was created inside the Office of Pesticides and Toxic

Substances (OPTS), which only covered the activity of the agency under TSCA and concerning pesticide products. It did not really contribute to coordination with other program offices. Third, a more original and shared activity of intermedia analysis was initiated. Groups were established that were tasked with experimenting with the integrated analysis of the toxicity problem, not at the level of individual substances, but on a chemical class basis by industry or by territory.

Toxics Integration as a Response to White House Supervision

In the short run, the impact of the Toxics Integration project was not substantial. But it did represent one of the first formal experiments in applying regulatory analysis as a decision-making discipline tentatively affecting the whole agency. Moreover, as regulatory analysis was being increasingly championed by the White House, the regulatory analysis personnel in EPA gained further support for their activities. Environmental and health regulation had emerged strongly in the 1960s, during a particular phase of American politics characterized by the strength of new public interest movements and greater political pluralism. But that period of intense regulatory change did not last beyond 1975 (Harris and Milkis 1989; Hoberg 1992). In the second half of the 1970s, environmentalism became less of a national priority and the control of regulation a much stronger one. In fact, systems to moderate regulatory interventionism were already in place. Richard Nixon, at the same time as he created the EPA, instituted inside the White House the office that would oversee its activity, the Council on Environmental Quality, and its process of quality-of-life review.[8] Regulatory quality later became an important leitmotif under Jimmy Carter, who embraced the theme of government efficiency and responsiveness during his presidential campaign (Graham et al. 2005).

After Carter's election, this materialized in the adoption of an executive order in 1978, requiring regulatory agencies to perform regulatory impact analysis, and the establishment of a Regulatory Analysis Review Group (RARG), comprised of representatives of the OSTP, the OMB, the Council of Economic Advisers, and the EPA, and tasked with reviewing regulatory proposals stemming from regulatory agencies) and of a Regulatory Council within the White House (with representatives of departments and regulatory agencies), chaired by Douglas Costle. The council had a variety of projects aimed at increasing the use of innovative approaches to regulation and at coordinating regulatory efforts among agencies. It started a calendar of federal regulations and issued a set of pamphlets on innovative approaches to regulation. In this position, Costle pushed the rationalization of regulations development through regulatory analysis.[9]

Providing support to toxics integration was another aspect of this effort. The RARG was already an increment from what was called "quality review," which had been instituted under Nixon, and a process that was already delaying many EPA initiatives significantly (Anonymous 1976). Dan Fiorino, who was the OMB interface with the EPA, found out that the office asked for the revision of EPA rules much more frequently than those of other agencies and departments, and that it extended the review time for EPA rules in about a third of cases—far more frequently than for other agencies.[10] The EPA staff started to get frustrated about the OMB withholding major rules for months in order to conduct a review. But that was nothing compared to what was to come once Ronald Reagan was elected president and launched his "regulatory relief" program.

During the 1980 presidential campaign, this became a stronger theme than ever, particularly under Reagan's rhetoric of government as the source of all ills in American society, and so it needed to be rolled back. A few days before Reagan assumed office, the *New York Times* published an article proclaiming, "OSHA, EPA: The Heyday Is Over" (Anonymous 1981), listing the scathing criticisms of agencies that had accumulated during the past years, and explaining how the whole new social regulation drive was in fact a terrible mistake. Ann Gorsuch succeeded Douglas Costle as Reagan's EPA administrator in May 1981, and her job was to do the dirty work of correcting that mistake: essentially paralyzing any strong regulatory intervention coming out of EPA. Until then an assistant district attorney, Ann Gorsuch had virtually no experience in Washington policymaking or public administration. However, she was a Reagan enthusiast and a good soldier for his "budget-cutting and rollback of government" message. Most of her leadership of the EPA was inspired by the intention to curtail the agency's capacity to make decisions and issue standards by drastically reducing staff and budget overall, closing the enforcement office, encouraging regional offices to settle issues instead of going to Court, and requiring more detailed review—by senior officials or by the SAB—of the standards prepared by the various offices, and facilitating review of proposed EPA rules by the OMB (Layzer 2012, 110).

Reagan's other move was to institutionalize regulatory and cost-benefit analysis in two ways: through Executive Order 12291, requiring regulatory agencies to perform cost-benefit analyses for all proposed regulations with an anticipated impact of over $100 million;[11] and through the creation of a dedicated office to review the regulatory analyses of agencies, the Office of Regulatory and Information Policy (established within OMB in February 1980). Jim Tozzi, the head of the office and a longtime leader on the White House's regulatory reform agenda, dedicated a considerable amount of his time to environmental regulations, which were particularly prone to controversy and disagreement (Anonymous 1982).

The OMB used the mantra of toxics integration to instill regulatory analysis further into the processes of the agency. Although the budget passback does not generally entail policy directives,[12] in 1981, it was accompanied by a request to come up with a more decisive strategy for toxics integration. This forced the agency to continue the integration effort. Taking stock and going forward with it, Dan Beardsley, then in the Office of Policy—sometimes dubbed the "mini-OMB" for its role in helping or stimulating regulatory offices across the agency to perform regulatory analysis—put out a new report on the subject of toxics integration in 1981. He noted that the problems that he had diagnosed in 1977 remained unsolved: Duplication of analytical efforts continued, leading to inconsistent application of policies and standards; and the priorities of offices remained heterogeneous, leading to a bigger problem for the agency—that of ensuring "effective or measured" efficiency. In that 1981 report, the autonomy of program offices was attacked directly. The staff of program offices was described as excessively loyal to the regulatory regime that they operated and to the administrative culture of their service. On the other hand, the authority needed to impose agencywide standards or procedures on separate scientific assessments was lacking. Managers and staff scientists were reluctant or unable to criticize work done in other parts of the organization.

One of the short-term consequences was the institutionalization of the intermedia analysis performed in the "Integrated Environmental Management Division," reporting to Richard Morgenstern[13] in the Office of Planning. The division was populated by people from across the agency, with a variety of disciplinary backgrounds. Its industry work groups focused on calculating the regulatory burden for a given industry and computing cost-effectiveness curves, taking into account production processes and control technologies, the transport and ultimate fate of pollutants, and the risk to exposed populations. The division was described as a stimulating place to work, where pioneering analytic work was performed, using such elements as nonlinear models and pollution geomapping. The integrated industry analysis demonstrated on several occasions that much money was spent chasing risks that appeared comparatively "trivial."[14] It showed that "few of the regulations under consideration at the time of the study lay on the cost-effectiveness curve for reducing additional risk" (Beardsley 1987, 144). In other words, the strict consideration of cost-effectiveness measurements would justify discontinuing many of the agency's actions. Gorsuch was enthusiastic about this promotion of joint analytic work, but she did not engage in any major reorganization, preferring instead to dedicate energy to implementing severe staff reduction plans.

Conflict over Guidelines: Assessment versus Analysis

Policy analysts and toxicologists found legitimacy in very contrasted kinds of bureaucratic designs. For the former, controversies surrounding chemicals surged from the uncertainty concerning the preferences of the public for tackling one or the other hazard. To govern those controversies, one had to render these preferences explicit, and show the process of ranking problems. From this perspective, policy analysts were to be at the center of the bureaucratic assemblage: they gathered all the necessary elements of information for the various decision criteria—from risk calculations to costs and benefits, making them available for the decision-maker. Proponents of quantitative risk assessment, as shown in the previous chapter, had a totally different design in mind: one in which uncertainty concerns the knowledge of a given individual risk, which needs to be reduced as much as possible, through various forms of data and calculation. Decision derives from these calculations and the choices among calculation methods. Toxicologists and exposure specialists come first; their calculations are provided directly to the decision-maker, who may balance calculations with other supplementary, contextual elements of costs and benefits to adjust the decision.

These two ways of assembling the agency and to produce credible decisions competed in the early 1980s in the agency through processes of reform and integration of the organization. The legacy of the Toxics Integration project and of economists was strong there, albeit unexpected and indirect. The Toxics Integration group engaged in what was the first survey of analytical practices across the agency. The agency leaders had no synoptic vision of what the various offices were doing; they only had the experience of inconsistencies between health assessments, as revealed directly by the press and the courts. The CAG in the ORD was probably the only group in the agency that had some understanding of what was actually happening in various program offices. At least it was in a position to compare its own approach with that of the Air and Water offices, which it served most frequently.

So the Toxics Integration staff undertook a survey of the practices of various offices, showing that the uptake of quantitative risk assessment was uneven at best: Many offices were working with cutoff hazard criteria, leaving dose-response statistical curves and exposure considerations to others—the Air and Water offices, typically. Most offices remained substance-centered, while the OPTS performed mainly multimedia assessments. Finally, most of these offices failed to consider costs and benefits analytically, alongside risks. A few other program offices balanced the costs of controlling some substances against evidence of carcinogenicity—the pesticide program being a case in point.

That survey belied Office of Policy economists' belief that jointly developed analytical methods, coupled with strong political impetus, would solve the integration issue. The report notes that the agency would best fulfill its overall responsibility for the protection of public health and the environment by "comparing (whether analytically or judgmentally) risk among substances and across media, and adjusting Agency budgets and actions accordingly." The report envisioned decision-making structures to link policy with scientific findings and economic impacts: "integration should take place as early as possible in the analytical process on which Agency decisions rest" (EPA 1981c, I–2).

The convergence with the CAG's efforts through its guidelines was unmistakable, and both the CAG and Toxics Integration personnel converged on using notions of risk to integrate separate regulatory and epistemic cultures in the agency. In support of that effort to develop the agency's integration, and hence the consistency and credibility of its evaluations of chemicals, the Toxics Integration staff developed a standardized taxonomy of analysis that could apply to all offices. They picked the notion of "risk assessment" to name the analytic and evaluative work that the EPA performed as a whole.

The notion helped to embrace the set of analytic operations pertaining to the analysis of health consequences of toxic substances across offices: "[T]he most important tool used for evaluating toxic problems is 'risk assessment', a three part process that combines (1) an evaluation of the inherent hazard of a given compound through a given route of exposure (air, drinking water, surface water, food), at various levels, with (2) a quantification of current exposure levels (prevailing levels times the population exposed to each level), to produce (3) an estimate of total risk related to current exposures to that chemical. Human health risk assessments in EPA have tended to stress cancer effects, but teratogenic and other chronic health effects can also be accommodated. EPA also has considerable experience at estimating environmental risks, such as risks to fish, crops, and other biota" (ibid., x). Thus, the policy analysts of the agency were pushing the codification of expertise in the agency even further than the CAG had dared to do. In the EPA thus far (as discussed up to now), risk assessment had been used alongside other notions of health assessment, hazard assessment, or exposure assessment. It designated the quantitative, mathematical part of the evaluation of health effects. In no way was it used as the generic, organizing notion subsuming all other parts, including the more judgmental and qualitative parts of an overall process.

The report outlined several options to further integration,[15] all of which drew on *analysis* as a mechanism of integration, definition of generic analytic tasks, and preparation of guidelines for making flexible judgment. A few years later, recounting the effort, Beardsley (1987, 144) again noted, "We decided to use risk analysis—rather than more predictable bureaucratic approaches like reorganization—as the integrating theme because risk reduction, of either human or ecological health, is the best working approximation we have to describe EPA's general mission. Within limits, we believe it can be a common denominator for measuring many different actions, a sort of Rosetta Stone for opening up communication among the separate programs."

Even though the policy analysts did not necessarily get involved in the development of further guidelines for risk assessment or exposure assessment, their involvement in the work of separate program offices, as part of the need to prepare for OMB reviews, came with a more structured representation of the knowledge that the EPA should draw on to shape decisions. Essentially, they advocated the integration of knowledge on health effects with economic analysis, with analysis embodying the most generic kind of competence in the agency. The Recommended Outline for Health and Environmental Assessment Documents of the Office of Drinking Water (EPA 1982) made it clear that a complete assessment had three parts: a health assessment, an exposure assessment, and a risk assessment. The document was meant for the associate assistant administrator of the Office of Water, a position invented by Gorsuch during her mandate to position more politically appointed people to supervise the work of regulatory offices. Like other assistant administrators, Merna Hurd was imposing on the staff a discipline of considering the risks, costs, and benefits of alternative regulatory options. The author of the document thus drafted an outline that, unlike most other guidelines, incorporated the consideration of control options for chemicals with negative effects into the very exercise of scientifically assessing those effects. That too was an exercise in assessing uncertainties—one that could justifiably be characterized as a risk assessment.

The same sort of forced consideration of cost and benefit knowledge in a standard, linear analytical process took place at the Air Office. A cancer policy task force there was asked to rethink the part of the guideline indicating the risk threshold that they considered significant. The group was led by Betty Anderson of OHEA and the CAG, with people from across the agency. It was to find a solution for such questions as: Should the agency decide on the basis of individual risk or total population risk?[16] Should it adopt a common "target risk level"? The 1976 cancer assessment guideline had put these questions aside, because so long as the OHEA developed guidance more or less alone, or in conjunction only with the Air and Water offices, in accordance with the precautionary statutes that they administered, these questions were irrelevant. But in a context where the agency as a whole was asked to consider the costs and benefits, as well as the proportionality of its regulatory decisions, they became unavoidable. The agency had to fashion a common response, matching the level of public contestation and OMB oversight that it had started to get in those years.

It soon appeared exceedingly difficult for the group to come up with generally applicable policy principles on these issues. The toxicologists leading the assessment of health effects kept in mind the uncertainty inherent in their assessments, and they needed to err on the safe side. To compensate for this controversial, de facto precautionary policy, they tried not to standardize or mechanize it too far, applying strict guidelines. This attitude contrasted with that of economists and policy analysts, who had confidence in their cost and benefit measurements. They found less uncertainty, and firmer ground for decisions, than health scientists. Anderson reported the difficulty to an administrator at the Office of Policy in the OPRM, asking him to arbitrate and to give directions on resolving the question.[17] The group agreed that it was necessary to use estimates of the magnitude of the cancer risk *along with* other factors to make sensible regulatory decisions, such as the costs and feasibility of control, benefits of the compound, and impact on the controlled industry and the economy—which is a major shift compared to the past, when generic cancer policies were restricted to considerations of health and science. But they did not manage to standardize a decision-making process using these various elements of knowledge, or to clarify the relative importance of health scientists and policy analysts in the forging of these decisions.

Conclusion

Economic analysis, under the guise of comparative analysis of the levels of the costs, benefits, and risks associated with actions on series of chemicals, and prioritizing these chemicals, emerged as early as other kinds of scientific expertise within the EPA. This rationale for decision-making, after all, was present in the group that designed the agency, including Alain Enthoven, the economist and systems analyst, and Terry Davies, the political scientist and advocate of integrated action on the environment. The agency they had imagined at the time of the Ash Council, for Nixon—featuring a central planning and management function—never saw the light of day. There were plans to organize the agency around this kind of decision-making mode based on a synoptical function of information collection and systematic comparison, turning the policy office into a central functional element of the agency. However, these plans did not materialize.

But the ideas underpinning such a design remained alive inside the agency. It reemerged during the mandate of EPA Administrator Douglas Costle because of a new political configuration. The many new regulations introduced in the 1960s and 1970s to regulate environmental problems and health risks started to be perceived as a problem. In the nascent era of regulatory reform, the agency was under pressure to speak with one voice, justify each new regulatory intervention, and avoid duplication. The problem of the impregnation of the environment with chemicals constituted the other facet of this configuration of evaluation of the agency's action. The effects of individual chemicals did not present the only issue. Their accumulation and dispersion across the totality of the environment constituted a new, more complex version of the problem. The focus of the controversies concerning the agency shifted. They were less about being wrong for one chemical than about being wrong about which hazard it was regulating in the first place. The orientations given by the administrator—addressing the chemical revolution across environmental media—encountered the thinking of economists in order to produce an emergent bureaucratic technology of risk-ranking. Still, giving greater power to economists in the definition of the agency's priorities meant placing them above toxicologists in the hierarchy of expertise of the agency, and above the leaders of individual regulatory programs. In a context in which the public continued to expect that the agency protects against chemical risks—rather than optimally distribute its resources between a variety of issues—this was a difficult organizational reform to make. For the time being, the economists' proposed design and bureaucratic assemblage was not fully instituted. The risk-ranking process did not become an obligatory, cross-agency process, involving the leaders of each program office. It did not lead to any formal reorganization of the agency. It was not used to project a new identity. The political imperative of optimizing regulatory intervention, as Reagan campaigned to become the president of a leaner, smaller government, only grew stronger from this moment onward, though. And so did the ambition to inform the decisions of the agency through economic analyses.

4 Codifying the Risk Assessment–Risk Management Framework

Ten years after its creation, the EPA embraced risk twice, but separately and to different degrees.[1] By the early 1980s, the notion of risk encapsulated a first design, according to which the agency produces decisions that are adjusted to the level of risk, determined by toxicological and statistical analysis and concurring policy assumptions about the nature of these risks. This design was embodied by cancer risk assessment methods. It was advanced by a fringe of the agency's scientists and adopted by a couple of its programs. It materialized in guidelines and improved communication between scientists, who were calculating the risks and estimating the hazards, and the various other actors involved in writing rules and standards--economists, policy analysts, lawyers, or office chiefs. Another design, a commensurative one, aimed at comprehensively reviewing agency activities against risk, cost, and benefit indicators, with a view to taking control over the agenda of the agency's offices in order to produce a more controlled and integrated image of what the agency was addressing. One design responded to industry judicial challenges against the ban of its chemicals, using the uncertainty surrounding carcinogenesis; the other tried to limit controversies stemming from the application of various risk criteria by separate regulatory offices to similar chemical conditions.

At the end of the 1980s, the risk assessment guidelines developed by the EPA—as well as other agencies—sparked more controversy. They displaced the legitimacy problem of the agency. It was no longer an issue of whether the agency had the authority to ban chemicals, but whether its mode of reasoning and making decisions about chemicals was right. For the chemical industry, these guidelines embodied an overly conservative and stringent regulatory philosophy of risk elimination, producing many false positives, and the action on cancer and chemicals was misguided. The chemical industry opposed these guidelines, and took on the agency's way to use science. It entered the game of design, suggesting that the proper administrative form of using science was an independent science panel, separate from regulatory agencies. This proposal was based on a different design than that of EPA's health scientists and economists: a more scientistic and predictive one, according to which science can eliminate uncertainties and establish future risks with precision in such a way that the space for autonomous judgment by a political official, between the walls of an agency, should be restricted as much as possible. Industrial action displaced the design process from the agency itself toward another influential site, the NRC, in which a panel of experts pursued the work of organizing disputes and came up with a new representation of necessary knowledge for bureaucratic acts. The NRC panel built the knowledge representation and technologies that were emerging in the agency—quantitative risk assessment, risk-ranking—in its resulting NRC report, *Risk Assessment in the Federal Government: Managing the Process*. It articulated a generic framework that combined these technologies and that instantly (and surprisingly) appeared legitimate to nearly everyone involved in the controversy.

Industry in the Game of Design

As part of its campaign against the EPA's conservative assessments of chemicals, the AIHC developed a full, positive institutional proposition that suppressed regulatory agencies' mission to assess risk, entrusting this scientific work to a new, separate science panel.

In late 1979 and early 1980, the AIHC stepped up its campaign to institute this science panel. In November 1979, the group published a first Recommended Framework for Identifying Carcinogens and Regulating Them in Manufacturing Situations (AIHC 1979). The framework highlighted several necessary reforms: a separation of scientific from social decision-making; giving an increased role to risk estimation in assigning priorities for regulatory action ("risk estimation includes both identifying potential carcinogenic hazards and determining the probability of adverse occurrences, essentially scientific functions"); the importance of risk estimation to select "approaches" (types of regulatory measures) for priority subjects; and the organization of a "sequential interplay" between regulatory agencies and an expert scientific panel (ibid., 6–9).

The AIHC devised a whole new controlled process for scientific advice that was based on a list of nine steps, involving the science panel or the agency in turn.[2] The boundaries between the panel and the regulatory agencies, and science and regulation, were drawn by two kinds of value judgments: scientific and social, respectively. The qualitative determination of existence of a hazard and the calculation of the most probable risk estimate (quantitative), "although judgmental, are basically scientific efforts requiring interdisciplinary, scientific expertise." But the determination of the degree of control "is a social-value judgment, which should be influenced by the scientific determinations but in the end is political" (ibid., 10). This determination should be left to agencies, and the rest left to a science panel. Here, the AIHC was enlisting powerful allies: William Lowrance's work, first (see chapter 1), but also the newly published report of the OSTP (Federal Register 1980).

On February 21, 1980, the AIHC circulated a supplementary proposal that outlined the organization of the science panel (AIHC 1980). It would be a new standing committee within the NAS. The rationale behind its creation was to recruit the most experienced scientists, to harmonize scientific advice, and to construct the scientific basis for a national cancer policy. The panel was seen as a real panacea: it would eliminate scientific inconsistencies in scientific evaluation by agencies, provide evaluations by the most objective professionals, reduce conflicts of interest, and speed up the regulatory process. Benefiting from the excellent reputation of the NAS and delivering high-quality scientific opinions (AIHC documents generally used the term *determinations* to refer to these opinions), it would also have the benefit of minimizing controversies around basic scientific issues. The AIHC insisted that the individuals in charge of regulation would benefit from assessments produced by competent, experienced, and objective professionals. It proposed that the NAS has complete latitude in nominating the committee members.

Much like Dan Beardsley in the agency's policy office, the AIHC drew on the argument that "risk analysis" could be a vector of greater objectivity in policymaking—one that could eliminate the "biases" of an agency that was structurally inclined to regulate. The AIHC promoted a sequential model of decision-making in which the analysis of benefits would work as a kind of embedded veto point in order to derail conservative assessments. During the public hearings on cancer guidelines, the physicist Richard Wilson

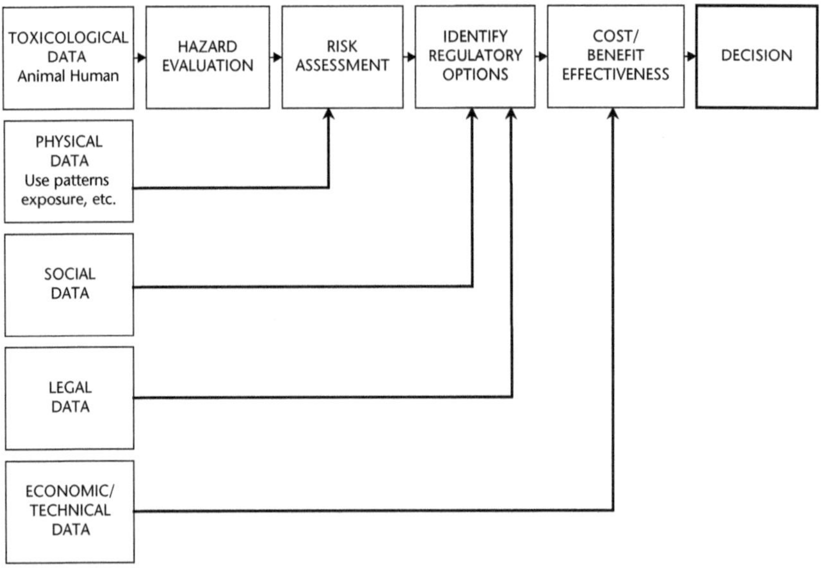

Figure 4.1
A linear risk analysis and decision-making scheme (adapted from Haight et al. 1980, 14).

maintained that the right approach to risk, as the former EPA administrator Russell Train said in his preamble to the 1976 guideline, consisted of first assessing the risk and then deciding what to do, taking into account not only the risk, but also the benefits associated with the substance: "It is important to start with a structure such as this because even though there are information gaps in the structure, without a structure illogical actions may be suggested" (EPA 1980c, 8).

This two-step design pleased Wilson, and more broadly the chemicals industry, as it afforded a procedural and systematic opportunity to overrule the estimates of the supposedly conservative scientists of the agency on the basis of benefits. The industry was also thinking in terms of this linear, step-by-step decision-making system, as in figure 4.1, where preliminary assessments of risk could subsequently be vetoed on the grounds of benefit or cost considerations before arriving at the final decision-maker. A report drawn up by the Chemical Manufacturers Association and the American Petroleum Institute for educating key industry representatives involved in the public debate on risk/benefit/cost analysis advocated this structure, with supplementary steps after the risk assessment to deflect the possible decision to regulate (namely, an "identification of regulatory options") (Haight et al. 1980).

The AIHC's discourse on uncertainty—where risks come from a lack of knowledge that future, supposedly "sound" science will compensate for—was coherent with this proposed design: removing risk assessment activities from regulatory agencies and entrusting them instead to a dedicated science panel outside these agencies. The "science panel" proposal of the AIHC broke the step-by-step linear process of decision-making into two large modules at the stage of regulatory option identification. The first steps of the process (comprising health risk assessment) were to be removed from regulatory agencies and entrusted to a science panel, while the rest was left to the agencies. This would take away from regulatory agencies the possibility of articulating and legitimizing health-protective regulatory decisions through science. Instead, it would make science more amenable to other methodological, epistemological pressures, and focus regulatory agencies on the task of devising optimal decisions from the point of view of their costs incurred on businesses, and of their benefits. In this particular assemblage of knowledges and expertise, a boundary is drawn around cost-benefit analysis, left to regulatory agencies, probably because this expertise is the least likely to generate strong regulatory measures for industries.

In the chemical industry's action, the model of separating scientific knowledge from policy decision-making was thus associated with an institutional model, based on the view that science is a particular type of content that should have its own dedicated institutions and space, to remain as free as possible from any possible source of bias. The "science panel" concept echoed a parallel, quite animated debate and similar institutional proposal for dealing with scientific contestation: that of the "science court." This institutional, expert deliberation design had been proposed by Arthur Kantrowitz, a physicist in fluid mechanics and gas dynamics, but also a public proponent of the separation of science from ideology. He articulated a proposal to set up an institution for forming scientific judgments in the context of disagreements among experts (Kantrowitz 1967, 1975).

The idea pushed by Kantrowitz (though not by him alone)[3] was that an organized confrontation of scientific arguments would be an ideal setting for producing truth and objectivity. Despite constant references to democracy, the proposal was strongly rooted in an instrumental vision of science—a belief in its capacity to produce facts, separate from the realm of values—and of overcoming excessively emotional debates. What Kantrowitz and other promoters of the scientization of public issues had in common was the belief that content-free procedures or settings would produce the right outcome in all cases. Following the right process of confrontation and deliberation among experts would produce objectification, regardless of the substance of the issue. Even though it proved to be illusory (Aakhus 1999), this institutional design appealed to those who defended a scientistic approach to the regulation of risks and technologies.

As a skilled and motivated promoter of his idea, Kantrowitz got President Gerald Ford to set up a task force under the Presidential Advisory Group on Anticipated Advances in Science and Technology, with the purpose of developing a workable scheme for such an institution. In those years (1975–1976), Kantrowitz published a paper in the journal *Science* that gave his proposal good publicity and sparked much debate on the nature of his proposal and its workability. A well-attended conference in 1977, sponsored by the NSF and the American Association for the Advancement of Science, concluded with an endorsement of the proposal by the leaders of the country's most prominent scientific societies—and of the AIHC.

There were notable differences among the designs pushed by Kantrowitz and the AIHC, however. Whereas Kantrowitz suggested bringing disagreements out into the open and forcing scientists to engage with opposing arguments, the AIHC embraced an altogether different method: it preestablished what the appropriate components of a consensual decision would be and defined a sequence of production, with standards applying at each step of the process. A strict corridor was defined that was presumed to lead to the articulation of consensual, fact-based decisions. In this process vision, which was quite fundamentally opposed to the forumlike idea of Kantrowitz's institution, the central expert panel was supposed to pour out authoritative formulations of facts—and hence it was not to be an arena for open disagreement. But the connection between the two enterprises certainly stemmed from the belief that there were institutional designs specifically suited to the production of scientific facts and truths.

The Committee on Risk and Decision-Making

The AIHC, Kantrowitz, and ongoing discussions inside the NAS showed that a particular phase of institutional imagination had opened. In the NRC, at about the same time as the AIHC was campaigning for a science panel, the mathematician Howard Raiffa's Committee on Risk and Decision-Making (CORADM) was struggling with categories to represent and order the kinds of knowledge and activities implied by decisions on uncertain collective hazards. The final report of this committee contributed to the thinking around the respective roles of assessment, evaluation, and control of risks, as well as the organization of their interactions in concrete regulatory practice (NRC 1982). It reflected Raiffa's perspective as a decision analyst, valuing analysis as a tool for making legitimate decisions. Analysis was helpful in decomposing the components of a decision, thus laying the groundwork for negotiations among actors making or being subjected to the rule in development.[4] Therefore, analysis should be preserved. But decisions in controversial contexts were obviously limited by what the report called "conflicts of interest," and with these, professional analysis provided no solution.

An additional solution, therefore, was needed to decrease the level of contestation surrounding risk decisions. Raiffa saw that solution as bracketing out the most controversial part of these decisions: the value that people assigned to the actual object of the decision. In a typically analytical style of thought (breaking down predecisional sequences of thinking), the report promoted the notion of a discrete risk evaluation stage, lodged between the assessment of the risk (pertaining to science) and decision-making or policymaking. It is unclear, when reading the report, whose job it was to consider values and how values could be decided. On the one hand, risk analysis was said to encompass risk assessment and risk evaluation, seemingly indicating that values should be an integral concern of the scientists assessing risks. On the other hand, the report occasionally folded risk evaluation into policy analysis, arguing that risk evaluation was connected to the identification and choice of policy alternatives, and as such was policymakers' responsibility.[5]

Raiffa believed this knowledge representation in terms of risk assessment and risk evaluation was concordant with the practices and problems of agencies, although no concurring evidence seemed to exist at the level of the formalisms developed by the EPA or of its organizational chart. It invented a risk evaluation step that was empirically nonexistent and prescriptively useless for risk assessors who were, de facto, faced with these problems of values and alternatives within the confines of their scientific exercise. The conflict that set organizational sociologist and political scientist Charles Lindblom against the rest of the committee members revealed the limits of this category of "risk evaluation" as a way to salvage a professional, objective analysis. Lindblom was intellectually and politically convinced that the knowledge underpinning decision-making and policymaking was not professional, specialized, formal knowledge. He argued that decisions were most often derived from and shaped by the articulation of ordinary and socially distributed knowledge. No knowledge stands outside or exists prior to decisions, independent of social perspectives and experiences. The whole demarcation of an intermediate step of risk evaluation to cloister the controversy-prone consideration of values, preserving science's "safe space" of facts, just did not make sense.

Lindblom fundamentally disagreed with the whole endeavor and thrust of the CORADM, which was to formalize procedures of analysis that would produce "superior policy alternatives" (NRC 1982, 30). Policies could not be rationalized on the basis of formal knowledge produced before they were made, precisely because they were not a sort of neutral enactment of predetermined data, but rather situations of active engagement with, and production of knowledge about, the problem. Lindblom maintained that the proper use of knowledge in action consisted of improving learning capacities through action, not of importing decisions from external knowledge, the objectivity and authority of which he considered to be mere myth (Cohen and Lindblom 1979).[6] On the page of the final report that listed the members of the CORADM, his name was followed by an asterisk with the following statement: "Charles Lindblom does not approve of this report." The limits of the science/evaluation/policy sequence as a representation, accentuated by the conflict between Lindblom and Raiffa, undermined Raiffa's entire effort. At the end of the day, the final report[7] did not put forward any concrete recommendations as to how to manage the interfaces among assessment, evaluation, and decision-making. No trace of this thinking can be found in the documents and practices of agencies today.

AIHC in Congress

Despite the failure of CORADM to institute a new design for science and making decisions, the AIHC considered it a good strategy to create its own proposal for a science panel endorsed by the various academies (of science, engineering, or medicine) in order to undermine agencies. The AIHC mobilized elite representatives of large chemical and petrochemical companies, who consistently made contacts and sent letters to various congresspersons to advance the collective industry proposal on risk assessment, specifically on its dissociation from regulatory agencies.

Several bills proposed in Congress echoed the concerns of these companies during that period. In February 1980, Congressman William C. Wampler (a Republican from Virginia) had introduced a bill suggesting the creation of an eminent scientific panel inside the White House's OSTP. The panel would effectively be akin to a court; anyone would be allowed to refer issues to it to adjudicate agencies' decisions. Another proposal was pushed by Congressman Donald L. Ritter (a Republican from Pennsylvania), which concerned the activities of the five health and safety agencies and foresaw the coordination of risk research by the OSTP, as well as the development of prototypical risk analyses by the five regulatory agencies, to strictly frame their capacity to develop these risk assessments.[8] The bill reflected the AIHC's ideas of restricting the scientific capacity of regulatory agencies and giving more control over risk assessment to the federal government.

The AIHC made another proposition to Congress: to have the NAS discuss, and possibly endorse, the science panel idea. To that end, AIHC representatives engaged in simultaneous lobbying of the NAS, the House of Representatives, and the Senate. In March 1980, the industry group sent a letter to Senators Thomas Eagleton and Ted Kennedy (both Democrats; from Missouri and Massachusetts, respectively). In this document, they required the senators and corresponding committees to include in the appropriation bills that were under way a proposal for an amendment on the need for agencies to avail themselves of the best scientific advice, as well as to give $500,000 to the NAS for a study on how best to achieve this. The NAS was in the process of working on risk assessment methods and on designs for distinguishing science, policy, and the gray zone of risk evaluation (NRC 1981, 1982). Having it give its imprimatur to the science panel idea, and even agree to form the panel within its walls, would decisively advance the attack on regulatory agencies.

Congress's oversight of the EPA and ongoing disputes about chemical cancer assessment meant that it was receptive to institutional proposals to examine new possible designs for science and policy. In November 1980, the Senate, supported by the NAS, approved the AIHC's proposal to allocate $500,000 from the FDA budget for a study of risk assessment. The objectives were formulated in almost the same terms as those used by the AIHC in its documents: to assess the merits of institutional sharing of scientific functions through the separation of the objective risk assessment process from the regulatory process per se; to consider the feasibility of unifying the various functions of risk analysis, and study the possibility of developing a coherent risk analysis methodology likely to be taken up by all regulatory agencies in the decision-making process; and to list the procedural and institutional questions that would arise from the interaction between the selected risk analysis methodology and the regulatory processes.[9] In short, the NRC was charged with investigating whether assigning risk assessment functions to a dedicated institution in the federal government, separate from specific regulatory agencies like the FDA, the EPA, and OSHA, would remedy the alleged distortion and politicization of science by regulatory agencies. Because this alleged distortion was embodied in the cancer guidelines, the question of guidelines was explicit in the charge. The committee had to examine the feasibility of setting uniform guidelines. The project was explicit: It aimed to reduce the autonomy of an agency that was perceived to overregulate, through a definition of the knowledge it could use and how.

The NRC's interest stemmed from the fact that risk was an increasingly hot topic, for which it was getting many report requests. A 1981 overview report on the activities of the NRC noted that 49 percent of the reports produced by NAS in the preceding years included some aspect of risk assessment (NRC 1981, 2). That review report, authored by a set of top academics in the nascent field of risk research and risk analysis (e.g., Robert Kates, Howard Raiffa, and Gilbert White), inaugurated a generic definition of risk assessment, encompassing a wide variety of analytic methods: "Risk assessment as its practice has evolved is a rubric covering a broad range of analytic activities including: the identification of hazard, the estimation of probabilities of occurrence of hazardous events of specific magnitude, the linkage of such events with various undesirable health, safety, environmental and other societal consequences, and the evaluation of risks by comparison with costs, with other risks, with benefits, with alternative ways of reducing risks, or with the risk of substitute activities. A complete risk assessment would involve most or all of these elements" (ibid., i). Reviewing a number of risk assessments done by the NRC itself, the authors learned that most of these exercises involved ambiguous definitions, paucity of data, or controversy around the main subject, and that the questions addressed in the reports were difficult to answer or intractable, with unclear modes of analysis. They reached the conclusion that in risk assessment, there should be a concerted effort to distinguish among scientific findings and personal or value judgments if these findings are to be translated into policy.

The NRC thus had some experience in the field of risk assessment, and this time around, it had a clear, feasible charge to answer—and one that promised an impactful report. Expectations were high.[10] The committee had a precise, feasible task, and a mission: first and foremost, to assess the AIHC proposal for a science panel (that is, the separation of assessment and decision-making into distinct institutions). Its work was very straightforwardly connected to events and debates that were just unfolding, and had attracted more attention only since the ILRG's work on cancer guidelines. In that sense, the pressure of events and of the discussions that were ongoing at just about the same time, outside the walls of the NRC across Washington, was probably stronger than usual.

The Risk Assessment Committee

Frank Press, the president of the NAS and the NRC, moved away from the solution of relying on the conclusions of the CORADM, or mandating Raiffa for this new study, and instead created a new committee named Committee for the Institutional Means for Risk Assessment (Risk Assessment Committee, or RAC). The Board on Toxicology and Environmental Health Hazards (BOTEHH)[11] of the NRC worked on setting up the panel to do the study. People were selected to represent the following subfields of expertise: risk assessment policy, risk assessment, law, economics, decision analysis, political science/organization theory, psychology/perceived risk, epidemiology, biostatistics, oncology, public, industry, and basic science/other. More informally, and in line with NAS's practice, the board was looking for people that would represent the various viewpoints on the subject: those of the agencies, of scientists, of regulatory bureaucrats, and of environmental or health activists and industry.[12]

Frank Press suggested taking William Ruckelshaus as an expert on risk assessment policy, by virtue of his previous position as EPA administrator. Ruckelshaus had also been a member of the CORADM. Bob Tardiff, a toxicologist and influential member of the NRC Board of Toxicology,[13] made several important recommendations, namely Morton Corn (to cover risk assessment policy), Richard Merrill (law), Joseph Rodricks (risk assessment), and Reuel Stallones (nominated as an "alternative," without mention of his particular expertise in cancer risk). Most of these people accepted the nomination, and they turned out to be key, influential drivers of the discussion. At this stage of the process, the NRC continued to be inspired by the experience of the CORADM, insofar as it aimed to attract first-rate scholars: Solow and Arrow were the nominee economists, and Charles Lindblom and Don Price the political scientists. None of these nominees were eventually selected, which left the committee without economic expertise. Political science was eventually represented by two political scientists who joined later in the process: Terry Davies and Ted Greenwood, both of whom would turn out to be essential to the committee's thinking, owing to their experience and primarily their understanding of the intricacies of knowledge and policy relations.[14]

As was often the case, the question of the committee chair proved to be a thorny one. In July 1981, NRC staff recommended that Press give the chair to Gil Omenn. It may be that the name of Gil Omenn was suggested to the NRC and to Press, who was personally involved in the selection of experts, by the AIHC. The AIHC had appreciated and defended the work of the OSTP and Omenn on carcinogen identification and evaluation. Omenn was also the deputy of Frank Press when the latter was the chief of OSTP.[15] Robbins, a member of the NRC staff, wrote to Press that "Gil is presently writing a monograph on air pollution and has been very interested in this question. I would consider him one of the best qualified people I can think of to do the job."[16] Omenn was given the chair just a few days later, but he occupied the position for a very short time. A full week after the first meeting of the committee in October 1981, Press asked him to step down. This change was motivated by the opinion of several members of the BOTEHH, that the position of Omenn and the OSTP for a "two-step" decision-making process--insulating science from regulatory control and leaving key choices to scientists in a central risk assessment program under the National Toxicology Program (Federal Register 1980)--seemed incompatible with examining all possible options. It seemed just too close to the chemical industry's preferred design of insulating a scientific stage and entrusting it to a science panel. Reuel Stallones, an epidemiologist and specialist of cancer and smoke from the University of Texas—and a consultant to the AIHC—was immediately appointed to replace Omenn.[17]

The Committee on the Institutional Means for Assessment of Risks to Public Health eventually comprised fourteen people. Besides Stallones, four members provided academic expertise in environmental health and toxicity testing, particularly with respect to cancer risks. Morton Corn was an industrial hygienist from John Hopkins University and a specialist of workplace safety risks related to aerosol particles. He was also the former administrator of one of the agencies of interest in the study, OSHA (1975–1977). Vincent Dole was a professor of medicine, a member of the Institute of Medicine (IOM), another branch of the complex of Academies, and a researcher in biochemistry. He was already mostly known for his research on heroin addiction, understood as a metabolic process. Kenny Crump was a mathematician and biostatistician who had developed a reputation in the field of environmental health and regulation for his method of extrapolation from low dosage of chemicals to high dosage—a method used by the EPA at the time. Elizabeth Weisburger from the National Cancer Institute brought her knowledge of oncology and carcinogenic substance identification.

There were several experts of administration of health and environment, most notably Richard Merrill, a graduate of the University of Virginia School of Law and former chief counsel of the FDA between 1975 and 1977. He was also a member of the NRC's Board on Toxicology. Ted Greenwood was an assistant professor in political science at MIT. He was on secondment to the OSTP between 1977 and 1979, dealing with energy policy, and since then had returned to his research, dedicated to agencies' use of scientific knowledge. Terry Davies, a political scientist and early researcher on environmental policy, was a member of the group that President Richard Nixon consulted to create the EPA in 1970 (Davies 1970). He was based at the Resources for the Future think tank at the time of joining the committee.

Mediating between the two groups was a smaller subset of two men who had a mixed scientific and administrative profile. One was Rodricks, a biochemist and toxicologist by training (PhD, University of Maryland), who was at the FDA between 1969 and 1980, ending his career there as deputy associate commissioner for health affairs from 1977 to 1980. Before joining the RAC, he had left the FDA to establish a private risk assessment consultancy. He had chaired the IRLG risk assessment working group (see chapter 3). The other was Omenn.

The rest of the group was less vocal, though they contributed to the overall balance of the committee and the variety and novelty of the recommendations it would make. Two members of the committee were identified as specialists in the emerging science of risk assessment. Warner North was a young up-and-coming decision scientist and an early developer of applications of this science to environmental and health issues. By his own admission, he was a protégé of Raiffa and other pioneers in decision science.[18] When the study began, he had already looked at such things as the process of making decisions about air pollution or chemical-related risks, both through his own research and also through his participation as consultant to NRC committees. He had founded his own consultancy, Decision Focus, after leaving the Stanford Research Institute.[19] The second risk research expert was Paul Slovic, a psychologist and pioneer in risk perception research, at the time a research consultant with his own company, Decision Research. Slovic and North knew each other since Slovic had worked with the behavioral psychologist and economist Amos Tversky, who interacted intensively with the Stanford Research Institute. Frank Mirer and William Utidjian had been selected to represent the public and the industry, respectively.[20] Mirer was a chemist and toxicologist who had spent time as a PhD candidate and postdoctoral researcher at the Harvard School of Public Health. He joined the RAC as the head of the health and safety department of the United Automobile Workers Union. He was mostly interested in occupational health chemical safety standards and the activity of OSHA. Utidjian was the corporate medical director of the chemical company Cyanamide.

Beyond the inherent diversity of backgrounds and differences in views on the science panel debate, the committee was unique in that a majority of members had some exposure to, or even experience in, administrative and regulatory affairs, either as political analysts or as consultants or staff members of these agencies or of the federal government. This probably gave the group a particular center of gravity. The director of the project for the NRC was Larry McCray, a political scientist by training, who had worked in NAS already in the 1970s, as well as at the EPA and in the White House. After Ronald Reagan's election, he returned to NAS to join the Commission on Life Science, under which the RAC had been formed. He played a decisive role in shaping the report. He recalls: "I think the thing I liked most was that we really knew what was going on in agencies. That's unlike other committees I have seen. I always say to my colleagues that we need practitioners.... This committee was different because people really knew about agencies. Rodricks, Merrill, Omenn ... these are people who really respected government."[21]

The Charge, Formal and Informal

The RAC first convened in October 1981. The group set out rapidly to discuss the AIHC proposal. The committee's collective position on the science panel idea crystallized early on. The members rejected the idea on the grounds that it was too impractical for regulators and too disruptive of their functioning.[22] But the committee had to take into account another contextual element, which played out in favor of *more* separation of science from policy: the ongoing scandals about the intrusion of the EPA's political appointees in the work of health and environmental effect assessors.

As the committee was meeting, the EPA was rapidly falling into an institutional crisis. Ann Gorsuch succeeded Douglas Costle as Reagan's appointed administrator for the EPA in May 1981. A district assistant attorney, Gorsuch had virtually no experience in Washington policymaking or public administration. The main reason for nominating her for the job was that she was a Reagan enthusiast and a good soldier for the budget-cutting and regulation rollback campaign that the president was launching against social regulation. Most of her leadership of the EPA was inspired by the goals of curtailing the agency's capacity to make decisions and issue standards, drastically reducing staff and budgets overall, closing the enforcement office, encouraging regional offices to settle issues instead of going to court, and requiring more detailed review, by senior officials or by the SAB, of the standards prepared by the various offices (Layzer 2012).

A sense of crisis very quickly emerged after she took office. Hearings held at the House of Representatives in October and November 1981 give a sense that the EPA was in the largest crisis of its young existence was already spreading in Washington, D.C., well beyond the federal government and through the business and environmentalist communities. At the hearings, an emphatic Toby Moffett (a Democrat from Connecticut) opened the session by stressing the exceptional nature of what was going on at the EPA at the time and the willingness of Congress to question and take on the leaders of the agency directly.[23] Moffett's concerns were quite widely shared. But the worries of the chemical industry were no less substantial, as an article in the October 21 issue of *Chemical Week* signaled.

News had emerged, in particular, of wrongdoing around the use of scientific arguments to prevent regulatory developments in the agency. John W. Hernandez, the deputy administrator, had held closed, unannounced meetings with industry representatives. These so-called science forums—which were called "science courts" by Hernandez in his letters of invitation to industry groups, betraying his proximity to the proposal floated by the AIHC (EPA 1981a, 1)—were held on six occasions in 1981 for high-profile chemicals considered under the TSCA—formaldehyde and DEHP—and served to cast doubt on the value of the scientific studies that established the risk and toxicity of the substances, including the EPA's own in-house estimates. Hernandez, who conducted the meetings alone, put the emphasis on the exercise of quantitative risk assessment, reminding everybody that "a thorough and independent scientific review of the evidence that relates exposure to pre-established dose-response effects must precede any regulatory decision" and that "scientific review, as a process, must not limit the policy-making authority of the administrator. Instead, good science, with ample peer review, must be the foundation for all appropriate Agency decisions." This authorized him to reject the "questionable assumptions" that underpinned the risk extrapolations of his own staff (EPA 1981b, 7).

Other direct interventions in health assessments included Hernandez sending a health assessment performed by agency staff to Dow Chemicals and deleting portions of the text arguing that Dow's manufacturing complex in Midland, Michigan, was the main source of dioxin pollution in the region. John Todhunter, the assistant administrator for toxic substances and pesticides, was no less bold: in the spring of 1981, his staff had come to the conclusion that, given a recent study by the Chemical Industry Institute of Toxicology showing a link between formaldehyde and nasal tumors in rats, the substance should be designated for priority review (Jasanoff 1987a). Todhunter reversed this decision in a memo to the EPA administrator from February 1982,[24] in which he single-handedly declared past exposure estimates to be "completely incorrect." Against the weight of evidence traditionally considered by the agency, Todhunter used the existence of one negative study to justify not regulating the chemical. He also overrode the existence of a positive carcinogen test in animals, arguing that the product was probably genotoxic, and thus exhibited a threshold. He lowered the EPA's typical acceptable risk threshold of one cancer in 1 million to one in 100,000 (US Congress 1982). Most of these interventions in science were judged illegitimate because they originated from political appointees, with a clear agenda of delaying or halting the regulation of major products. The interventions did not go unnoticed, precisely because these products and environmental problems were in the news anyway. These cases were widely publicized through congressional oversight hearings and media inquiries.

These widely decried political interventions in the science of chemical risks constituted the public backdrop in which the RAC was due to arbitrate for or against the proposal of the AIHC to disempower regulatory agencies to perform risk assessments. In a sense, the problems caused by Hernandez and Todhunter lent support to the AIHC's proposed institutional scheme. Their behaviors were breaching a well-accepted norm of independence of science, and of mutual autonomy of science and decision-making, on which the AIHC was capitalizing to deploy its enterprise of limiting the powers of regulatory agencies.[25] But the RAC quickly took distance with the science panel idea, because it was a clear attack on and a potential disruption of an institutional system for which the RAC saw no need.

From a Linear Scheme to a Modular Framework

To develop its recommendations, the RAC first clarified the components of the design of science-based decision-making on risks. The group had quickly realized that it was missing definitions that were key to simply mapping out what regulatory agencies were doing when they were dealing with uncertain hazards, as well as where lines could be drawn between science, policy, and other aspects. According to McCray, the construction of the definitions that eventually composed the risk assessment–risk management framework, of which RAFG is so often given as the source, "was really an internal thing."[26] At no stage did the group seek to draw up the formal framework that RAFG is now credited with providing, and for which it is now famous. Furthermore, there was no intention (let alone a mandate) to standardize the practice of regulatory agencies based on that scheme. Ironically, what was later to constitute a renowned framework were the definitions that the group designed for itself, to give some order to the set of terms that had emerged in recent years to characterize risk knowledge—from *risk assessment* to *risk analysis*, through *risk evaluation* and *health assessment*. The definitional work required was a way to manage the various viewpoints inside the committee, where contradiction, or conflict, was evident. People with a background in biological sciences were not accustomed to considering that scientific practice, even for regulatory purposes, might be informed or underpinned by values, whether implicit or explicit.

Stallones, Corn, Crump, Dole, and Weisburger were designated by Davies as the "science camp": those people from biology, broadly speaking, who were most likely to defend the idea that risk assessment is science and that science is objective and not influenced by values (Davies 2003). The "social science group," or the "values side," were considered to include those who initially would be more inclined to defend the view that risk assessment is a mix of scientific and political values. But others in the group did think along those lines, and rather than sweeping the issue aside, they addressed it head on.

The particularity of that committee was that it acknowledged that there was a need to analyze risk analysis—that is to say: to unpack it, clarify what it was made of, schematize it, repackage it, and define what would be a proper institutional shell for it. Larry McCray did an essential part of this work in what he termed his "anatomy of risk assessment" (McCray 1983). The anatomy of risk assessment was really an X-ray of uncertainties and "assumptions"—not "values"—intruding on the process of health and effects assessments, which listed the thirty-six points at which choices are made in a risk assessment process. In that forensic study of the exercise of scientifically assessing risk, McCray showed that multiple choices were being made at all levels based on professional judgment, experience, and, possibly political preferences. Indeed, risk assessment is nothing but a series of "discrete decisions," a sort of scaffolding of judgments that allow one to produce the final estimate (ibid., 83).[27] The anatomy was essential to define the utility of guidelines: These were tools to explicate the numerous choices, many of which were extrascientific, that were made in the course of analyzing cancer data. It was decisive (Jasanoff 1987) to inspire the group's conclusions on the need to keep risk assessment and risk management in interaction, and to reject the science panel idea. These assumptions are the ground where scientists responsible for calculating hazards and bureaucrats defining the standard meet.

The group also benefited from the many hours of presentations by members of the EPA, FDA, and other early practitioners of risk assessment. Elizabeth Anderson, from the EPA's CAG, in many ways the precursor of risk assessment in the federal government, spent a great many hours with the committee outlining the meaning of the terms in use in the agency, or at least in certain parts of it. Therefore, the RAC did not start from scratch. The EPA was already in the process of self-codifying and schematizing its work, defining for the purpose of its guidelines the meaning of *risk assessment, health assessment, dose-response assessment,* and the like. The RAC invested even further in this conceptual work, drawing from Lowrance's 1976 book *Of Acceptable Risk: Science and the Determination of Safety*, which Slovic distributed to the group to consider. The committee also surveyed how various agencies were organized to produce and use scientific data, identifying the pros and cons of each organizational model.[28] Those who worked on the survey were not the natural or medical scientists of the committee, but a lawyer and two political scientists: Merrill, Davies, and Greenwood. In parallel, the project director, McCray, a political scientist, surveyed the mechanisms for institutional separation and the current agency practices.[29]

Whatever the source of the inspiration, the RAC showed particular agility in assembling the set of terms to formulate a cohesive, integrated scheme that would describe the whole of the regulatory process. They assembled and simplified the representation of environmental scientific assessment and decision-making. Traces of this work are apparent in the draft frameworks shown in figures 4.2 and 4.3.

From this sequence of figures (the graph on page 21 of RAFG, reproduced in figure 0.1, being its final form), the following process can be inferred. The first move was the semantic *generification* of risk assessment, under which four clearly defined and distinct operations were progressively subsumed. The notions of risk assessment, risk assessment policy of the analytic process, and others already existed in the bureaucratic or scientific discourse.[30] But a diverse range of definitions and schemes coexisted (Lave 1982; OTA 1981). In documents of the EPA predating RAFG, risk assessment was used alongside other terms, but with considerable variations and confusion. For people in the OHEA, risk assessment covered the quantitative work of computing a dose-response curve. Combined with exposure assessment, it produced the synthetic estimate of the excess risk. By 1982, however, things were changing in other parts of the EPA, but the actual place of quantitative extrapolation and calculations in the procedure was not completely clear or free of contradictions. In the 1982 document of the Water Office (EPA 1982), the term *health assessment* defined the hazard that may potentially be derived from being exposed to a chemical, and included what the OHEA termed "risk assessment" ("the quantification of the carcinogenic risk, the threshold for acute and chronic toxicity, the impact on sensitive populations, the ADI ..."). Exposure assessment reviewed not only the data on the

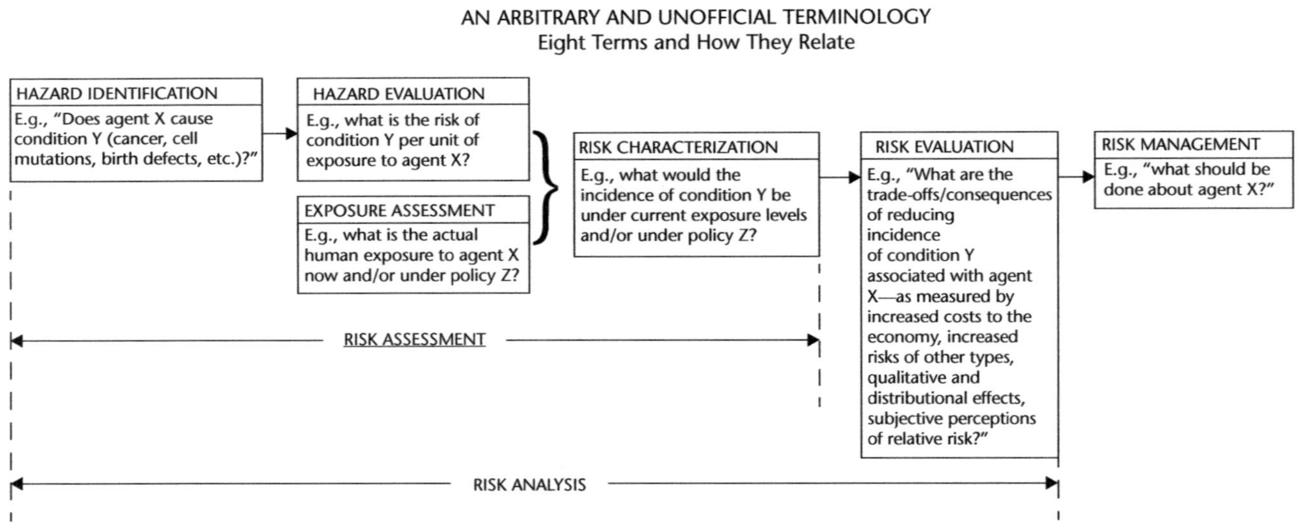

Figure 4.2
First draft of the institutional framework for risk decision-making composed by the NRC's RAC.

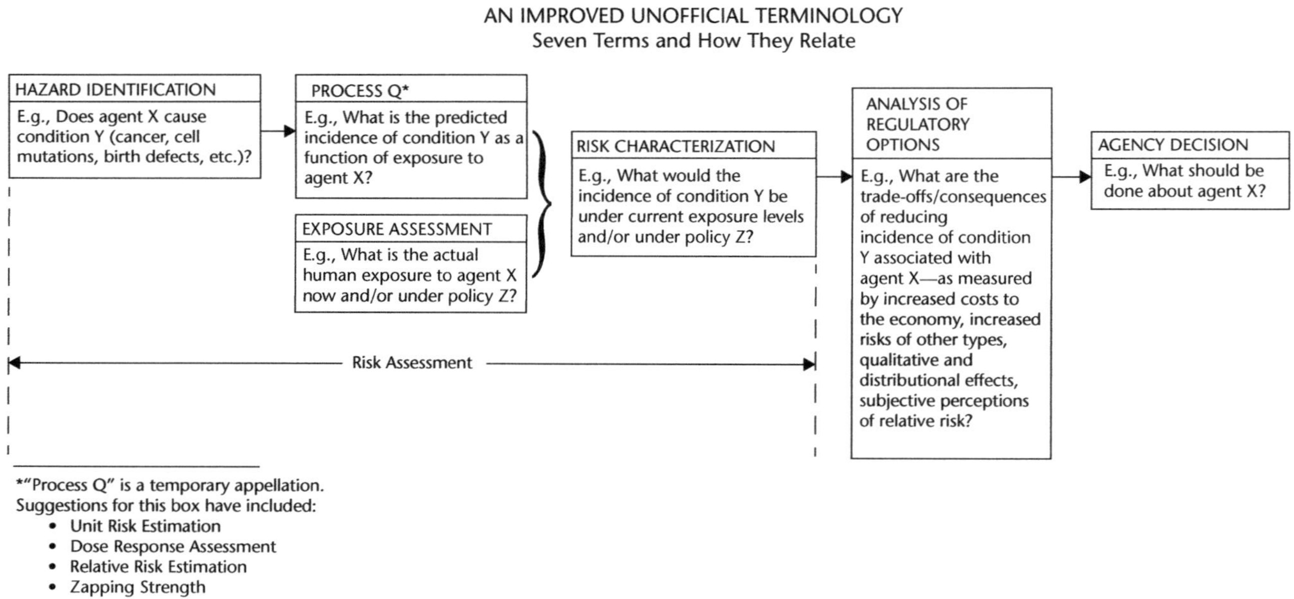

Figure 4.3

Second draft of the institutional framework for risk decision-making composed by the NRC's RAC.

size and nature of the population and species exposed, but also the technology for controlling this exposure. The risk assessment part combined the first two components through an estimate of the likelihood that the hazard would materialize in the indicated population. It also included references to regulatory and nonregulatory control options. Throughout the document, risk assessment was used in lieu of the generic health assessment notion.[31] In the report of the Raiffa committee (NRC 1982), assessment was always twinned with evaluation (as discussed previously). So risk assessment was not a generic category under which exposure and hazard assessments could be subsumed. Figures 4.2 and 4.3 show that the specific contribution of RAFG is to have promoted, or generified, the notion of risk assessment, not by inventing it from scratch, but rather by promoting it and dropping other parallel terms.

The stage 2 graph mentions "temporary appellations" and displays the options that were being considered. The committee was explicitly in the business of forging workable, collective, and one could even say conventional appellations, as opposed to exhaustively and objectively describing the practices of agencies. In fact, an appendix to the report ("Anatomy of Risk Assessment") was dedicated to this sort of description of the thirty-six steps of risk assessment. The report went further in the direction of modeling risk assessment through a set of parsimonious (and, by necessity, more generic) categories. Four clearly delineated operations, corresponding to well-identified disciplinary expertise, were defined specifically as the four steps to populate the rubric of risk assessment. Genericization also provided a way to hold together the necessary emphasis on quantification of risk and to preserve the narrative-based interpretation of risk in risk characterization—something for which Omenn pushed.

The second move (which was somewhat surprising for a committee that was meant to focus on risk assessment) was its invention of the generic notion of *risk management*. Until then, that term either referred to an engineering exercise involving the design of systems to render them less risky, or it was so broad that it was an ineffective technology to use to make a decision (e.g., NRC 1982). The committee focused the definition of risk management on the notion of policy alternatives, based on various analyses and evaluations (of benefits, costs, social values, and other elements). It became "the process of weighing policy alternatives and selecting the most appropriate regulatory action, integrating the results of risk assessment with engineering data and with social, economic, and political concerns to reach a decision" (NRC 1983), much in the way the AIHC or members of the nascent risk analysis community were starting to use it. One of the members of the AIHC, Shell's Vice President for Health, Safety and Environment, Paul Deisler, had introduced the term *risk management* in just this fashion, through the linear scheme shown in figure 4.4.

As a consequence, risk management became a broader category than *risk control* or *decision and action,* and also a recognizable, autonomous activity. The notion of policy alternatives also precluded the impression of the definitiveness and authoritativeness of the concept of *decision,* and also of a mechanical move from science to decision. Risk management was a weighing of options as a relative, revisable judgment, with an inherent temporal dimension. It was very well suited to an agency that worked through controversial decisions for decades, without ever reaching closure, and that wanted to evolve toward more fine-tuned prioritization and comparison of risks, as in the Toxics Integration initiative.

The third move, which resulted from the first two, was the elimination of the controversial notion of *risk evaluation.* In previous NRC reports, as well as in public discourse, values had consistently been a bone of contention and the main problem in the concrete allocation of tasks and responsibilities for decision-making. Of scientists, economists, lawyers, or administrators, the question of who would be the evaluator designated in previous expert reports, such as that of the NRC Committee on Risk and Decision-Making, was fundamentally ambiguous. The work that the RAC did on risk terminology unlocked this problem by simply dropping the term *value.* In the proposed architecture of terms, structured by notions of risk assessment and risk management, it simply became unnecessary. It was replaced by a more straightforward language of choice among scientific assumptions.[32]

The political dimension of risk decision-making was covered by the symmetrical notions of risk assessment and risk management. Both activities were conceptualized as similar, insofar as they grappled with uncertainty. In previous reports and proposals, such as the one by the AIHC, uncertainty was supposed to have been dealt with and eliminated during a prior scientific stage. Now, risk assessment and risk management covered the political aspects of dealing with risk, which took away altogether the need (at least provisionally)[33] to speak of values. Risk evaluation was replaced by a much more positivistic notion of *risk assessment policy.* In a clear act of

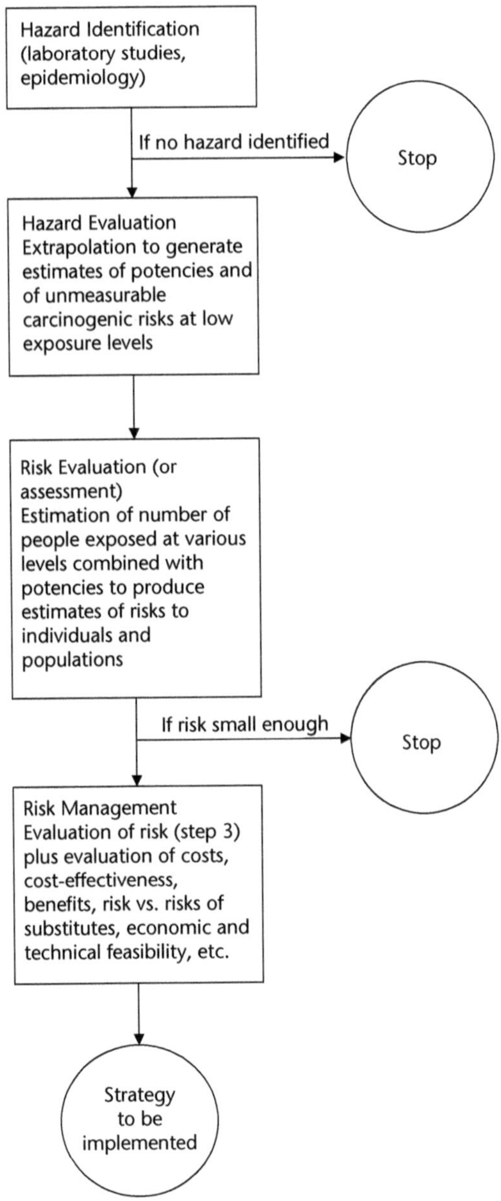

Figure 4.4
The graphical framework for risk decision-making of the Office of Technology Assessment (adapted from OTA 1981), based on Deisler's notion of risk management.

simplification and deconflictualization, that facilitates the packaging and commodification of this knowledge of decision-making (Suddaby and Greenwood 2001), a term was coined to embrace all the judgments made by assessors in the face of uncertainties and gaps in their knowledge, as listed in McCray's important annex. Interestingly, uncertainties, assumptions, and risk assessment policies were neither represented in the graph nor included in the set of highly cited definitions on page 3 of RAFG. As a result, the idea did not travel much with RAFG and was mostly traded for an even more positivistic idea of science policy (Jasanoff 1995); more on this topic is covered in chapter 5.

The combination of the two generic terms of *risk assessment* and *risk management*, which took the place of *science* and *decision* that had been used more frequently until then, made it easier to envision a variety of ways in which science and decision-making could be articulated in practice. The report made it clear that the distinction between risk assessment and risk management closely parallels the distinction that the OSTP report had made, three years earlier, between a first stage of identification and characterization of chemical substances and a second stage of evaluation of regulatory options. But in the work of the RAC, the linearity of the sequence was somewhat minimized: A regulatory decision was depicted as the product of two activities of risk assessment and risk management running in parallel, not the end sequence of a linear process starting with assessment. The RAC thus effectuated an important and innovative terminological construction in which regulatory decision-making in the face of risk ceased to be represented as a linear process of science informing policy. Instead, it was seen as the assemblage of two concurrent activities—namely, risk assessment and risk management—one of which was rooted in environmental and health sciences, and the other in economic and regulatory analysis.

The group effected a move from a representation of risk calculation as a "rigid formula"[34] to a depiction as "concomitant processes." These terms form what could be called an institutional thesaurus. In that lexicon, the terms had either an associative relationship (risk assessment/risk management) or a hierarchical one (*risk assessment* being the generic term for the four specific operations of hazard identification, dose-response assessment, exposure assessment, and risk characterization). The terms could be combined to shape the process of risk regulation. The thesaurus served to index, flexibly, different types of policy content, as well as to distinguish how and when to employ them to shape a regulatory decision. This was a very effective scheme, not just for making decisions, but for making sense of risk politics, comparing different agencies' organizations, highlighting the importance of guidelines, and demarcating zones of ignorance and judgment from zones of knowledge and fact. The whole design, one could say, is generative rather than merely representational (Drucker 2014).

The subtly intertwined and abstract definitions of risk assessment and risk management were an alternative to the science panel. The committee refrained from imposing a single organizational template and becoming involved in the design of local applications of a general scheme. It adhered to a higher-level, somewhat codified (if not coded) discourse on the associative relationship between risk assessment and risk management. The concepts conveyed the idea that it was not advisable to remove risk assessment from regulatory agencies.

Reading *Risk Assessment in the Federal Government*

By September 1982, people in the NRC were convinced that the committee's work would have a greater impact than other NRC reports had had. First, it had aroused the interest of many in Washington, and expectations had increased. Moreover, the press had covered the work already, as the committee was progressing. On January 15, 1982, the *Environmental Health Letter* reported: "Regulators Urged to Separate Facts from Values in Risk Assessment." A month later, the newspaper *Food Chemicals News* published an article titled "NAS Committee Urged to Consider Guidelines for Risk Assessment," reporting on a public meeting held by the RAC on February 10.[35] When the report reached reviewers, it soon seemed that the new formalism would have palpable effects on the understanding of science and policy conundrums in regulatory agencies.

The introduction to the report pointed out that the context in which the charge emerged was one of controversy and conflict. Then it moved directly to the presentation of those now highly quoted and institutionalized definitions:

> Regulatory actions are based on two distinct elements, risk assessment, the subject of this study, and risk management. Risk assessment is the use of the factual base to define the health effects of exposure of individuals or populations to hazardous materials and situations. Risk management is the process of weighing policy

alternatives and selecting the most appropriate regulatory action, integrating the results of risk assessment with engineering data and with social, economic, and political concerns to reach a decision.

Risk assessments contain some or all of the following four steps:

- Hazard identification: the determination of whether a particular chemical is or is not causally linked to particular health effects.
- Dose-response assessment: the determination of the relation between the magnitude of exposure and the probability of occurrence of the health effects in question.
- Exposure assessment: the determination of the extent of human exposure before or after application of regulatory controls.
- Risk characterization: the description of the nature and often the magnitude of human risk, including attendant uncertainty. (NRC 1983, 3)

This definition drew from the work of McCray, appended to the report as a working paper on the anatomy of risk assessment, which defined the concept as a combination of "a hazard identification or a dose-response assessment with an exposure assessment" (McCray 1983, 84). This meant that it compounded knowledge of how hazardous the risk object was in different conditions (meaning, for a chemical, at different levels of exposure or dosage), with knowledge of how many people lived in those conditions. The outcome turned out to be increased knowledge of when and where (or for whom) the harm was likely to materialize. In RAFG, these definitions were represented in a general diagram depicting key operations and mechanisms in shaping regulatory decisions (see figure 4.1, earlier in this chapter). The committee spent very little time on this diagram, but it was the form under which the report traveled by far the most.

The definitions were laid out in chapter I of the report, drafted by Warner North. Armed with these terms, the group then focused on what it called the "tricky issue" of how to handle the relationship between risk assessment and risk management, as well as between science and policy. The remainder of the report thus addressed the interplay between risk assessment and risk management, notably through the use of guidelines (chapter II, drafted by Rodricks) to clarify the respective ambit of politics and of science in the process, and the organizational arrangement of these interactions (chapter III, drafted by Merrill).[36] Chapter IV detailed all ten recommendations.

The report argued that the main political problem in risk decision-making was not whether and how science should inform policy (or vice versa), but a practical issue of making sure that choices pertaining to the analysis of hazards and uncertainties on the one hand, and to regulatory strategies on the other, were made not only transparently and separately, but also in full awareness of one another. Risk assessment and risk management were both in dialogue, and yet transparently distinguished from one another. This rule was most important for what the report called "risk assessment policy"—that is, the models and assumptions that the various scientists performing the calculation of the risk chose to apply. In the words of recommendation A in the report: "[R]egulatory agencies [should] take steps to establish and maintain a clear conceptual distinction between assessment of risks and consideration of risk management alternatives; that is, the scientific findings and policy judgments embodied in risk assessments should be explicitly distinguished from the political, economic and technical considerations that influence the design and choice of regulatory strategies" (NRC 1983, 7).

In the rest of the report, and in the appendix in particular, risk assessment was clearly presented as a mix of science and policy. Originally, the report argued that there was politics in risk decision-making throughout the process, in both risk assessment and risk management. This was stated several times in the report, notably on page 7 (recommendation A), and was also noted in subsequent presentations of the report around Washington. In his 1983 testimony before the Committee on Agriculture of the House, Rodricks expressed it thus: "One central point [in the report] is that, while subjective judgments that have policy implications—particularly on the degree of conservatism that should be applied in areas of scientific uncertainty—may be an inherent part of risk assessment, certain other types of policy considerations, such as the economic importance of a substance, should not be permitted to influence risk assessments" (US Congress 1983c, 360). In short, there were politics or elements of policy throughout the regulatory decision-making process, but some were classified as ingredients to assess risks, and others as ingredients to manage them. Both, however, were equally political, which is why neither could claim to have the upper hand.

Admittedly, there was residual ambiguity in RAFG as far as uncertainty was concerned.[37] The definitions on page 3, in the summary of the report, seem to exclude uncertainty, defining risk assessment as the use of "the factual base" to define risks. Nowhere was it mentioned that it incorporated inferences or assumptions to compensate for missing data. Recommendation A presented risk assessment as a compound of "scientific findings and policy judgments," but again emphasized very little the uncertainty of the whole exercise. "Risk characterization" was the only module in the whole exercise that seems to involve uncertainty. So the summary was expurgated of almost all uncertainty. These first pages were written by Stallones, the chair of the committee, whom Davies placed in the so-called science camp (Davies 2003). Those who defended the view that risk assessment was a mix of science and policy were rather unhappy when they discovered the content of those pages—but it was too late to make any amendments to it before publication.[38] Another ambiguity lay in the general design of the process: Was risk assessment a sequence of necessary and logical steps—as in the stylized decision tree depicted on the cover—or not? Page 3 spoke of "steps," but also mentioned that a risk assessment might contain "one or more" of the components of the framework. On page 21, the graph contained "elements of risk assessment," but described a converging flowchart, in which most elements seemed interdependent and necessary. That graph, which arguably circulated more than McCray's appendix (in which uncertainties and corresponding inferences were detailed), made no mention of uncertainty, or even of the crucial category of "risk assessment policy" that denoted the space where policy and science were hybridized.

The other two general recommendations essentially concerned the issue of guidelines. Recommendation B recommended that agencies develop what was termed "uniform inference guidelines"—namely, a document listing, as comprehensively and in as much detail as possible, the choices that risk assessors might make in the course of a risk assessment. Recommendation B derived from chapter III of the report, where a highly informative and clear history of the effort of creating guidelines in regulatory agencies was recounted. These lines of the report were also extremely helpful to make sense of what the guidelines actually were and why they were an important instrument. The report spoke of "guidelines," not to refer to any official legal pronouncements of procedure satisfying legal requirements, but rather "the principles by which risk assessments are to be performed." The report used this definition "because that is the term Congress used in the legislation that authorized this study" (NRC 1983, 52), and because the term had gained currency since its use by the EPA and the IRLG.

Recommendation C advised the federal government to create a central board for risk assessment methods, whose function would be, among other things, to review the guidelines adopted by regulatory agencies, and to ensure that they were in keeping with the advancement of scientific knowledge in the area. That recommendation read like a balancing act: The report clearly accredited the need to change something to the current institutional organization of science-based decision-making and of regulatory agencies, but it also went against the AIHC proposal to take risk assessment away from agencies, which were perceived as too disruptive.

The reviewers of the report noted how the report furthered understanding of the processes and politics involved in regulating controversial substances. NAS generally organizes a very strict and demanding review process for each of its reports.[39] The review of RAFG was no exception. In fact, Al Lazen of the NRC made sure that the review would be even more stringent than usual. Six months later, the reviews arrived on the chair's desk, and they were excellent. One reviewer was of the opinion that the report was first rate and provided "an excellent analysis of the subject, stated with exceptional clarity. The clear separation of risk assessment and risk management is particularly helpful." Another stated simply that "the overall report is superb ... well-written, very balanced, extremely timely, and offers lucid, realistic and intelligent conclusions and recommendations." Two others stressed that the "workable" and "reasonable" recommendations would definitely improve the situation. Stallones commented to his fellow committee members, a few weeks before the report's release: "The general comments of the NAS were so good that I have included them as an appendix to this memo. When your car won't start, your students won't study, your rats won't die, or your boss just won't, then you pull out these pages and read them again, and bask in the warmth of well-deserved praise."[40]

The analytical quality of the group goes a long way to explaining how and why this integrative, simplified outcome could be reached. The NRC staff had found that the job was of relatively high quality, despite an uncommon and complicated institutional mission. Lazen, the head of the Life Sciences division of the NRC, who oversaw the committee, wrote to Frank Press: "The committee is one of the most intellectually gifted I have had the pleasure to deal with. The chairman, after a slow start, is now a very strong and effective leader who will do very well in front of the press or a congressional hearing. The report's acceptance and impact will depend in large part on the credibility of its authors as individuals, and here we are on firm ground."[41] The members of the panel, in retrospect, readily praised one another's intellectual quality and breadth of knowledge about risk assessment and agencies. McCray was described as a "superb analyst," Rodricks as someone who was able to grasp the science and yet understand the institutional and political big picture as well, and Merrill as a first-class mind and legal scholar.[42] The decision-theoretic perspective brought in by Warner North was also recognized as a crucial component in the committee's collective capacity to articulate an effective knowledge representation. The intellectual and political diversity of the group, combined with the effective direction given by the NRC staff and the chair, contributed to turning the committee into a place where things that were conflicting outside could be discussed inside.

Thus, there was a fine line between success and failure in the work of creating frameworks. In this particular case, the committee could very well have failed due to the inability of the members of the group simply to communicate with one another and to engage in the elaboration of a common view. There was actually a real possibility that this would happen. Lazen, ex post facto, wrote to the chair with the following words of gratitude: "How you got a consensus out of such a vocal and diverse group will always be something of a mystery to me—but I'm not knocking it!"[43] The immediate, instrumental action of the new EPA administrator, with the proposed design, soon showed that he was exactly right.

Conclusion

The episode of the writing of RAFG richly illustrates how and why, in their diversity, experts of science and of its use in decisions can foster the design of legitimate administrative forms. It demonstrates how this form of science can bring order to an area of governance otherwise steeped in public controversy, interest group conflicts, and interinstitutional disagreement. The moment that a report on the legitimacy of regulatory agencies to analyze and regulate risks was requested from the NRC is one of intense controversy around particular chemicals and the risks they cause, but also and more broadly, around the legitimacy of regulatory agencies to intervene to control these products. It is also a configuration in which scientists with an experience in administration, with growing social capital and reputation in Washington, D.C., as well as a common belief in the possibility to invent acceptable decisions thanks to science, were brought together. A configuration, finally, that sparked multiple processes of abstraction, formalization, and standardization of new designs for decision-making, underpinned by different representations of what comprises legitimate bureaucratic knowledge and technology.

What is particular, in the case of the RAC, is that it concentrated a collective capacity to commodify knowledge, deriving in this case from the application of a decision-aiding perspective and an ability to think in terms of process; an ability to insert one's contribution into a longer history of codification and to build on elements of already genericized and tested bureaucratic knowledge; finally (and perhaps more important), the internal diversity of the group functioned as an initial trial for the framework, which made it resistant, presumably, to subsequent trials and controversies in the wider world. Frameworks, this case shows, result from the modeling of intraorganizational diversity, its components, and their assemblage. They successfully emerge where the diversity that is modeled in the framework is represented in the very site in which it is designed and in its content.

It is frequently argued that the RAC constructed a new framework, almost ex nihilo. This chapter illustrates the concrete work involved in creating an organizational and administrative order, by renaming and articulating the elements of decision-making that were already emerging from the practice of agencies, and their own, initial design work. The committee was less inventing new elements of a framework than working to articulate already-identified elements, renaming them so that they become part of a process that appeared to respond to the expectations about what people in an agency should or should not do. Its most original product is also the one that is less often cited: it is the formalization of a notion of risk management. The notion allowed rearticulating science and policy in a way that administrations such as the EPA—an agency that the experts knew of and in some sense defended—could recognize and appropriate. It asserted the decisionistic design emerging in the agency, combining the calculation of risks using medical, toxicological, and statistical expertise, and the use of criteria of cost, benefit, and policy opportunity, which help compel a decision and end a dispute. The framework uniquely recognizes the place that health scientists and economists were in the process of carving for themselves in the agency. The posterity of RAFG was in no manner written. But the configuration in which it emerged made it a potentially influential design effort. What happened, nearly as and when it came out, confirms this point.

5 Ruckelshaus and Risk: Representing the EPA

March 1983 is an important date in the history of the institutionalization of risk-based decision-making more generally. That month, two events took place almost simultaneously in Washington, D.C., the conjunction of which defined and installed a particular design at the heart of federal policymaking and regulatory practice. The first was the awaited publication of RAFG. A day before it came out, the report was presented to an assembly of regulators, risk professionals, industry representatives, and congresspeople at a well-attended dinner at NAS on February 28, and publicly released the next day. The second event occurred eight days later, when EPA Administrator Ann Gorsuch resigned. President Ronald Reagan had appointed her in March 1981 to apply his deregulatory, antienvironment policy agenda at the EPA. Gorsuch did so with a vengeance, severely cutting the staff, implementing drastic budget reductions, confiscating important dossiers to make decisions alone, and slowing and discouraging the development of regulations and enforcement efforts at her own agency. She gradually corroded not only the staff morale, but also the effectiveness and image of the EPA in the industry, as well as in the wider public. She resigned after the House of Representatives cited her for contempt of Congress for refusing to hand in agency records that congressional committees wanted to investigate to clarify allegations of mismanagement of the Superfund program. On March 17, 1983, Reagan reappointed William Ruckelshaus to the position he had held a decade earlier, with the goal of restoring the image of the agency and the morale of its staff. Ruckelshaus returned to the agency with revived energy, vision, and fresh ideas to make it effective and credible again.

The governance of risk and its constituent risk assessment–risk management framework are overwhelmingly sourced to RAFG.[1] But the NRC's report would probably have been just another well-regarded but confidential text if Ruckelshaus and his aides had not picked up on the architecture of categories and processes that it advanced and people in the EPA had not already started to conceptualize what risk assessment and risk analysis comprised. The particularity of what happened then was that Ruckelshaus implemented—and solidified at the same time—a knowledge representation that RAFG had distilled, because it provided an image of EPA's processes and outputs that concurred with the expectations of various audiences of the agency simultaneously. Indeed, various designs were included in RAFG and in the interpretation that Ruckelshaus and his staff made of it, including the decisionistic logic of deriving decisions from set cancer theories, the commensuration logic of risk-ranking and prioritization, and a third, deliberative logic of constructing images of public concern interactively with the public (i.e., *risk communication*). The categories emerging in RAFG, the political crisis surrounding the existence of the latter, and Ruckelshaus's enterprise of legitimation of the agency formed a unique combination of factors leading to the material and symbolic reorganization of the agency and its way of making decisions.

The Dissemination of RAFG in Washington, D.C.

Anticipating a larger-than-usual impact for RAFG, the president of the academies, Frank Press, decided to hold a dinner the day before its official launch. More than 100 guests attended this highly successful event at the NAS in Washington, D.C., on February 28. Having collected impressions from around town at the dinner, Larry McCray, the NRC staff director for the project, wrote to members of the panel on April 18 that it had been deemed a success by most of those who were present. Press himself characterized the dinner as "probably the best the Academy has had,"[2] with high attendance by members of Congress—twenty-seven of them—and the heads of affected regulatory programs, as well as high-quality presentations by Reuel Stallones, Gil Omenn, Richard Merrill, and Joseph Rodricks. Omenn also considered the dinner an effective event.

The report was published the next day as a neat, readable, accessible document with a vivid red cover (hence the nickname it will later be given, "the Red Book"; see chapter 9). The *Washington Post* signaled its release in a short article. The NRC distributed hundreds of copies of the RAFG, but it hardly needed to: the report instantly found its readers. This was soon confirmed by the sales numbers: the Academies Press reported that it had sold 600 copies as of April 8, and 1,250 copies by June 24. By all appearances, then, the report was a hit. Less than a month after the release of the report, Philip Smith, the NRC's executive officer, received a letter from Joe Penick, senior vice president of Mobil Oil Corporation and a key player in the AIHC. On the whole, Penick approved of the report and congratulated himself on the fact that it both supported the AIHC's intention in proposing a central risk assessment panel, and was more constructive, offering "a better fit with current procedures of the regulatory agencies." In short, it had a better chance to succeed politically and institutionally. Penick also acknowledged that the report brought about a key conceptual shift: "We agree to the need to distinguish the scientific 'risk assessment' process from the social-economic considerations for 'risk management' required in regulation and standard setting."[3]

The work of disseminating the report was not limited to standard editorial marketing. A significant share of the members of the panel sent copies in person to key contacts or traveled to present the essence of their recommendations to various organizations in Washington, D.C., and in the country more generally. Immediately after the launch, Press wrote to various congresspersons, to the director of the National Institute for Environmental Health Sciences, and to Ruckelshaus, who had just been appointed by Reagan as EPA administrator. Press's executive officer, Phil Smith, sent the report to Jim Tozzi, deputy administrator of the White House's Office for Information and Regulatory Affairs (OIRA), and to Douglas Costle, former administrator of the EPA and a member of the Council for Environmental Quality under President Jimmy Carter. In May 1983, Joe Rodricks testified before the House Committee on Agriculture, Subcommittee on Department Operations, Research and Foreign Agriculture. He testified again before the Senate Subcommittee on Natural Resources, Agriculture Research, and Environment in May 1984. Stallones summarized the report before members of the Toxicology Forum in Aspen that same year.

Those multiple contacts left no doubt about the general reaction to the RAFG. In Washington, D.C. at least, regulators, bureaucrats, lawyers, industry representatives, and risk professionals—that small, nascent risk regulation community—broadly accepted the analysis presented in the report. Most of the reactions that were recorded show that it was interpreted as a report dealing with the problem of the relation between science and policy—though it replaced these categories with risk assessment and risk management, each embedding a kind of policy content. Other members of the panel, who were defending the more sophisticated "risk assessment policy" concept, were frustrated by this perspective (North 2003).[4] The interpretation of RAFG as saying science should be separate from policy, however, was not completely germane to what the author said in the report. The first few pages introduced risk assessment as a "factual base," not mentioning the role of assumptions and policy judgments involved in choosing parameters to calculate risks at low doses, for instance. And the graph "Elements of risk assessment and risk management," on page 21 of RAFG (figure 0.1), so often reproduced, did not convey the role of uncertainties in a seemingly unstoppable, flowing process of learning from research to produce an estimate of the risk. It does not feature risk assessment policy either. Even Gil Omenn, though a full participant in the panel's work, considered that it was essentially saying the same thing that he had argued in the OSTP report of 1980: that science and decision-making were two separate elements.

Closing the Crisis

A report or a study, however well written and authoritative, generally has very little impact on politicians' strategies and decisions because they simply lack the time to consider ideas and pore over long texts. But the contrary may be true when the study comes at just the right time and place, when and where the politicians need it. In this case, the Academy "got the timing nearly perfect" (Barnes 1993, 8). The report came out just as Reagan was searching for Gorsuch's replacement. Reagan soon decided to bring Ruckelshaus back to the agency to solve this full-blown, open crisis.[5] By his own admission, Ruckelshaus did his best to find alternative people for the job, but eventually decided it was his duty, and challenge, to accept the mission. He finally accepted the mission offered by Reagan on the condition that the budget of the EPA be restored to the 1981 level and that he keep control over the choice of political appointees. Reagan made an exception to his "administrative presidency" strategy (Golden 2013) and agreed to supervise the actions of the EPA less closely than in the preceding years, confident that Ruckelshaus was the man for the situation and that a depoliticization of the agency would help in deflating the crisis surrounding the environmental issue. He secured an increase in the agency's budget, particularly that of the ORD. While the White House held a discussion of each and every proposed candidate, there was no veto on any of the people chosen by Ruckelshaus and his chosen deputy administrator, Alvin Alm.

Ruckelshaus's persona was that of an effective, dutiful, and impartial civil servant of the highest capacity, whom Reagan trusted to restore the image of the agency and the morale of its staff, as well as the overall credibility of that administration with regard to environmental problems. Many of the people who commented on his nomination used the word *integrity* to describe his reputation (Shabeccoff 1983). Environmental groups, though not opposed to Ruckelshaus's renomination, did not explicitly support it either (Lippincott 1985). They were not reassured by his intention, expressed at confirmation hearings, to hardly change the substance of Reagan's policies but rather focus on adapting the EPA's management style (Layzer 2012).

Ruckelshaus knew he was on a mission. He had little time to restore the reputation and functioning of the agency, and indeed had given himself a tight deadline (he wanted to leave the job after only a couple of years). Before officially taking office in May, he spent time consulting with a wide range of actors of environmental policy, from environmental activists to various industry groups to EPA staffers, to get a sense of what they viewed as the most important and urgent problems to be dealt with when he returned to the job. What he gathered from these conversations was that the agency needed to demonstrate that it was active and effective in protecting the environment and people's health—a point on which Gorsuch had truly instilled doubt in the population. According to Ruckelshaus, starting to intervene again "very clearly and vey publicly" was also in the interests of industry, where there was indeed even a demand by some segments (Anonymous 2008). Ruckelshaus took his time to reflect on the best language to use to reshape the agency's image among its staff and the public. Thinking in terms of "risk assessment" and, more innovatively, "risk management," was how he framed it.

Ruckelshaus already had been given draft copies of RAFG by the president of the Academies in February. He was well prepared to consider its ideas because he was immersed in the intellectual climate of the time and the ongoing conversations about science-based regulation and risk assessment. He had refused to serve on the RAC but had served on Howard Raiffa's CORADM. He knew several members of the panel. In his position, he could not ignore the debates surrounding the competences and legitimacy of agencies to employ science, as well as the AIHC's campaign for a science panel—a proposal that made little sense to him.[6] Most of his take on risk was inherited from intellectual work around the EPA, in the newly created SRA, in the NRC, and in intellectual references used there, such as the notion that safety is a judgment rather than a calculation (inherited from William Lowrance's influential book *Of Acceptable Risk*), and the distinction between voluntary and involuntary risk. In the words of his future assistant administrator for ORD, Bernie Goldstein, he "jumped on the Red Book."[7] Ruckelshaus's subsequent reconstruction of this choice was the following:

> I was looking for a way of sort of calming the controversy down that existed with my predecessor at EPA, Ann Bedford, and to put the focus of the issue in front of the country in a reasonable way, and the Red Book was a very good tool that helped me accomplish that. Because it allowed me to talk about a lot of issues involving the assessment of risk, which as I say, should be a process that should not in any way be subject to political intervention, and the policy judgments, which are political, with a small p, they should not be a big p political, which society decides for itself what the risk is and what to do about it. And that is a difficult problem, under the best of circumstances. And I was trying to get that whole debate or discussion of the issue of risk assessment and risk management out of the political debate and this Red Book allowed me to do that.[8]

Ruckelshaus was recognized as an effective and skilled public communicator. In the first few months of his second term as EPA administrator, he gave multiple speeches in a number of prestigious locations, addressing a variety of audiences. The first full representation of the EPA as a risk agency was in a speech he delivered before an audience of scientists at NAS on June 22, 1983. The speech opened by depicting a public full of panic and fear, concerning its natural environment, public health, and economic survival. It rapidly moved to name the "idea of science" as one of the fundamental answers to these fears. Science alone could not provide the answer, though, because of its dissonance with public policymaking in a democratic system: "Nowhere is this more troublesome than in formal risk assessment—the estimation of the association between the exposure to a substance and the incidence of some disease, based on scientific data" (Ruckelshaus 1983, 1026). Ruckelshaus then brought in the new notion of risk management:

> Scientists assess a risk to find out what the problems are. The process of deciding what to do about the problems is risk management. The second process involves a much broader array of disciplines and is aimed toward a decision about control.

> In risk management it is assumed that we have assessed the health risk of a suspect chemical. We must then factor in its benefits, the costs of the various methods available for its control, and the statutory framework for decision. The NAS report recommends that these two functions—risk assessment and risk management—be separated as much as possible within a regulatory agency. This is what we now do at EPA, and it makes sense. (Ibid., 1027)

This speech was probably the first public declaration in which the EPA *as a whole*—all of its statutes, offices, and modes of intervention—was redefined using the language of risk: a sort of official birth of risk-based governance. Being delivered at the NAS, to an audience of scientists, and featuring many poignant lines about science, the speech was noted as a commitment to the use of science in federal policymaking.[9] It figures in several books and essays on environmental policymaking and risk regulation (e.g., Dietz and Rycroft 1983; Levenstein and Wooding 1997), and is frequently cited in the academic and policy literature.[10] But there were other reactions to it. The *New York Times*, for instance, emphasized those parts of the speech where Ruckelshaus announced a new cross-government, interagency initiative on risk management. He wanted other agencies to embrace risk and its dilemmas, to relieve pressure from the EPA, and convey more effectively to the public the difficult problems that the federal government in general was facing (Shabecoff 1983).

The main objective was to project in public the internal dilemmas and difficulties of the administrative and scientific processes that the EPA had to operate, perhaps much more than the actual development of risk assessment science at the agency. The redefinition of the EPA's identity in these terms was a direct, transparent response to the accusations against Gorsuch and the concerns of EPA staff. Whereas Gorsuch and John W. Hernandez were accused of manipulating the process of evaluating scientific data in closed meetings with industry, Ruckelshaus gave autonomy to the process of risk assessment, separating it from risk management. Whereas Gorsuch had severely curtailed the agency's budget for research, Ruckelshaus defined science as its most critical resource and announced that the budget of the ORD would be increased. He also stated that he would work to develop long-range research in the agency, thus responding to a plea by many in Congress, industry, and academia. Whereas the policies of Gorsuch seemed to be motivated entirely by the idea of reducing federal interventions, Ruckelshaus focused on considerations of safety, scientific robustness, effectiveness, and costs. Finally, whereas controversies arose from the misunderstanding of how the EPA assessed risks and derived regulatory measures from it, the risk assessment/risk management architecture was in place to convey to the public a much more direct and approachable image of the agency, one that spoke to expectations about its action in Congress, in regulated industries, in the media, or in the environmental community.

The EPA as a Whole

The risk representation contained, first, a clear, integrated agenda. The administrator insisted that the nature of the issues that the agency was dealing with had fundamentally shifted. While the pollutions the agency took on in the past were material, with identifiable offenders, toxics now represented a more diffuse, evasive, and general threat.[11] This appreciation, arguably a very political judgment about the issues that comprised its agenda, was compounded by internal qualitative and quantitative surveys. One of the reports that Ruckelshaus used was a document called "Trends Likely to Affect the EPA in the Next Ten years," which contained the graph shown in figure 5.1.

Risk assessment, second, not only established the EPA as a scientific agency, but also highlighted the dilemmas involved in using science. Ruckelshaus spent time in his speech outlining the conflict between the work of science and the work of making laws on the environment and health. He reminded the public of this in a sentence that strongly resonated with most recent expert reports on risk assessment, including RAFG: "In assessing a suspected carcinogen, for example, there are uncertainties at every point where an assumption must be made" (Ruckelshaus 1983, 1027).

In the June 1983 speech, Ruckelshaus distorted what RAFG meant, as well as this emergent common wisdom among risk researchers. He spoke about science as much as about risk assessment, betraying his belief about the objectivity of knowledge on risk. He defined risk assessment simply as knowledge about "the nature of the risk," and risk management as the question of "what to do about the risk," as if the risk were easy to capture. Accordingly, this speech called for a clear separation of the two aspects, arguing that this was what RAFG had recommended and seemingly claiming that a neat separation between the objective world of facts on the nature of the risk and the world of choosing what to do about it, based on other, more political considerations, could easily be found (Jasanoff 1987). In fact,

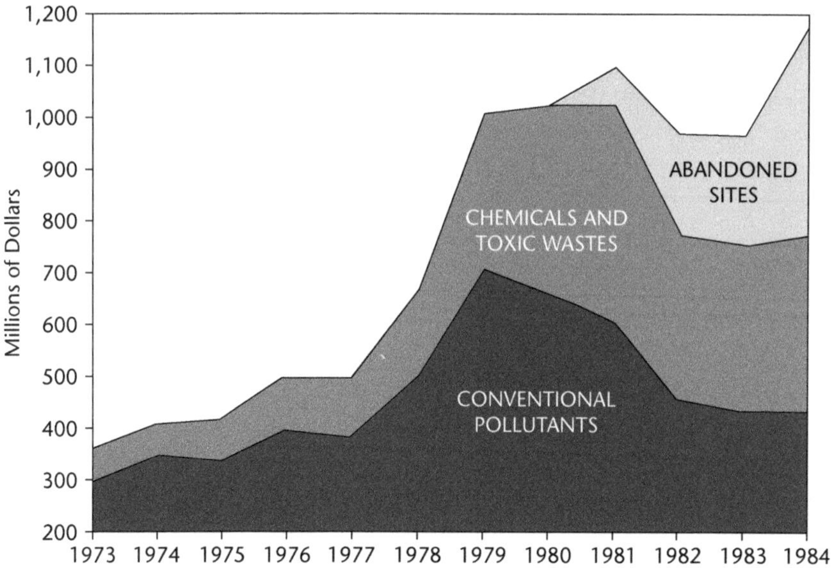

Figure 5.1
Funding history of EPA operating programs by class of problem (adapted from EPA 1983c).

several members of the RAC were embarrassed about the way the report was used by Ruckelshaus as well as others in order to stress this separation of science and policy. Warner North, for instance, recalls that he was troubled by the call for separation, which he considered a misinterpretation of RAFG: "We had encouraged conceptual distinction, but not organizational separation, and we had stressed the need for ongoing communication between the risk assessors and the risk managers. Ruckelshaus went on to give more speeches subsequently published in leading journals.... But he repeated the language that encouraged misinterpretation of what the Red Book actually said" (North 2003, 1150–1151). Some of the very same scientists who were advising the agency were also thinking, contra Ruckelshaus, that uncertainties were inherent to science as and when it dealt with risk, and that risk assessment was inherently limited, in its precision, accuracy, and credibility, by the choices at the level of parameters that were made to compensate for data gaps[12] that constituted risk assessment policy.

But there are reasons to believe that North misinterpreted Ruckelshaus's misinterpretation of RAFG. While Ruckelshaus was speaking before an audience of scientists, his speech was clearly addressed to other audiences of the agency. He was there to save the agency from accusations that its political appointees manipulated science to legitimize the controversial choice of leaving the chemicals in use. It was sensible politically to proclaim and demonstrate, through concrete descriptions of administrative processes, that EPA top officials would not tinker with the scientific data to impose their preferred regulatory options or dictate to risk assessors what their calculations had to be.[13] To be sure, the conceptual refinements and organizational recommendations of the NRC about inference options and risk assessment policy were lost in his public presentation.[14] Ruckelshaus was not indifferent to these sophistications, but the benefits of the framework in terms of explaining to the public the problems that the agency was dealing with, and its use of science, far outweighed the cost of localized internal critique.

This time, an internal reaction to his representation of the new EPA concerned the emphasis on health risk. Stressing the health agenda, Ruckelshaus was putting forward the action of certain parts of the agency. That focus seemed inappropriate for those who remained primarily interested in dealing with environmental pollution and contamination, as under the hazardous waste and Superfund programs. To correct the imbalance, Lee Thomas, the assistant administrator in charge of these programs, had to point out on the occasion of the establishment of a cross-agency report on risk assessment and risk management, that risks to flora, fauna, and natural resources were part of the agency's mission, as much as risks to human health.[15] By stressing assessment of risks to health as the agency's central commitment and resource, the administrator gave the impression of forgetting some of the fundamental roots of the agency's research programs: namely, the work of its engineers in designing and testing technologies for pollution control and reduction. This was an area in which the agency had amassed a few important successes by developing technologies that would not have existed otherwise—the main example being the catalytic converter.[16] In September 1983, a group of engineers from various labs of the ORD wrote a joint letter to Ruckelshaus to indicate that, by stressing health issues and risk analysis, he was in essence giving up on a fundamental part of the identity and defining source of legitimacy of the agency. They asked him for guarantees that research funding would continue to accrue to them. Those budgets overall declined, as did the ORD's budget over the years, but the EPA's investment in the area declined under a conjunction of trends: the privatization of much of these technological developments and the concomitant shift of regulatory agencies toward the instrument of performance-based standards in the area (Vinsel 2012).

While Ruckelshaus stressed risk assessment and its separation from policy in many of its public discourses, the most innovative part of this discourse was the use of the notion of risk management, to speak about policy and government of the environment. Risk management nicely bridged different goals and modalities of what the agency was supposed to be doing. It could easily be interchanged with the notion of "risk reduction," which Ruckelshaus readily employed in communication with the general public, notably through the press. He argued that if risks needed to be measured and assessed, then it followed that levels of risk could be acted on, and the EPA would be acting to move these levels downward—not to eliminate them entirely. The notion of risk reduction became one of the central themes in the agency, and lastingly so (EPA 1988). At the same time, risk management, as defined in RAFG as a weighing of different considerations, including costs, was appealing to industrial audiences. It helped demonstrate to the public that the agency was keeping economic and industrial questions in mind. In the first lines of his June speech, Ruckelshaus hinted at the problem of "economic survival," along with the challenge of safety and environmental protection. He also explicitly spoke of costs and benefits as elements of risk management decisions. Ruckelshaus was not an environmental zealot; he was a seasoned Republican, with the environmentally conservative sensitivity that defined a substantial portion of the party back in those years. He also had a good track record in terms of environmental decisions made since his first stint as head of the EPA. But he was not indifferent to the issue of industrial development.

In his view, environmental protection was to be weighed up against industrial development and economic growth—which count as benefits in the language of risk management. This was particularly important with regard to air pollutants, in which the EPA had to face some of the most complicated cases of decision-making, such as ozone. The Clean Air Act was a pure risk statute, with no reference to any consideration of costs and benefits. Ruckelshaus, early on in his mandate, wrote to Vice President George H. W. Bush, calling for a less restrictive Clean Air Act (Layzer 2012). He clearly used the language of risk management to give a place to the consideration of costs and benefits, and more generally to regulatory analysis, in line with Reagan's regulatory reform agenda, and to weigh in on this legal issue. The separation of risk assessment from risk management was not simply a way of insulating the science, or a promise not to cover up his decisions in technical language. It was also a way of isolating a space of risk management in which he could freely weigh scientific considerations—and the conservative options that often came with them—against other motives of decision-making, such as costs, benefits and political feasibility.[17]

In doing so, Ruckelshaus was executing the kind of engagement with audiences that he thought was crucial to his job. He was in fact instituting a clear division of work between himself and his deputy administrator. His main job was to be active in communicating to the various audiences and interest groups the agency's decisions, operations, and knowledge, not to manage it day to day. As his successor, Bill Reilly, noted, "Ruckelshaus had a very clear concept of the need not only to ensure integrity in public service, in government, as a government official, but also to communicate to the country what a government agency was doing and why" (EPA 1995b). At a moment when the public at large, Congress, and environmental groups seemed to react primarily to issues of health risks and to the problem of the quality and integrity of the science that the EPA used in its decisions, it seemed entirely rational, from a political communication point of view, to stress these aspects. The configuration of political crisis around Gorsuch and other EPA managers created an opening for new discourse about the organization. The fact that the representation that Ruckelshaus used originated in the work of a prestigious, highly credible scientific body and carried its imprimatur made it a natural political choice.

Representing the EPA's Action

The renewed legitimacy that the framework gave to the EPA was so strong, in fact, that Ruckelshaus also felt authorized to put the agency at the center of a governmental branch, redefined as risk management. In the summer of 1983, he contacted the heads of agencies and departments involved in making decisions about toxics and risks to health. Some were in a similar situation as the EPA, applying statutes that made it mandatory for them to measure the risks and safety of chemicals, and to forge regulatory decisions, such as the FDA (food additives), OSHA (chemicals present in the workplace), or the CSPC (risks from exposure to chemicals in consumer products). But the three departments of health, of transportation, and agriculture were also included in the plan. The idea developed over the summer by Ruckelshaus and his aides was for a council chaired by one of the heads of agencies, rather than by the White House, to exchange information or coordinate decisions on chemicals of common interest, harmonize risk assessment methods, and more—provided that all the agencies agreed. Ruckelshaus built on the IRLG (see chapter 2), but because the risk assessment–risk management framework had tied science to regulatory decisions, regulatory work was now explicitly included in the mandate of this new interagency group as well.

Ruckelshaus had the green light from Reagan to take such an initiative, even though the OMB little enjoyed the enterprise.[18] But in any case, the fact that Ruckelshaus thought that he had the power to launch this initiative so soon after the White House had demoted the previous cross-agency coordination group showed that there was a new intellectual ground from which to observe Government (with a big *G*, as in Ruckelshaus's speech) as a whole. The initiative demonstrated how generic the fact of being faced with controversies, uncertainties, and challenges to decisions had become for many administrations, and how risk assessment and risk management redefined it.

Ruckelshaus argued for the new coordination initiative in the following terms in a letter to his counterparts in September 1983:

> Our agencies are becoming ever more involved in the most difficult kind of decisions: those where government must decide how much society is willing to pay to reduce health or environmental risks. Government always had to rule on such issues, but never before as explicitly or as frequently. Science's ability to detect harmful substances has increased so dramatically that it is no longer possible to suppose that risks can be wholly eliminated: trace amounts of chemicals appear everywhere. Despite substantial improvements in health care and longevity, polls show that the public believes that life is getting riskier, not safer. In my recent discussions with you and with others, I sense a recognition that we need to regain control of the terms of this important public policy debate—not let it become more polarized and destructive.[19]

The statement of purpose tabled at the inaugural meeting on December 15, 1983, also stressed controversy, contestation, and conflictual relations with the public as the new conditions with which governmental action had to deal. They demanded greater coordination and harmonization than ever attempted before. Many decisions of federal agencies in regulating chronic health hazards were controversial; the roots of the controversy lay in changes in public expectations and concerns about health protection, as well as the fact that the costs and benefits of regulatory policies often fell unequally on different groups in American society: "There is therefore a serious need to articulate and clarify—internally and to the public—the necessary differences in how the agencies deal with controversial topics of risk management" (Anonymous 1983).[20]As the quote illustrates, it was the notion of risk assessment and risk management, and the whole scheme for producing decisions that they comprised, that enabled him to think that the action of these agencies, and of the federal government overall, despite its heterogeneity, could be streamlined and organized.

Design, Organizational Image, and the Deflection of Criticism

The recodification of EPA's actions in integrative terms of risk, assessment and management, was extended to defend the agency against a potential threat from Congress. Since 1979, in liaison with the AIHC that lobbied for it,[21] Congressman Don Ritter had been pushing legislative propositions to address some of the criticisms of risk assessment as it was carried out at the time.

The Ritter Bill had already been rejected in March 1983, but Ritter forged an alliance with Representative David Martin to push a new bill on risk assessment. In terms of that bill, the White House was to designate a regulatory agency to coordinate joint research projects with other regulatory agencies to improve the value of risk analysis. Title II of the bill included a proposal to create a Central Board of Scientific Risk Analysis under the NRC, with the role of establishing guidelines to be applied by agencies and reviewing specific analyses by agencies in view of a regulatory decision. The bill differentiated between "risk analysis" (quantifying probabilities of a risk) and the ambiguous task of "risk evaluation" (determining the acceptability of that risk to individuals and society). [22] This design, contrasting with RAFG and emerging knowledge representations inside the EPA, involved a reduction of the autonomy of the agency, to produce the science necessary to advance decisions. It denoted the altered legitimacy of the agency among Republicans in the House.

Since RAFG had been published, and the EPA had resurrected the IRLG under a new form, involving managers of agencies and not only its scientists (thanks in part to the platform articulated in RAFG), the agencies were able to coalesce around the EPA to deflect Ritter's and Martin's ideas. Nearly everyone opposed the idea of creating greater supervision and dependence in the agency's regulatory decisions, including the NAS. In May 1983, Press wrote to Martin, warning that the governing bodies of the NRC had not, as yet, considered the establishment of a new, standing risk assessment board in the NAS or NRC, and that it seemed inappropriate for the risk assessment board to undertake specific risk assessments at the request of federal agencies. The procedure would be cumbersome and lead to prolonged delays.

Former members of the RAC also viewed the bill negatively and were sufficiently reassured by the ideas put forward in RAFG to oppose it. Ted Greenwood wrote to McCray to say that he was "appalled" by the use that Don Ritter and David Martin had made of the report in Congress, precisely because their proposed board for scientific criteria and principles denied the most important point of the report: that risk assessment is a mix of science and policy. Worst of all, the bill "seem[ed] to be trying to use the NAS committee's legitimizing ability for a set of concepts totally contrary to what we wrote and intended."[23] Most of those who were consulted by the House on this proposal regretted the fact that the congresspersons did not use the categories coined in RAFG. The risk assessment/risk management twosome was much more appropriate to capturing the political challenge involved in using scientific estimations of risk. Better than "analysis" and "evaluation," it conveyed the potentially controversial nature of the ties between science and decision, as well as the fact that science had to be carried out independently, yet also had to be performed in close connection with the exercise of making a decision. By insisting on this scheme once again, most of the people who came to the congressional hearing helped to demonstrate the amount of disorder, delay in regulatory decisions, and intractable conflict of scientific authority the bill would recreate.

Of course, the EPA was not the slowest to respond. Elizabeth Anderson, the toxicologist chief of OHEA, who was then heading the efforts of the EPA on cancer risk assessment and championing the use of cancer guidelines, had scanned the bill and developed a list of counterarguments, which she shared with key people in the agency.[24] She claimed that the creation of a central risk assessment panel would cause delays and would not bring closure to controversies because it would just be another point of discussion during risk assessments. It would deprive agencies of very important means, resources, and competence. It would, furthermore, give the NAS a quasi-policy role (as under the language of "criteria"), inappropriate for an academy, the stature of which derived from its uncorrupted adherence to scientific excellence. Staff in the Office of Water analyzed the bill and its likely effects on the EPA in similar terms,[25] as did the scientific advisor to the chief of OPTS and the assistant administrator for OPPE[26]: elimination of some of its key powers to perform risk assessments, creation of more delays in delivering decisions on high-profile chemicals and, essentially, paralysis by analysis.

Finally, Ruckelshaus wrote to the OMB and to Representative James Scheuer (a Democrat from New York who was a supporter of the EPA) a six-page letter opposing the bill, and Bernie Goldstein presented the agency's argument at a hearing before the same congressperson. Given the amount of "semantic confusion" in the matter, the ability that this language offered to explain to the public what agencies knew (and did not know), and how this knowledge factored in final regulatory decisions, Goldstein made a plea for the bill to be aligned on the RAFG's scheme (US Congress 1985). The representatives of the FDA, the CSPC, and OSHA did likewise, representing the way that their agencies approached health and environment, and succeeded in producing slightly less controversial policies, now that they explicated uncertainties inherent in the risks considered, and took into account all other priorities and motives to balance this knowledge. The hearing ended up being an effective education and defense of regulatory agencies and of their actions, in the words of the chair of the subcommittee, James H. Scheuer, a Democratic representative and consistent supporter of environmental policy (ibid.). The idea of a science court and the industry's project to eventually curtail the power of regulatory agencies in the area of science seemed to have been halted.[27]

A Risk-Communicating Agency

Ruckelshaus conceived of his role in communicating to the public and engaging with audiences and constituencies of the agency as being of paramount importance, and he chose an experienced and effective manager to run the agency so that he could concentrate on this role. He believed that regardless of how important communication was for this role, it did not depend solely on him, and it needed to be designed within the organization.

Ruckelshaus asked Milton Russell, the new assistant administrator for policy, planning, and evaluation, to develop activities around communicating with the public about the complexity and uncertainties surrounding risks and their reduction. Russell was an economist by training, with specialization in the study of energy markets and policies. In the 1970s, he had spent time both at the think tank Resources for the Future and at the White House. From 1974 to 1976, he was the senior economist for energy issues on the Council of Economic Advisers, where he met Al Alm, the agency's former head of policy (1973–1976). In March 1983, Alm had been chosen by Ruckelshaus to be his deputy administrator. He looked for dedicated, competent, and politically neutral administrators to reinvigorate the agency, and he offered Russell the job of assistant administrator for policy. The latter accepted despite his lack of experience on environmental issues, and soon went to work on ways of adapting the agency's routines to what Ruckelshaus thought was a new defining condition of legitimacy and authority for regulatory agencies: their dependence on the level of information in the public and the latter's understanding of the particular dilemmas facing the agency when dealing with uncertain issues.

Russell worked on explicating Ruckelshaus's initial hunch—that the EPA should be positioned as an agency aimed at reducing risks for the population, and that it should "do more to increase the public understanding of environmental risks and the considerations that must be taken into account in making risk management decisions."[28] In a memorandum on "Communicating with the Public on Issues of Environmental Risk,"[29] he showed that the question of the public, and of its understanding of the science and other components of regulatory decisions, was not one of the problems of the agency pertaining to external relations and public engagement. It was transversal and concerned the agency's scientists, rule-developing lawyers, and field-level officials in regional offices alike.

In this memo, Russell explained that one of the central problems in the area of risk assessment, for the agency as a whole, was that "DMs [decision-makers] and [the] public [are] unclear about [the] nature of estimates—how uncertain, how conservative?" When it came to risk management, the problem was that decision-makers did not base their decisions on information that the public could best understand, especially scientific information about the number of people who were actually exposed to the hazard in question; the dominant discipline in the agency for assessing effects on health was essentially experimental, calculating dose-responses in animals, not information about people—and information about the actual benefits for the population's health, accruing from the decisions that the agency made. Finally, as concerned communication strictly, the public and Congress had trouble interpreting risk estimates and did not understand that absolute safety was not possible and that some trade-offs were inevitable.

At a noticeable distance from the claim that people's perceptions of risks were different from those of experts and were wrong by the logic of trained probabilistic judgment (e.g., Fischhoff et al. 1978), Russell worked from various assumptions: that uncertainties abounded; that the agency had its own particular ways of framing and interpreting risks, sometimes orthogonal to how people saw those same risks; and that disagreements between various segments of the public and the agency were there to stay. In the agency's efforts to reduce risk, the objective should be to frame and inform controversies, not to eradicate them through supposedly authoritative and objective calculations: "informed disagreement would be preferable to the present situation in many ways" (Russell 1984, 2). Accordingly, the operational problem of the agency should not be "how to convince people we are right," but rather how to capture the "environmental values" that different "public(s)" were most interested in seeing protected, and "how much environmental quality [these publics] want the country to buy" (ibid.). Russell went on to sketch out answers to the question of what decisions to communicate on, who should do the communicating, toward which publics, defined in what way, and through which channels and networks. With Roger Gale, Ruckelshaus's closest advisor at the time, he also began to standardize the messages to use in all EPA staff communication and in Ruckelshaus's communications outside the agency:

> There are a number of basic messages that we feel it is essential to emphasize. Among them:
>
> - We seek to reduce risk.
> - We will always have some risk.
> - We realize that issues are complex and that there is an element of uncertainty.
> - We attempt to anticipate problems before they bite us.
> - We distinguish between scientific assessment of risk and the management of risk.
> - We balance risk and benefit.
> - We enforce the law; the mighty are not above us.
> - We listen.
> - We protect the public.
> - We tell you everything we know.[30]

It was clear, thus, that risk communication first emerged as a discipline of communicating on the constraints, achievements, and overall legitimacy of an administrative organization dealing with uncertain risk, not simply about transmitting scientific information to the public. The 1984 report *Risk Assessment and Management: Framework for Decision Making,* which summarized and institutionalized the new image of the organization for the public, and in many ways was a testament of the transformations initiated at the time of Ruckelshaus's second term, endorsed the point in its final lines:

> The point can not be made too often. In one sense, risk management *is* a form of communication. Technical analysis of the costs and benefits of a proposed action is not a device for coming up with the "right" or "rational" answer: all such analyses are far too sensitive to subjective values and far too dependent on uncertain data for us to pretend that they are. Risk management, and the technical analysis that contributes to it, is largely the *exposition* of the information we believe is reliable, the values we wish to apply and the way that these two are linked to produce a set of policies ... Obviously, not everybody will agree with the values so expressed, but in order for the debate about values to begin and for the democratic processes that ultimately establish values to take place, everyone has to know what the values underlying our decisions really are. (EPA 1984a, 35, emphasis in original)

Those messages served to anchor the image of an agency that was responsive to the public and to what it experienced of the agency in particularly controversial situations—those of the Gorsuch years and the still-frequent controversies that erupted here and there during 1983. The emerging message allowed Ruckelshaus to go toward audiences with which relations had been complicated or inexistent in the past. Russell's memo pleaded for engaging with a number of constituencies that the agency had not considered enough: "risk-oriented constituencies," media managers, state governors, and Congress (Russell 1984a). The risk assessment–risk management framework provided the structure for engaging with the public not as an agency that knew everything, but as one making the best possible decisions in the face of uncertainties.

Ruckelshaus put it in practice himself, in Tacoma, Washington. Taking the opportunity to make a decision on a high-profile, controversial case—arsenic—he initiated a new kind of public event: a town meeting for direct interaction between top EPA officials and an unselected public.[31] In July 1983, he proposed a mandatory pollution control technology for Arsaco's Copper Smelter, plants that represented the country's biggest source of arsenic pollution. The technology was supposed to reduce the level of arsenic emissions by 17 percent. He announced that he would directly consult the residents of Tacoma, where Arsaco's Copper Smelter, the country's biggest source of arsenic pollution, was located, to collect comments about whether that level was acceptable and what other decisions could be made to address the situation. Ruckelshaus did not get away easily with that announcement. He was accused, first, of being lenient on a polluting industry: the proposed restriction on emissions was lower than expected, at least by environmental groups, particularly for a substance that caused cancer and, by convention, was believed to do so at any dose. The press and environmental groups claimed that Ruckelshaus had compromised health protection with job protection—particularly jobs in Washington, where he lived with his family during the 1970s. His initiative of a local public and open workshop on the risks of arsenic was hardly understood. The assistant attorney general of New York State, behind the 1978 lawsuit, argued that Ruckelshaus was giving up on the difficult task of arbitrating between health and jobs, placing the communities in Tacoma before this "artificial" choice. On July 23, Ruckelshaus replied to a *New York Times* editorial depicting him as Caesar, shying away from shouldering tough decisions. In this letter to the editor, he replied that by proposing the said standard (Ruckelshaus 1983):

> [W]e are proposing precisely what your editorial suggests we propose: that ASARCO installs controls on its Tacoma smelter to reduce arsenic emissions to the lowest level we believe is technologically achievable, and thus further reduce the cancer risk to the citizens of Tacoma ... The people of Tacoma are not being asked to make the decision; they are being asked for their informed opinion. They know that the right to be heard is not the same thing as the right to be heeded. The final decision is mine.

The agency did not replicate the Tacoma experiment.[32] However, along with other complicated cases, this issue taught Ruckelshaus the need to have options in mind, present them, and demonstrate his decision-making skill and controlled judgment over the science by applying one option. That is precisely what the entire set of projects deployed in the agency on communicating risk to decision-makers, and on risk management and risk assessment, were about. That bureaucratic design was intended for an agency that would generate options for a central decision-maker, such as Ruckelshaus or Alm, to make decisions and carry them into the public space, with increased levels of potential acceptance.

This particular experimentation was not deemed to be a great accomplishment, but still, "the improvement in the Agency's ability to explain risk to the public" remained a key priority in those years.[33] This concern of the EPA about communicating with the public was not foreign to the risk assessment–risk management framework. The framework that emerged later, in 1988, which designated risk communication as its third pillar (NRC 1989), actually codified what the EPA had initiated. Those two categories were understood as the ideal way of producing communicable and understandable decisions. This concern was incorporated right from the start in various "risk projects" through which the agency was redesigned (including the training of regional staff to improve community relations; development of a strategy by the Press Office to advance the understanding in the media of the agency's approach to risk assessment and risk management and increase journalists' awareness of the dilemmas and difficulties involved in public health decision-making; identification of live cases on which to work to develop risk assessment/risk management techniques; and town meetings with the EPA administrator, to get the concepts of risk assessment and risk management introduced into the popular press).[34] In other words, the results of Tacoma and of Ruckelshaus's sensibility for direct communication with the public were what the EPA staff later called "live cases" and "town meetings."

Risk communication was soon systematized, thanks to a team formed by Milton Russell. Russell and the Office of Policy team on risk communication worked to ingrain this understanding of the public in the agency in several ways. They worked with Ruckelshaus directly, feeding him ideas and knowledge about the emerging field of risk perception. Russell organized "breakfast meetings" for Ruckelshaus, himself, and scholars versed in the philosophy, ethics, or sociology of risk. According to Russell, this was part of establishing "the milieu and the understanding at the highest level," so that people could deal with these issues and explain them to the public.[35] Ruckelshaus and Russell thus met Paul Slovic on January 25, 1984, which led to a formal proposal by Slovic to the OPPE, for activities pertaining to risk communication (monitoring and evaluation of efforts to increase public participation; communication of state-of-the-art knowledge of risk communication to EPA staff; and development of methods to communicate risks in a way that reduced conflict and to incorporated views elicited from the public, into regulatory decisions). Slovic confirmed that the agency should change the presentation of risks that health assessment documents typically used: a single, synthetic figure expressing excess levels of cancer in broad populations (such as, "there is a one in a million chance of developing a cancer from exposure to that substance over a period of fifty years"). As he put it in a letter to Russell: "[A] robust conclusion of risk perception research is that one cannot simply present statistics and assume that people with different life experiences, training, etc., will understand them as intended."[36]

The agency entered into close cooperation with NSF's Vincent Covello, who was in charge of the National Science Foundation program on Technology Assessment and Risk Analysis, and one of the prominent scholars studying the social construction of risk and risk communication. A joint national conference on risk communication was organized in January 1986, where ample space was made for the EPA's experience through interventions by Ruckelshaus and Lee Thomas (who succeeded Ruckelshaus as EPA administrator in 1985) and discussions of prominent controversies handled by the agency, such as hazardous waste and ethylene dibromide (EDB). Later, Russell was contacted by the NRC to join the Committee on Risk Perception and Communication. With this panel, the NRC had decided to engage in an effort focusing on risk communication. The result of this work, the report *Improving Risk Communication*, is frequently cited as the initial recognition that risk communication forms an important institutional practice in health and environmental policy. The first lines of the preface of the 1989 report on risk communication noted: "This report [RAFG] focused on improving risk assessment and risk decisions within the government. However, a major element in risk management in a democratic society is communication about risk" (NRC 1989, ix). *Improving Risk Communication* reflected a high level of thinking and experience in dealing with communication with the public in situations of uncertainty, inconclusive scientific evidence, and controversy, and it took many strong positions on the need to factor in the public and the reception of knowledge by large audiences in risk assessment and risk management itself, as opposed to simply aiming to refine the communication of already-established calculations or regulatory decisions. This was pretty much what Ruckelshaus, Russell, and other EPA staff had experienced and conceptualized.[37]

Risk communication has produced a number of recipes. One example was the chemical EDB, from which the EPA learned a lesson through a case study (Sharlin 1985, 1987). The EPA's recipes for risk communication soon stabilized and spread. The associate director of the Office of Policy Analysis, in the OPPE, Derry Allen, worked, first, to understand the research on perception and communication of risks, and then on assessing what could apply in the EPA's operations, developing material and trainings for the agency. The accumulated experience in risk communication translated into a short document entitled "The Seven Cardinal Rules of Risk Communication" (Covello and Allen 1988). The rules, given as follows, were the basis of a subsequent training program initiated in 1989 and were translated into program-specific guidance by regional staff:

- Accept and involve the public as a partner.
- Plan carefully and evaluate your efforts.
- Listen to the public's specific concerns.
- Be honest, frank, and open.
- Work with other credible sources.
- Meet the needs of the media.
- Speak clearly and with compassion.

Risk communication, as well as the deliberative rationale it embodies, continued to be an important point of reference for regulatory practice at the regional level, for the members of staff in regional offices that engaged most directly with communities.

Conclusion

In 1983, an unlikely encounter between three things occurred: a bureaucratic entrepreneur, ready for and capable of implementing new objectives and ways of working for an embattled agency; a set of notions constitutive of an integrated administrative design for managing risks thanks to science—the original risk assessment/risk management structure crafted by the authors of RAFG; and a political context (a crisis, really) created by the decisions of William Ruckelshaus's predecessor, and her resignation. The fact that the EPA leadership was caught in such turmoil, and that Ruckelshaus was picked by Reagan to return to the EPA and accepted this mission, constituted a totally unlikely series of events. That the RAC concluded its work, apparently successfully, exactly as and when Ruckelshaus needed a solution is also utterly random.

Altogether, still, this design configuration was not so unlikely. The crisis that the EPA found itself in, and the problems that Gorsuch caused, were known. The distortions of scientific assessment by political appointees, which heavily contributed to causing the crisis, were the kind of issues that courts, industry groups, scientific advisers, and EPA bureaucrats themselves had been reflecting on since at least the second half of the 1970s. These were the issues that motivated the formalization of logical decision-making processes in order to capture uncertainties and make credible decisions. In other words, what happened at the EPA in 1983 was as much the reflection of a long-lasting controversy about the administration's scientific legitimacy as the effect of a sudden, deep political crisis forged by the exceptional behavior of Gorsuch and her aides.

The redefinition of the EPA as a risk agency during the second term of Ruckelshaus as administrator also reflects the configuration of those days, particularly the dense set of relationships that emerged between officials and scientists interested in the administration of uncertain environmental and health issues. Ruckelshaus was a member of the loose network of bureaucrats and scientists that were then reflecting on the best, legitimate ways of making decisions about risks, inside and outside the agency. He was one of the bureaucratic leaders of the country that knew of these risk ideas, he knew the emergent methods of risk-based decision-making, and he even knew about the ongoing work of the NRC committee that was working on the institutional means of risk assessment. Most of the people whom he picked to return to the agency and restore its authority were in some manner aware of these notions. Some helped articulate them as well.

When March 1983 came, a full design, assembled from diverse notions of risk assessment, risk-ranking, and risk management, had emerged. And all the people who could reproduce it in the EPA, in its diversity, were available to do so then and there. They applied the scheme of risk-based decision-making in the operations of the agency and in communicating about the agency. Those years were special, in that a group of bureaucrats and advisers only just assembled by Ruckelshaus and Alm to take control of the agency shared a common way of speaking about the organization and its goals and modes of action. For a moment, the enduring political controversy about the use of science in policy and the power of experts transformed the EPA into an uncontroversial agency that used risk to define its image, the forms of knowledge that it used, and its concrete decision-making operations.

6 Risk Management: The EPA as a Decision-Making System

Risk Assessment in the Federal Government: Managing the Process (RAFG) is often presented as the source of the risk assessment–risk management framework and the inspiration for William Ruckelshaus's policy of publicly separating science from policy in the agency—but that is a misreading of the report (North 2003) because it explicitly recommended distinguishing these things intellectually, and forced thinking about their interaction. But the single most important material consequence of the risk assessment/risk management knowledge representation was the capacity that it granted to the administrators of the agency to design a joint decision-making process involving the various programmatic and functional offices of the agency, and assembling the expertise and specific conceptions of uncertainty of toxicologists, economists, and policy analysts. In other words, the framework supported a redesign and integration of the agency, in the form of a decision-making process that attended to the various dimensions of an environmental issue, and concerns of EPA's audiences. All of this was evident in Alvin Alm's decision to pursue the Toxics Integration effort of the late 1970s, the gradual acculturation to the knowledge representation borne by RAFG across the agency, the ensuring design of a sophisticated "options tracking system," and the creation of multiple analytical guides and formats for application by all the agency staff.

The Agencywide "Risk Projects" of 1983

Given the loss of credibility of the EPA and the diminishment of staff morale during Ann Gorsuch's stint as the administrator, Ruckelshaus and Alm wanted to reenergize the agency quickly. Alm came up with the formula of the "task force" in his first week back at the agency in May 1983. He believed that the swift launch of several task forces was important for the agency to be able to deliver quickly in a context of high internal and external expectations. The task force formula was instrumental, the manager thought, in getting people involved and revitalizing the EPA. It was also a way of demonstrating, without delay, that the days of Gorsuch's autocratic style were at an end.

The day before he delivered his NAS speech, Ruckelshaus signed off on the decision to create ten task forces. The first concerned dioxin and other complicated cases of pollution, such as nonpoint-source pollution and groundwater pollution, and others tackled structural institutional difficulties concerning the enforcement of EPA decisions. Each of the task forces had ten to twenty members drawn from around the agency and mobilized more staff in the different offices as necessary in order to produce memos and reports. But the most generic task force was the so-called Toxics Integration Task Force (TITF), designed to continue the efforts initiated in 1977 to establish agencywide processes for comparative analysis of risk, ranking, and prioritization across all offices.

There was no splitting of the Toxics Integration effort among offices this time (see chapter 3): the entire exercise was located in the OPPE and chaired by Richard Morgenstern, the director of the Office of Policy Analysis there. Its agenda became more ambitious. Its mandate now was to address various problems "that can be grouped under the general heading of 'risk assessment' and 'risk management.'"[1] RAFG and its proposed definitions helped the managers and policy analysts in the agency to align the various offices on common descriptions of bureaucratic processes and problems. As in RAFG, all other terms—from risk evaluation to hazard assessment, in passing by technology assessment—were obliterated in this simplified picture of what the EPA should deliver. The risk assessment and risk management categories were defined in task force documents as they were in RAFG: as generic processes summarizing all actions aimed at controlling risk. And while until 1983, the Toxics Integration work groups had had limited agendas, this time around, the task force had an extensive reach across the organization and its subjects of interest.

The objectives of the TITF were to create consistency—a favorite goal of the economists of the OPPE—at all levels. The agenda of the first meeting of the task force had the following points: "consistency in risk assessment" (developing methods and guidelines for application across the agency);

"consistency in risk management"; "high-visibility chemicals"; "intermedia inconsistencies" (improvement of coordination between offices managing water, air, contaminated lands ...); "interagency coordination."[2] The fact that the risk designs were already en route to becoming normal bureaucratic knowledge and technology across the EPA, beyond the pockets in which it had first been tried such as the CAG, enabled Morgenstern to advance this streamlined, integrated agenda affecting the entire organization. Not only was most of the Washington, D.C., regulatory community abreast of this new representation of agencies as structured around scientific risk assessment or risk management, but the EPA itself was structured by it. Notions of risk assessment were already employed in the agency, notably at the CAG.

RAFG, given its high profile and recognized quality, further helped to structure this knowledge and spread it across the EPA. The report had penetrated the agency through the various offices and at the level of political appointees and program managers.[3] John Todhunter, the head of the OPTS (then in his final weeks at the agency) was at the dinner event for the launch of the report. He soon wrote back to Gil Omenn, whom he had been seated next to, that he had instructed his staff to develop options for implementing the recommendations of the report. He stressed the need to act on the recommendations with regard to peer review—recommendations that were "meritorious and fully consonant with policy positions that I have expressed in the past."[4] Don Clay, also of the OPTS, wrote to Lee Thomas when the latter was acting EPA administrator (March–April 1983) to stress the merits of the report. Clay considered the report "a fine piece," and went on to suggest that each office in the EPA comment on it and submit their evaluations to a single source in the agency.

Several senior scientists that Ruckelshaus and Alm had picked for the agency in 1983 concurred to promote the RAFG framework and further install risk assessment and management in the agency. John Moore became the assistant administrator for pesticides and toxic substances in October 1983, but he had worked as a consultant for the agency starting in May 1983. He had a doctorate of veterinary medicine and was a certified toxicologist. Before joining the EPA, he had spent fourteen years (1969–1983) at the National Institute for Environmental Health Sciences, as head of toxicology research and testing and deputy director of the National Toxicology Program. He fully understood the benefit of RAFG's redefinition of science-for-regulation as risk assessment. He was called to the agency to attempt to

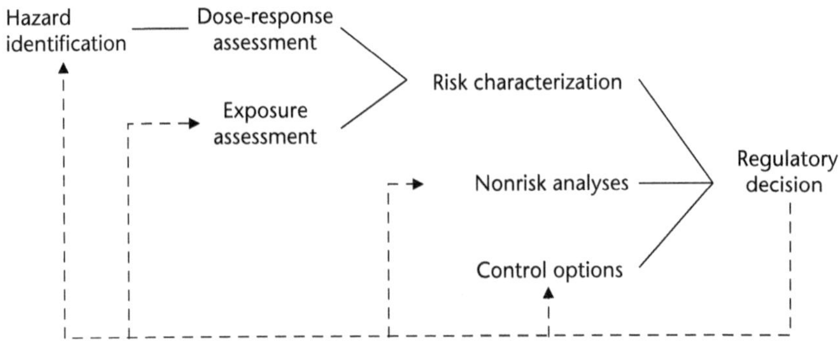

Figure 6.1
Graphical representation of a regulatory decision-making nosology proposed for the EPA.
Source: Richard Hill, Office of Toxic Substances, memorandum to John Moore, assistant administrator for the OPTS, "A Regulatory Decision-Making Nosology," US EPA, January 12, 1984, Milton Russell Special Collection.

institute risk assessment in the programs, and RAFG laid out the path to do just that. The document had the advantage of being "not this huge document, it was readable, it could be used as a guide," so Moore told people to get it and read it.[5]

In his office, he circulated a memo by one of his staff members, entitled "Regulatory Decision-Making Nosology," based on the definitions of RAFG. The document was meant to develop common definitions of risk terms because some confusion was noted within the agency as to what *risk assessment* meant. A graph converted RAFG's proposals into a decision scheme (see figure 6.1). Richard Hill, the author of the paper, sent it to Moore and to the ORD's OHEA with a note saying that "in the carcinogenicity guideline work-group [Alm's Toxic Integrations A-1 effort] it was apparent people were using terms in different ways. The appended is my draft to help demystify matters. There may be merit in getting general agreement across the agency."[6] Moore agreed to this and sent the paper to Milton Russell, thus supporting the ongoing process in the EPA of acculturation to this new knowledge representation and technology of decision-making. This design work, prompted by RAFG, shows that the report did not impose a new framework on the agency, as much as it helped disambiguate a knowledge representation that was already emerging in the agency, and that its officials were ready to apply.

Finally, Bernie Goldstein,[7] who became assistant administrator for research and development at the agency in October 1983, was also very much alerted to what was in RAFG, even before it got published. He knew several panel members and frequently interacted with them professionally.[8] Much like Moore, he was perceived as a person of high intellectual stature, a "solid thinker"[9] who quickly saw the interest of this framework for constructing a bigger picture of what the EPA was actually doing and could deliver to the public. His office, the ORD, was also aware of these notions. The OHEA, of course, was already partly organized along some of these lines. But RAFG emerged there as a new and authoritative manner of defining regulatory work.[10] This was especially the case for the version that stressed the scientific potentiality of risk assessment and its neat separation with policy, through a well-demarcated exercise of clarifying the science policies of the agency.

So, perhaps for the first time since the EPA was set up, the whole team of assistant administrators was renewed. Special talents in science, law, and management were brought in. They were picked by the EPA administrator, synchronized to bring about major changes to the agency, and aligned on a similar set of intentions for the development and application of risk assessment techniques in order to improve the science and policy interface. All of them were aware of the RAC's efforts and had read the report. RAFG showed many of these people the directions in which to take the agency, both to consolidate its expertise in risk assessment or risk ranking and to eliminate the controversy-prone practices of the recent past (conflicts between offices, intervention by program managers into the science, etc.). It helped to align people from different offices, as well as the people in these offices and top leaders of the agency, to a common set of notions that had started to be diffused throughout the organization and corresponded to its key operations.

Designing an Options-Generating System

Perhaps the most decisive change that the risk assessment–risk management framework brought about is the unified representation of the agency as a machine to produce decisions and of its internal structure as an organic system to develop these decisions. How risk assessment and risk management effectively became a framework embracing the activities of offices across the agency in order to become more integrated can be seen in the establishment of an options tracking system.

Like Ruckelshaus, Alm had worked at the EPA in the early years of the agency. After leaving the EPA in 1973, he went on to work for President Gerald Ford on energy policy, based in the OMB, and he stayed on under Ford's successor, Jimmy Carter. During the second half of Carter's term, he became assistant secretary for policy and evaluation in the US Department of Energy. When Ruckelshaus asked him to come back to the EPA to be his deputy director, Alm was the director of Harvard University's energy security program. Everywhere he worked between 1973 and 1983, Alm introduced further policy analysis and economic and regulatory analysis to design and deploy public policy programs effectively, and he received credit for that.[11]

Coming back to the EPA after the years of Gorsuch's politicized and error-prone management, Alm introduced his own personal management philosophy, essentially a mix of discipline, stimulation of personal involvement, and intensive analytics. In a memo dated August 18, 1983,[12] addressed to all the agency's political appointees, and particularly the assistant administrators and regional administrators, he clarified that his philosophy was to set clear goals and schedules and to hold people accountable for meeting them once they were agreed to. He wanted "professional commitment and motivation to get the job done." His expectations of the EPA staff were high—and he noted that they were one of the best and most professional teams. On the other hand, he announced that he would apply a more inclusive and participatory decision-making system, thus giving staff opportunities to use their experience.

In his approach, it was analysis that brought people together in a disciplined decision-making system that respected their autonomy and experience. Alm wanted environmental results to be monitored, based on reliable data, and to be periodically evaluated to avoid succumbing "to day-to-day demands." More important perhaps, one of his stated principles was that "[a]long with good data, we need high-quality scientific and policy analyses. As the Administrator recently stressed to the National Academy of Sciences, we must make the scientific and policy assessments behind our decisions clear so that the public understands what we have done and why."[13]

Alm worked hand in hand with Milton Russell, whom he had recruited to become head of the OPPE at the agency (as discussed in chapter 5). Upon his arrival at the agency in May 1983 as a consultant to the administrator,[14] Russell received a copy of RAFG from Ruckelshaus and Alm, with the mission to create a new approach to decision-making for the environment—a more structured and integrated one: "The issue was, how could we in effect create a new approach, a new image, a new analytical, objective basis for regulation and environmental issues in the USA? And consequently, Mr. Ruckelshaus was looking for, my office was supposed to, impress and create an analytical framework, which was risk analysis, risk management, and so forth. And my job was to promote that, not only within the group there at EPA, the office, but across the EPA and outside."[15] Ruckelshaus and his deputy Alm pushed Russell to search for the elements of that generic and harmonized decision process in the risk assessment–risk management architecture and in the generic thought processes articulated in RAFG. The report and its architecture of terms were used to continue the effort of agency integration that was attempted at the end of the 1970s, but that did not lead very far. In 1983, the EPA looked very much like it did in 1981 and in the early 1970s, as a compound of offices dealing with separate matters (see figures 3.1 and 3.2, chapter 3).

Starting in the summer of 1983, Russell and Alm worked on a process that could help track each and every dossier and decision in-the-making in the agency, filtering out the most important and politically problematic decisions so that they would rise to the deputy administrator and administrator levels for them to focus on. The options-tracking system was very much a political process because its function was to detect and channel policy issues that had the potential to become controversial or litigious, either inside the agency or toward the industry, environmental groups, or at the OMB. The options review and the risk assessment–risk management framework were closely linked: "[T]he options selection process was the implementation of the risk analysis/risk management operation. It was the way we organized, structured, enforced, and regulated if you will, made it part of the normal flow of the system."[16]

So the system was there not only to track operations routinely, but also to create the conditions to address issues that were in the process of becoming controversial and politically salient. In more operational terms, the goal was to make sure that everything flowed smoothly, to be able to track delays caused by staff offices, including the twelfth floor where the administrator was, and also to make sure that the administrator and deputy administrator were on top of issues and ready to make decisions, with all information in hand. The system, furthermore, was intended to ensure that someone was there to question the decisions favored by program offices. The OPPE had the role of scrutinizing the decisions shaped in the program offices and being a "third pair of eyes"[17] on decisions in preparation.

In practice, the system was run by the Office of Management Systems and Evaluation inside the OPPE. It would track on average 200 to 250 operations at a time. More important operations, designated level I, were those that were likely to have an economic impact of over $100 million, and were subject to cost-benefit analysis as per the 1981 Presidential Executive Order 12291, enforced by the OMB. The decision of how important an operation would be was made by a workgroup involving people from the program offices. A level II or level III operation issue was handled directly in the program office and signed off by Russell. For operations of greater importance, discussions were held in a meeting with Alm every month.

The staff would meet with Alm and the various program offices to go over any of the problems in the procedures and to deal with delays and controversies. Following these troubleshooting meetings, an option selection meeting would be held. The meetings brought together around twenty people, including members of the OPPE, people from the Office of the General Counsel, the Enforcement Office, a representative of the ORD, and the assistant administrators of program offices. The day-to-day logic of individual program offices could not simply be carried into these meetings in which the functional offices played a key role—something that Alm ensured would happen. A representative of the OPPE would often lead those meetings, or Alm himself. A protocol for conducting the meetings was put in place, institutionalizing cross-office debates on decision options, thus diminishing again the possibility of one program office or another having a unilateral influence on a dossier. Recommendations from each office would be discussed, explicitly considering options and alternatives that were compared using notions not only of cost, risk, and benefit, but also of alignment of the criteria of each legal statute. Ruckelshaus, Alm, or Russell would then select the preferred option. A final decision meeting in the administrator's office would be held before the final sign-off on the regulation.

What was being resumed there was the older attempt to resolve internal conflicts stemming from irreconcilable statutes, thanks to a transversal process of analysis. According to that plan, the OPPE would finally receive the sort of mandate that it should have had under Douglas Costle's third phase of EPA structuration, along functional lines (see chapter 3). The options review system was fitted to the previous institutionalized practices of the steering committee and the so-called workgroup. Both had been gradually introduced during the 1970s, not least because of the Toxics Integration efforts. This time, the political circumstances allowed Ruckelshaus to stress analysis and management as central functions of the agency, and thus to put the OPPE in a central position. The fact of placing a political appointee at the head of this nonstatutory office reflected the novel political status accorded to policy analysis in this EPA. This office, now under the direction of Russell, was to become "the premier analytical office."[18] Its task would be to develop, with the contribution of all offices, a preparatory basis for making decisions—and a reliable, robust one at that. The thing that Ruckelshaus and Alm, as the top decision-makers in the agency, wanted from the OPPE was to ensure that they would never have to question an analysis or a set of recommendations submitted to them. What was necessary to achieve this was a controlled, uniform decision-making process across the agency.[19]

However well institutionalized it may have been, the options system still relied on the political importance that the leaders of the agency granted it. Between 1983 and March 1985, while Ruckelshaus and Alm were there, the system functioned effectively. The fact that it worked meant that it forced those people in a position to release standards and rules to the public to detect whether the decision they contemplated would be challenged outside. It prompted them to perform a sort of political opportunity analysis of upcoming decisions and participate in the construction of an internal agenda that would reflect external controversies. For such issues, the options selection meeting was a kind of preliminary test of the proposed decision—an internal peer review that anticipated the deconstructive pressures that the decision and its scientific foundations would undergo after public release. John Moore, head of the OPTS at the time, recalls that options tracking was "not an automatic system for generating decisions, but really a way of asking people 'tell me how much you thought about this really' … rather than accepting the automatic 'one in a million cancer risk'. It was not necessarily a system to challenge the risk assessment; instead, Alm was saying 'I accept that the risk is such, but what are the management things you defined that lead to this, what did you give credit to … ?'"[20]

So long as there were leaders in the various offices that understood and agreed with this point, the system worked. Even after Ruckelshaus and Alm left the agency, the system lived on, notably because key political appointees would still be there to defend and run it.[21] Moore understood that better than anyone else. His office was the source of many potentially controversial decisions: The Pesticides Office made many decisions every year, on widely used products, and had to run the complicated cancellation procedures; the toxic substances office, at least for the "existing chemicals" part of its activity (Boullier et al. 2019), made few decisions, but it dealt with controversial, embattled chemicals, many of which came under more than one office. Moore duly worked within the framework of options selection to identify the issues that should be elevated to level I interest—those that had a large "magnitude of effects and costs," and hence a "high degree of controversy," and could potentially become a "policy precedent,"[22] such as formaldehyde.

Those meetings also helped to discuss regulatory procedures under particular statutes, such as the special review procedure run by the Pesticides Office for withdrawal of pesticides from the market.[23] As mentioned previously, the special review procedure was complicated because it was difficult to find proof that the product in question was definitely the cause of a hazard, and the burden of proof to reverse a decision to authorize a product was high. These procedures often lasted several years, which created another difficulty: As the procedure unfolded, more scientific data appeared, changing the basis of the decision. The OPPE developed routines for these special reviews that frequently involved touchy legal questions. The options meeting was an occasion to work out how to make these procedures more collective, without automating them through formal rules and procedures. On all these aspects, the particular interest of Alm and the OPPE was for extra clarification of the motives and basis of the office's decision, with a view not to develop more "automatic decision rules," but rather to put in place a series of screens to detect upcoming political problems.[24] Again, this meant that the process was dependent on professional engagement and political support because the meetings were moments of collective analysis of political, regulatory, and scientific issues, not a mechanical decision point in a set procedure. It belonged to a team model and a bureaucratic-pluralistic style of decision-making (McGarity 1991; Furlong 1995) that continued to inform the way the EPA was working, even after the options selection process lost the energy that Alm and Russell had injected into it.

Risk, Costs, Benefits, and Regulations: The EPA Analyzed

The options tracking and selection system was the product of a particular way of seeing the agency: as an organization plagued by internal discrepancies in regulatory cultures and political agendas, and thus was unable to collect, share, and integrate information about the problems that the public expected it to treat. But it was also the product of the sets of categories that surfaced in 1983–1984, and which allowed for the almost-ontological redefinition of what the agency was and how it should operate to respond to these external demands and legitimacy criteria. In the framework of risk assessment and risk management, the agency was an organization that turned out decisions by forcing those who knew about risks and those who knew about costs and feasibility of regulatory interventions to confront one another. Alm and Ruckelshaus, effectively, were making decisions by arbitrating among them. They had the help of OPPE economists and regulatory analysts, who were there less to force offices to count, measure, and calculate everything—indeed, OPPE people frequently reminded others that not everything could be measured and that there was no point in establishing automatic procedures for decisions based on data—and more to prompt them to clarify the political judgment, or science policies, underpinning the proposed rule. The risk assessment–risk management framework, the design that then best embodied the legitimate way of evolving decisions using various scientific inputs, served as the template and was reflected in a series of tools.

Numerous tools came out of the office of the deputy administrator and of the OPPE at the time to "allow the agency to apply the risk assessment/risk management discipline in a more orderly way."[25] Alm first developed his criteria for an acceptable "decision package" to bring to his and Ruckelshaus's consideration, particularly for level I regulation.[26] Thus, the chosen regulation should provide the greatest net benefits to society; allow flexibility in compliance; induce innovation in the regulated community; be consistent and complementary with other federal programs and agencies; impose the least possible burden on the public in terms of reporting, information collection, and recordkeeping; and disrupt competitive markets as little as possible. These criteria, overall, implemented the regulatory reform program and took an economic view of regulation. But they were meant to inform one overarching public principle and element of agency reputation: "The Agency's primary goal is to develop standards and rules that protect human health and the environment." The package was to include analyses showing that all options had been considered on the basis of the best data available, and "clearly present: the costs, risks, and benefits of the options"; present a range of assumptions, including a "best estimate" as well as a "conservative case"; and give a comparison of the action's cost-effectiveness with that of similar regulatory actions. In terms of presentation, the memo required the decision package to include "a clear formulation of the problem" and "conclusions flowing logically from basic information"; "a clear, consistent, and logical rationale for the proposed action"; and a "presentation of risk analysis information in a clear and consistent format."

In the memo, risk analysis was the generic rubric for all information, data, numbers, and calculations that had been produced through an analytic effort, be it about costs, benefits, or risks. Risk analysis, in other words, was the overarching process encompassing risk assessment. And while the tools for risk analysis still implied a large degree of trust in the competence and professional judgment of the health risk calculation experts, they also explicitly rejected reliance on one preferred mode of analysis and too much monodisciplinarity. Natural and social scientists, biologists and economists, needed to work at constructing decision options. Risk analysis pushed for a greater number of analytical options, thus introducing more flexibility into the process, and partially replaced implicit and contained professional judgment by a collective, organized process of options selection. This generic notion of analysis helped the at least nominal integration of the organization, the assemblage of various internal expertise.

The OPPE complementarily revised and extended the use of regulatory impact analysis guidelines, absorbing the OMB's own guidelines developed in the aftermath of President Ronald Reagan's adoption of Executive Order 12291. This new Regulatory Impact Analysis (RIA) guidelines (EPA 1983c) had timelines, methodologies, criteria, and procedural sequences for performing RIA and introduced a number of methodologies that would remain in place in the agency for a long time. The guidelines stipulated that the problem should be formulated clearly, with all options outlined, including nonregulatory ones.[27] Benefits and costs should then be examined in turn. Its novelty was to incorporate, under the term *benefits*, prescriptions for when and how to perform a quantification of health effects. For the very first time, the guidelines recapped the bureaucratic knowledge available in the EPA to address standardized regulatory objects: risk (defined in the guidelines as "the probability of experiencing an adverse health effect from the pollutant under consideration"—this was essentially the agency's first document ever to define the term *risk*); cancer risks; and noncancer risks. Noncancer risks were to be assessed using the conventional methodology of computing a NOAEL, divided by a "safety factor." Cancer risks were be evaluated following CAG's approach: a weight of evidence assessment to define how dangerous a substance may be and a quantitative risk assessment based on an extrapolation of dose effects observed in animals, combined with whatever existing exposure data. Those flexible instructions were soon reasserted in the OPPE's "risk documentation paper," which was an optional paper to be prepared when the assistant secretary wished to review the significant risk information developed by the staff.

All these analytical prescriptions were reflected in Alm's final tool: the risk management information reporting format, introduced in May 1984. The purpose of this format was, according to Alm, "to present risk data in a manner allowing comparison across rules and programs in a way more accessible to decision-makers. Because this form will accompany regulatory packages, it will provide the Administrator, me, and other reviewers with easy access to this critical information." It should be used for "all 'major' rules, and all 'significant' rules which specifically involved the management and control of environmental and human health risks." The office responsible for filling it in was given the status of "Lead Office." A format had to be made available to Alm with options selection review packages, completed to the best possible extent, "as part of the steering committee review." The format was prepared by the OPPE's Office of Policy Analysis, and an OPPE analyst checked that it was completed correctly.[28] The tool was meant to promote coordination across the agency, again, but also to prepare for the OMB's review of these rules. The OMB created a lot of uncertainty by keeping EPA rules under review for long periods of time, without clear timelines. It also had the tendency to talk to various EPA officials across the agency, sometimes with different information about their review of one given decision, thus creating more confusion among offices. To counter this, Alm had selected the OPPE to be the OMB's interface. The

OPPE was supposed to become the agency's internal "mini-OMB,"[29] helping the offices prepare for OMB review and coordinating the responses to the White House office.[30] At the end of the day, most of the rules developed by the agency were accepted, given the underlying process for developing them and the procedural guarantees put in place by Alm to consider costs and benefits properly. If he had an issue with the OMB, Moore (who managed the pesticides and toxic substances program) simply held his ground, waiting for the administrator to take his position at the White House. In his experience, the OMB generally backed down.[31]

Taking Options on High-Visibility Chemicals

Several decisions in the mid-1980s showed that the overall organizational structure for analysis, where risk, cost, and benefit were formalized as logically interdependent items of decision, naturalized cost-benefit analysis. In July 1983, the issue of the control of arsenic emissions in the air was on the agenda of the EPA again—the subject of the now-famous Tacoma meeting (see chapter 5). The agency revised its rule in light of a balance between the excessive number of cancer cases (believed to be one per year), and the reduction of the risk that the depollution technology would allow. Imposing such technologies at a cost, for reducing this small number of excesses cases seemed unbalanced. So nine out of the twelve smelters initially subjected to the rule were exempted from it (Pasztor 1986). Environmental groups were massively critical of that cancer calculation and of the weighing of costs and risks.

Ruckelshaus had little time to save the agency, and a complicated legacy to handle around a few hot, complicated dossiers, notably those issues that had embodied the failure of Gorsuch and that symbolized its faults: dioxin, benzene, formaldehyde, and EDB mainly. The scheme in RAFG was not just a good way to present to the public what the EPA was doing; it was also a good key to interpret what had been done wrong in these cases. The problem was not so much that politicians had distorted the science—for, what would undistorted science mean, and by what standard would a pure, objective science for risk assessment be judged? What had gone wrong was that Lavelle and Todhunter pretended that the science was straightforward. They grounded their actions on this supposed, deceptively objective state of the science in order to advance their preferences, not anticipating that their actions would be discredited by the questionability of the science, its fundamentally disputable nature.

Again, RAFG provided the key to understanding this by making it clear that risk assessment was shot through with assumptions, that program managers necessarily carry into their standards, unless such assumptions were made explicit at some point in the decision-making process. Ruckelshaus understood that scientific uncertainties were in fact quite difficult to overcome, and impossible to do away with in the short time span the EPA had to issue a regulatory measure. But he could reduce the political uncertainty associated with those measures made on the basis of uncertain science—that is, the potential that these decisions had to be contested through the science employed to buttress them—if uncertainties were assumed. What that translated into, in decisionistic language, was: options. Scientific uncertainties would not preclude acceptable decisions if they appeared transparently as a factor in the choice among options.

In 1983 and 1984, Ruckelshaus took a number of decisions on the controversial, convoluted dossiers that his predecessor had failed on, many of which had been on the agency's agenda since its establishment. One was dioxin. Ruckelshaus and Alm, reviewing all the hot issues that would have to be dealt with quickly with other leaders of the agency, came to the conclusion that an agencywide task force was needed for this topic alone. By December 1983, the task force had concluded its work, not with a decision to launch more studies or revise the existing health assessment, but with an announcement of a multimillion-dollar program for the assessment of health risks linked to dioxin contamination, with sampling or monitoring in a hundred new sites across the country. If doses higher than one part per million would be found, then a clean-up would be ordered and executed under the Superfund program.

EDB was another such long-lasting controversy that embarrassed the agency. In 1975, the Environmental Defense Fund had petitioned the EPA to do something about this product used as a fumigant, which a National Cancer Institute study in rats and mice had found to be carcinogenic. The agency had followed suit, initiating a so-called Rebuttable Presumption Against Registration (RPAR) process, or special review, to assess whether to ban the product. The preliminary decision to ban EDB was issued in December 1980, but that process was stalled by Ann Gorsuch, while the head of the OPTS was redrawing the health effect assessment with his own back-of-the-envelope calculations.[32] After Todhunter resigned from the Office of Pesticides, his successor, Don Clay, acted quickly on the matter, issuing an emergency suspension of the registration of the soil fumigant, justified by the recent discovery of high levels of groundwater contamination in California and Hawaii. The case became more complicated as a public panic set on, after discovery of residues of EDB in food products, with consumers calling EPA staff directly to know whether they would die from eating this or that product, and asking for advice on what to do. States pressed the EPA to develop a specific guidance on use of the product—to which Ruckelshaus first responded that he did not want to act in emergency. When the Natural Resources Defense Council threatened a new lawsuit on the EDB decision, Ruckelshaus changed his mind and held a press conference to launch an accelerated study, leading to an emergency suspension for fumigant in grain, and a six-month phase-out of treated fruits (Anonymous 1984c).

Formaldehyde was another controverted case of chemicals regulation in which the EPA was embroiled. It was one of the chemicals for which John W. Hernandez, Gorsuch's deputy administrator, held a so-called forum with the industry, which created one of several scandals during Gorsuch's tenure. Hernandez and Gorsuch had limited themselves to creating a research clearinghouse for studies concerning the chemical, even as the CPSC was banning the use of formaldehyde foam insulation in February 1982 (Mosher 1983). The official rationale behind the EPA not pursuing the regulation of formaldehyde was that animal tests demonstrating carcinogenicity should not be given too much weight as evidence of risk in humans—against the established guideline and WOE methodology. In July 1983, with Ruckelshaus back at the EPA, the Natural Resources Defense Council and the American Public Health Association sued the agency, accusing the EPA of setting an unreasonably high standard for meeting health hazard criteria in TSCA and of violating its own conservative scientific principles in assessing the chemical's carcinogenicity. The lawsuit pressured Ruckelshaus to put his new approach to the test: letting scientists work, on the basis of established guidelines, to construct a set of possible regulatory options. The EPA did just that: Ruckelshaus reverted to the application of its cancer policy, considering substances found to be animal carcinogens as being carcinogenic in people as well, and announced that the agency would consider regulatory options on formaldehyde, ranging from ban to partial ban or no ban. The move seemed bold and was greeted in just that way by the press (Pasztor 1983, 1984). The agency was seen to be performing its mission of protecting the public's health, following a graduated scientific and regulatory response that seemed credible.

On all these issues—a March 1984 proposal to establish a rule applying to particulate matter, blocked since 1978, offered yet another example—Ruckelshaus capitalized on the dual advantages of a flexible and generic framework for decision-making: transparency concerning the criteria on which the agency was making its decisions, including scientifically, but also the right and legitimacy to develop and adopt different regulatory options for variegated chemicals, in a way that was comprehensible to the public, even when those decisions allowed greater exposure and risks for the public (Shabecoff 1983f).

Anticonsistency

The development of these tools for risk analysis had important consequences—namely, that of enabling the administrator, deputy administrator, and assistant administrators overseeing functional offices (research, policy, legal affairs) to take control of the development of major rules and standards. The creation of a cross-office process covering risk, cost, benefit, or options analysis was a soft, legitimate intrusion in the business of program offices. But now that there was an internal, transparent, and informed process for taking into account costs and benefits alongside risks, legitimized by a global knowledge representation defining the act of making decisions at the EPA in terms of a sequential consideration of numbers and uncertainties concerning risks, costs, and benefits. The agency was organized to produce these considerations and also to produce arguments justifying them. With differences among offices, regulatory analysis in the Ruckelshaus years was fully institutionalized (McGarity 1991), to the point that it was becoming a standard part of the decision-making and options generation process, including in legislation that (whether formally or by interpretation) discouraged taking cost considerations into account, such as the Clean Air Act.

There was some resistance to the introduction of more formal regulatory analysis. Alm and Russell were not unaware that most offices would resist requests to perform such analysis. For many offices, these innovations were too numerous in 1983–1984. The many guidelines introduced, on top of the options selection process, were resented, and complaints were becoming more frequent by the end of 1984 that the review process was causing delays, and that the OPPE did not demonstrate enough flexibility in implementing it. Russell denied the problems, emphatically defending what had been initiated in the past year: "My point is simply this—that under your (Alm's) leadership during the past year we have established a decision system that is the best in EPA's history, and the best for an agency of this kind in the government, and I don't want us to undermine what we have achieved."[33] A retreat with his senior staff to reflect upon these issues led to the conclusion that the OPPE was spending less energy on developing rules to be applied at a distance than on helping offices to improve and working with them directly. Still, the OPPE was not abandoning its vision of expanding analysis in the agency, getting more done by program offices, and "pushing risk assessment/risk management."[34]

But the language of "consistency" was frequently called into question by people outside the OPPE, as were the benefits of the forceful creation of central analysis and decision-making channels across the EPA. In the early days of the interagency risk management initiative, driven by the OPPE and largely centered on common analytic methodologies, Don Clay, the head of the OTS, wrote to Dan Beardsley, one of the leaders of the interagency effort, that "optimization of costs and benefits may not and should not in all cases define the ultimate goal for all agency activities related to chemicals."[35] To which Beardsley replied, as sanguine, that "I know you folks in OTS [Office of Toxic Substances] think OPPE wants to optimize every environmental problem which passes before our eyes, from personnel decisions to TSCA amendments. We suspect you would want us to coordinate the Second Coming … No matter how unique the specific chemical problem, it seems reasonable to try to respond to that problem in as organized and structured a manner as possible."[36]

While the acrimonious correspondence focused on the program of interagency coordination, it illustrates program offices' resistance to the OPPE's strategy of promoting integration through analysis. In the summer of 1984, the assistant administrator of the OPPE and two of his aides were busy writing the flagship report *Risk Assessment and Management: A Framework for Decision-Making*. The report was envisioned by Ruckelshaus, Alm, and Russell as the first major agencywide document explaining how the managers of the agency planned to create a presentable structure of the EPA and demonstrate its capacity to produce credible decisions on uncertain, controversial issues. It was projected as a "potentially significant statement of further policy direction of the Agency."[37]

The report was eventually endorsed by all assistant administrators and subsequently published. But in the process of writing it, Joe Cotruvo, the head of the Water Office, found that it was "overly optimistic about the changes that would come about in decision making when the risk assessment/risk management process becomes universal in the agency."[38] Moore, the assistant administrator for toxic substances and pesticides, noted that the report did not clearly define what was meant by consistency—"why it is an improvement, and how it will foster a strong base of public support as the report asserts"—and strongly recommended that the agency carefully evaluate whether consistency was desirable and under which circumstances it should be applied, given the differences between regulatory statutes and offices.[39] Much of the discomfort among the reviewers of the draft report came from the fact that the people in the OPPE described the methods of risk assessment and policy analysis as too standardized and mechanical. They did so to uphold the image of the EPA as a unified agency identified by its transparent procedures. But people involved in the development of regulatory decisions in program offices found that the economists misrepresented their work and the role of professional judgment in decision-making.

John Moore and the managers of the OPTS were particularly resistant to the way that the economists of the OPPE used science as an instrument to integrate the whole agency in a cross-program decision-making system. For the OPPE, risk assessment was a transversal process that could lead to differentiated risk management decisions in separate offices. But the OPTS had always produced its own science and performed risk assessment in house. At the very moment when Alm and Russell were describing risk assessment as a cross-office process, the OPTS was busy organizing itself along the lines of RAFG too, defining its own internal scientific and regulatory decision-making stages. In the offices that were implementing the TSCA, the people in charge of both existing and new chemicals were busy outlining a process based on RAFG, with scientists and types of data assigned to the three key steps of hazard assessment, exposure assessment, and risk characterization (Zeeman and Gilford 1993; Boullier et al. 2019).

The report was finally accepted by all the managers of program offices once it was rewritten to look like a proper framework: a set of generic principles that described to the outside world how the agency functioned, not a set of instructions to be applied uniformly by the professionals in the various program offices.

Risk Assessment and Risk Management as Culture

The guidelines and tools developed by Alm and the OPPE would not be sufficient to eliminate potential disputes inside the agency on chemical uncertainties so long as they continued to work in isolation, applying the criteria written down in their legislation. The staff of program offices needed to realize what they were part of—namely, that the standards they were developing would eventually appear to the outside world as decisions of the agency. The OPPE embarked on the design and delivery of a training course on risk assessment and risk management for all agency staff, "trying to get that whole framework deeply understood inside the agency."[40] This training policy was inaugurated at a major conference on risk assessment in early 1985, with a list of prominent speakers from both inside and outside the EPA, attended by a chosen list of participants from across Washington, D.C. Soon after this first public event on risk, the OPPE tested its new "generic course on assessment/management/communication"[41] in June 1985 with about 100 staff from the EPA headquarters and regional offices.

Goldstein and Moore taught risk assessment, while Russell took care of explaining risk management and regulatory analysis. All drew from the graph in RAFG, which provided the new informal structure of agencywide operations: a sort of mental organizational chart that was disseminated to replace the policy towers that people had in mind. The graph had changed into a Venn diagram and flowchart hybrid, perhaps to accommodate people's affiliations in the organization (see figure 6.2). The large circles and the space offered to every subdiscipline seemed to indicate the self-sufficiency of these various jobs and forms of expertise in the organization. But the arrows inside the circles drew a decision flow. They characterized the interdependence among the types of expertise and redefined their action identities with regard to an institution- and interest-free decision-making system (specific offices were not represented).

The training included hands-on session learning through case studies. Over time, thousands of EPA staff became acquainted with the famous—albeit entirely fictitious—controversy over the "widely used chemical

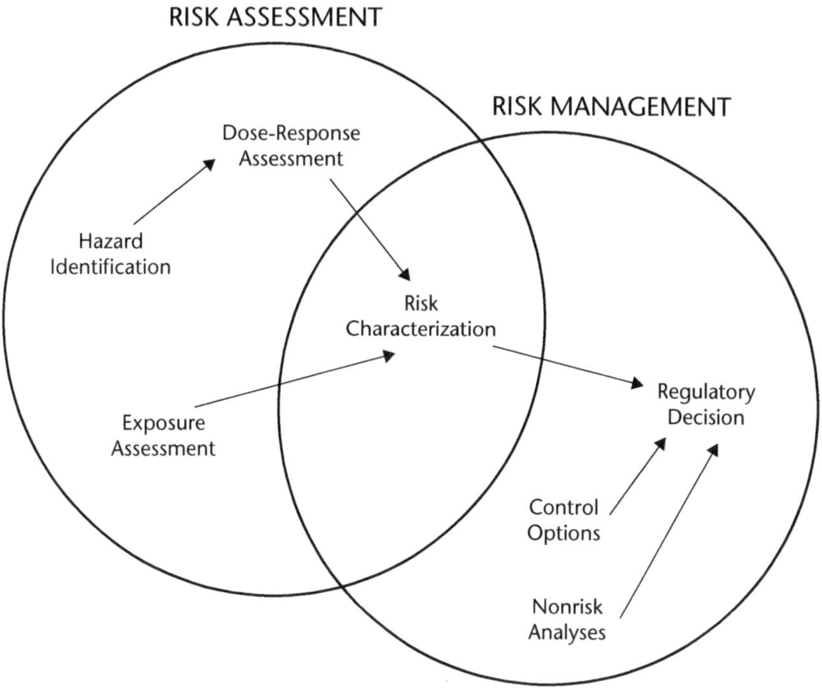

Figure 6.2
The NAS paradigm, as represented in EPA training material.
Source: Bernard Goldstein, "The Elements of Risk Assessment," EPA Workshop on Risk Assessment, March 17–18, 1985, Easton, MD; Milton Russell, "The Elements of Risk Management," EPA Workshop on Risk Management, April 13–14, 1986, Easton, MD. Milton Russell Special Collection.

dinitrochickenwire," which participants in the training had to approach in the role of either risk assessor (being instructed, as appropriate, not to ask themselves questions about what to do with the situation from a regulatory point of view) or risk manager. By the end of 1985, the OPTS and ORD had trained fifteen facilitators to conduct the "famed RAW" (Risk Assessment Workshop) involving the "notorious dinitrochicken-wire."[42] During the months of October and November 1985 alone, the OPTS and the ORD conducted fifteen separate RAW sessions at the EPA headquarters for the enforcement office, the OPPE, the ORD, and the OPTS.

The training was offered several times a year, to around 100 participants every time, leading to thousands of staff being exposed to an integrated knowledge representation of what the EPA as a whole was supposed to produce. By 1992, more than 18,000 people in the agency had attended a risk assessment training of some kind (Anonymous 1992). Information about the many workshops organized across the country on risk assessment, characterization, communication, or management was disseminated through a dedicated internal newsletter that started to be published in 1986 by the OHEA. This bimonthly publication, *Risk Assessment Review,* had been initiated by the newly formed network of regional risk assessors. It was an important instrument, too, as it circulated information across a wide range of offices concerning risk studies performed in any office or region about any substance or risk. It decreased, by a small proportion, the number of duplicated studies and contradictory assessments in the agency. It also gave form to a "risk assessment community" across the agency.[43]

Conclusion

The problem of the use of science to forge environmental decisions, insofar as it resulted in the articulation of an original design termed *risk assessment and risk management,* fostered the birth of a new kind of administration. In those years, a set of bureaucrats drew on the notion of risk management as the consideration of policy options to adjust decisions to the knowledge of risk in order to design a more integrated EPA—an entity that could smoothly involve scientists, policy analysts, and lawyers from various offices in processes of risk assessment and risk management. These processes were generic: They applied across the offices that comprise the agency and link the bottom of the organization with its top—the political leaders of the agency, recast as the ultimate decision-makers. These processes, furthermore, were scientific in style; they were defined in terms of the knowledge and information that was necessary to construct policy options and shape good decisions. In those days, it was the risk decision-making of economists and policy analysts—in which risk decisions can be optimized, so long as an adequate, homogenous level of information is reached on all aspects and objects of a decision—that prevailed.

It was their particular discipline of making decisions sequentially, through explicit criteria and completeness of information, that was disseminated across the EPA: a particular sort of bureaucratic science that incorporated and redefined the utility of the scientific calculation of health and environmental risks of toxicologists, biologists, health scientists, and statisticians, in a rare moment of collective redefinition (both symbolic and material) of the objects, knowledge, and processes of the agency. The most notable aspect of this transformation is that science was, at this moment, brought under the purview of the leaders of the agency. Risk assessment, now closely coupled with risk management, became an object of explicit interest and policy for them. A framework linking risk assessment with risk management meant, in essence, the possibility of managing risk assessment.

7 Designing a Science for Decisions

The framework for risk assessment and risk management that William Ruckelshaus articulated in 1983 emphasized the separation of the two activities because these notions helped the administrator to construct a public image of the internal operations of the EPA and its legitimate mode of using science against uncertain events. It helped to show that the agency was both legitimately using the best possible science and assuming its responsibility to make protective decisions, even where and when the science, given remaining uncertainties, could not conclusively indicate what the exact risks were, and how to reduce them. In this chapter, I show that, in those years at least, proclaiming the separation of risk assessment and risk management to construct a general, acceptable image of the agency for the public was counterbalanced by a closer political management of risk assessment. By defining the conditions of the legitimate articulation of science with policy, the risk framework enabled greater administrative steering of the science. This policy influence was not felt in individual risk decisions, nor in the involvement of political appointees in the actual work of risk assessors. But it materialized via the involvement of higher-level managers of the agency in the manufacturing of guidelines for risk assessment.

As the risk framework integrated assessment with management, science became a tractable object for the managers; scientific research and scientific assessment of risk became an object of action for the managers of the agency and the architects of its decision-making process, particularly Alvin Alm, Milton Russell, and other policy analysts. It was this way precisely because science was becoming an element of the design that the agency wanted to project to its audiences. The reforms of the ORD and SAB, the institution of a risk assessment forum, and the revision of the 1976 guidelines for cancer risk assessment—three major reforms aimed at limiting internal disputes in the agency, with enduring effects—clearly illustrate this.

Reorganizing the EPA's Science and Research

In their first month at the agency, Ruckelshaus and Alm decided to mandate the SAB to review the work of the laboratories and define a possible reorganization for the ORD. The group of five scientists included Bernie Goldstein, not yet confirmed by the Senate, but already at work as a consultant to the administrator. Ernest Cloyna, the chair of SAB, sent the conclusions of the expedited review to the administrator at the end of July 1983.[1] After having visited a number of laboratories across the country, the review group concluded that the assistant administrator should be a scientist with great stature in the scientific community, and that professional qualifications should be the primary criteria for recruitment of both the assistant administrator and the deputy assistant administrator. The group also noted excessive centralization in the ORD and a need for more involvement of research laboratories in the research planning process. The two recommendations seemed contradictory, but they were not, in the context in which the ORD was reorganized into five megalaboratories. Given the scope of their research and their internal planning processes (each had a research planning officer), these laboratories would ensure that greater decentralization did not mean fragmentation and irrelevance. The final report of the group (SAB 1983) pushed for more peer review of ORD's research and temporary recruitment of academics through the Intergovernmental Personnel Act. The number of staff at the EPA's ORD was high, in spite of massive cuts in the ORD budget and declining morale, but in some laboratories, the Reduction in Force policy pursued under Ann Gorsuch was threatening expertise, which remained thin. The report called for a fresh injection of new personnel into the laboratories.

In the report, the OHEA occupied a central place. Indeed, the OHEA and its components, the CAG and the Exposure Assessment Office, among others, had in recent years assumed "a key role in coupling science and regulatory decision making in the Agency," developing evaluations that required critical consideration of large amounts of scientific and technical data on health and environmental effects, while being attuned to "the particular needs of the regulatory office" (SAB 1983, 33). The report suggested supporting this newly acquired role of the office, turning it into a center with the same status as the four other megalaboratories. Most of the OHEA teams would be relocated to the Research Triangle Park in North Carolina in order to ensure close interaction with the Air Office, specifically its Office of Air Quality Planning and Standards (OAQPS), its main client, as well as several leading universities in environment and health research and the National Institute for Environmental and Health Sciences. The report suggested that the CAG and the director of the OHEA maintain their location at the EPA headquarters in Washington, D.C. It thus capitalized on the experience of the CAG and of direct coordination between large laboratories and program offices rather than through the ORD chiefs. That situation was a legacy of the past: Program offices enjoyed direct relationships with labs, which in turn were directly attached to their former departments. The CAG was a new EPA-created lab that had developed a close relation with a program office, the OAQPS. It seemed to have evolved a better coupling with regulatory work than the previous research committees of the 1970s, operated by the ORD hierarchy.

The changes decided by Ruckelshaus and Alm went in the direction advocated by the SAB report. They identified research as a critical area. Ruckelshaus secured an increase of the EPA budget from Ronald Reagan's level of funding (up to $214 million in fiscal year 1986, the last appropriation that Ruckelshaus negotiated). They recruited Goldstein to revive the prestige and political importance of the office in the agency. They did confirm the OHEA's central role in the articulation of research and regulatory decision-making, as well as its presence at the EPA head office in Washington, D.C.

The risk assessment–risk management framework, as well as the internal assemblage between people, expertise, and offices that it helped institute inside the agency, reestablished an image of independence of EPA's science from arbitrary policies. In the aftermath of the then-well-publicized actions of former political appointees on the assessments of dioxin, benzene, or formaldehyde, Goldstein picked up on the language of risk assessment and risk management promoted by Ruckelshaus, to draw a line between the ORD and regulators: The ORD would cover risk assessment, while the program offices would take care of the risk management. This was definitely not intended to redesign entirely the work of the program offices. Many of these offices had scientists to make the kinds of health and environmental assessments needed for their decisions. But the dualism meant that the ORD would certainly not get involved in risk management and that, conversely, program offices were not supposed to ask the ORD for data and studies that it already had in hand and that might help to support their preferred decision. Goldstein very publicly asserted this new generic characterization of EPA research and the ORD during his confirmation hearing both in Congress and in the media (Anonymous 1983).

The risk assessment–risk management design was also there to inform how far the ORD could concretely work to coordinate research for regulatory needs. Goldstein was particularly active on that front, suggesting to Alm that all research programs initiated by regulatory offices go through the same research committee as all ORD research programs.[2] This did not materialize. However, what did happen was that, from the mid-1980s onward, representatives of the ORD were included in the interoffice workgroups. For Goldstein, this was an essential mechanism to ensure the quality and relevance of ORD research. When properly structured in risk assessment/science policy/risk management terms, the meetings helped to identify, and agree on, the uncertainties during regulation development. Goldstein wrote back to all the ORD directors to explain how excited he was about this new ORD role, and that he would give it full priority.[3] Later, as Goldstein left the agency in 1985, he supported the creation of a new group, the Office of Technology Transfer and Regulatory Support, managed by Peter Preuss, formerly with the OHEA. The staff of this new office was "the focal point for the program offices' interaction with ORD" (EPA 1990a, 8). Its main function was to analyze and integrate scientific and technological information in the development of regulations. The group was later disbanded and replaced by other similar services. It was an important first attempt to establish a risk assessment–risk management routine to bridge the gap that existed between the ORD and regulatory offices.

The risk assessment–risk management combination similarly supported the role that the SAB acquired in the agency in the 1980s. The board aligned on this design in order to demonstrate its usefulness and become more mechanically involved with the agency's regulatory work. The staff director for the SAB,[4] Terry Yosie, actively designed a role for it, in close interaction with Ruckelshaus. Yosie had joined the EPA in July 1978, after obtaining a PhD at Carnegie Mellon University in Pittsburgh, where he studied the development of technologies for water pollution control. He joined the EPA staff for the SAB directly, moving up quickly to become staff director for the SAB. In that position, he reported directly to Ruckelshaus and Alm, with direct and continuous access to them. These new strategic advisory functions were inaugurated by working on the recommendations of the SAB group that reviewed the organization of ORD labs. The chair of SAB and other senior members of the SAB, along with Yosie, actively pushed for the recommendations of the review group (promoting the role of the OHEA, setting both applied and fundamental research as two concurring objectives of the ORD, planning research programs better, etc.) and folding the ORD into the risk framework much more than the review group had,[5] along with the SAB as well.

Yosie argued for a more routine involvement of the board in the review of the scientific work of program offices. He outlined the plan in a memo to Ruckelshaus at the end of 1983, offering the administrator ways of "making effective use of the SAB," [6] demonstrating the benefits, for the administrator, in relying on this independent yet allied source of advice. What Yosie outlined was a system to define an agenda for the SAB that would be more independent from the variable, ad hoc needs of program offices, and more reflective of the agency's strategic knowledge needs. With greater planning of SAB review work at the highest level of the agency, the board could contribute to increasing coordination among offices, police the border between risk assessment and risk management, identify ways to upgrade research quality, and act as a special sounding board for new strategic ideas tested by the administrator.[7] In effect, he was positioning the SAB as a kind of review panel for risk assessment, as recommended by the authors of RAFG (SAB 1987), to counter the AIHC's proposal to create an external, independent, and perhaps policy obstructive science panel.[8]

In numbers, the increase in the reviewing activity of the SAB was unmistakable, with formal reviews going up from ten to seventy-seven between 1981 and 1987 (SAB 1987). In substance, it played an increasing role in the review of proposed risk assessment guidelines, as well as in the analysis, document after document, of the methodologies and interpretations made by various program offices. The reviews of the board were on several occasions instrumental to changing the orientations of the risk assessments, forcing offices to abandon their risk estimates, considering omitted scientific factors, testing nonlinear hypotheses where these appeared credible, and so on. The complicated dossiers of asbestos, ozone, arsenic, and dioxin all bear the traces of these interventions by SAB panels (Jasanoff 1995).

The Problem of Accuracy and Consistency: Toward the Risk Assessment Forum

Alm's managerial vision, predicated on RAFG, took a much firmer hold of the agency's internal scientific work. In a sense, risk management meant a management of risk assessment, and hence stronger intervention on the scientists in the agency. This did not necessarily involve day-to-day supervision of data choices and interpretations, but rather pressure to explicate analytical choices, models, and data sources, with their attendant uncertainties, so that managers could discover the nuts and bolts of decision-making and intervene if necessary. In the very first months of his new mandate, Alm kept talking about the necessary accuracy of the procedures of risk assessment. He wrote to all assistant administrators and regional administrators of the agency—whether or not they were concerned with risk assessment, like parts of the Air and Water offices at EPA headquarters in Washington, D.C.—that "The Administrator, as part of his overall goal of improving the scientific bases for Agency decisions, has taken a particular interest in improving risk assessment procedures." Therefore, everyone in the agency was required to use "the most accurate possible procedures for risk assessments."[9]

Embracing accuracy was daring and could certainly provoke and frustrate the proponents of risk assessment in the agency. First, risk assessors took it for granted that these calculations could not be accurate and precise. Ruckelshaus compared risk assessment to a spy: If you tortured him long enough, he told you what you wanted to know (Ruckelshaus 1984, 157–158). In the early 1980s, at an interagency meeting in the IRLG, a US Department of Health official had denigrated risk assessment as being as accurate as a five-year weather forecast (Barnes 1993). Risk assessors themselves knew this: Minute changes in the assumptions made, such as in the level of estimated exposure of a fictitious individual to a substance in an exposure model, could lead to enormous differences in the final estimate. Goldstein, who managed large platoons of risk assessors in the ORD, acknowledged this.[10] Risk assessment is accurate only insofar as one trusted the analytical choices and expert assumptions articulated and applied by risk assessors. Going for accuracy meant challenging this capacity and authority to make agency decisions based on their judgments.

But accuracy was also a challenging theme for Alm and policy analysts more generally. The usual approach in policy and cost-benefit analysis (i.e., Alm's presumed perspective on decision-making, as well as that of the economists in the Policy Office) matched a culture of regularity and consistency, not accuracy. In the "mechanical objectivity" that is typical of these analytical exercises, something gains the value of being true if and when it is found to be commensurable with past and future calculations, not by correspondence to a supposed natural state (Porter 1992, 1995; Gill 2009). In approaching accuracy, Alm and the OPPE people were going beyond their usual epistemological territory to bring science into an organization for decision-making purposes.

The head of OPPE, Milton Russell, was in charge of advancing this agenda, together with Bernie Goldstein, the ORD's chief. Arriving at the agency in May 1983, Russell wrote to all assistant administrators in the agency that science was problematic in several respects. First, estimates developed in the science part were of "dubious accuracy" and sometimes "grossly exaggerated"; decision-makers and the public alike were generally unclear about the nature of the EPA estimates—how uncertain, how conservative were they? The public and Congress had trouble interpreting risk estimates and failed to understand that absolute safety was impossible. There was a great lack of accuracy in risk assessment, and the nature of the estimates (uncertainty, conservatism) used was unclear.[11] This intervention inaugurated the close, but respectful, involvement of the managers and economists of the agency in everything relating to the science underpinning regulatory decisions, and in choices in the face of uncertainty.[12]

The TITF was the channel through which projects of reforming risk assessment practices and rules were developed. Based on the discussions held in the risk assessment subgroup of the TITF, Alm announced three direct centralized actions on risk assessment: audit of risk assessment practices, creation of a risk assessment forum, revision or creation of guidelines for risk assessment. In his words, "These three actions demonstrate the Agency's commitment to scientific quality in regulatory decision making. I believe they will confirm that our scientific work has been done well in the past, and help prevent inconsistencies and deficiencies from occurring in the future. I thank you for your input to date and I look forward to working with you as we continue to improve our risk assessment process and products."[13]

In February 1984, the EPA contracted with Joseph Rodricks, formerly of the FDA, the chairman of the IRLG risk assessment group, and a member of the RAC, to perform an audit of risk assessment practices in the agency.

Rodricks had just launched a consulting business specializing in risk assessment,[14] and he performed the three-month, $15.000 audit in that capacity. His conclusion, unsurprisingly, was that the offices were not performing risk assessment evenly, at least judged against the background of the emerging orthodoxy of RAFG. Some were performing only the first steps of the full assessment, and some offices were strong on cancer while others ignored that aspect. The audit also made it clear that offices were treating genotoxicity in different ways, which might have had a major impact on how a carcinogen was assessed.

Cases of contradicting assessments among offices remained rare, fortunately. Most of the offices' work concerned routine assessment of regulatory issues, with no need to involve any other office and no outside scrutiny. The problem of inconsistency would appear for subjects of greater public attention, where the administrator would have to choose one assessment for the whole agency among competing ones, and where contradictions or hesitations would immediately be picked up by opponents in legal challenges. So the bigger problem that the audit identified was one of risk: The agency was not protected from subsequent conflicts among offices because different offices used different guidelines, or no guidelines at all, and because there was no common review mechanism for all offices.

A more long-term effect of the codification of the agency as risk assessment/risk management was the creation of a new cross-agency institution, authorized by Alm, called the Risk Assessment Forum (RAF). The forum was expected to fulfill four functions, some of which referred to the suggestion by the RAC to create a board to review risk assessment policies and guidelines published by agencies: "(1) review risk assessments upon the request of the Administrator, Deputy Administrator, Assistant Administrators, or Regional Administrators; (2) provide a mechanism for interchange on science issues in risk assessment; (3) advise the Administrator and Deputy Administrator on precedent setting cases and important risk assessment issues; and (4) recommend revisions or updates to the risk assessment guidelines, as appropriate."[15] It would be chaired by the main risk assessment group of the ORD, the OHEA, comprising representatives of all risk assessment groups from program offices, as well as representatives from the Office of General Counsel and the OPPE. It was topped by a Risk Management Council in 1986 by Ruckelshaus's successor, Lee Thomas, and populated by the institution's top "risk managers" to check on science-policy positions decided in the RAF.[16]

The RAF soon demonstrated its usefulness, in coordinating the process of constructing agency-wide methods of assessing risks, and of constructing shared decisions in the agency. The forum was the site of invention of a bureaucratic technique that was long used to demonstrate the agency's way of establishing risks to the outside, the uncertainty factor, but only because it was a place in which controversies surrounding uncertainty and its perception could be safely translated.

As mentioned in chapter 3, there was no formal guideline for determining safety levels for noncarcinogenic chemicals until that time. The area was simply too broad to be amenable to standardization: too many chemicals, causing too many different adverse effects. The result was that the methodologies adopted by the various offices of the EPA were even more diverse than in the closely framed exercise of carcinogen risk assessment. Some offices determined their rule for a chemical based on the evaluation of its most critical health effect (e.g., an effect on the kidneys), defined as the observable effect occurring at the lowest dose. Other offices picked one effect that they deemed more important or worthy of attention than another, regardless of the levels measured in experiments. Some offices measured these effects over the lifetime of a person to determine the daily dose at which one could safely be exposed, and others at less than the lifetime, considering that lifetime measurement produced overly conservative results. There were many more variations as well.[17] The result was that on a single chemical, there could be as many as thirty or forty risk estimations and so-called acceptable daily intake (ADI),[18] which was the source of an organizational conflict, as well as a major risk for the credibility of the agency.

That conflict started in November 1984, when the Office of Toxic Substances complained that scientists in the ORD duplicated evaluations that it had already performed. The problem surfaced on several occasions that year, leading to a rather hostile exchange of memos between the two offices, with the administrator and deputy administrator copied in. The OPPE stood in between, appalled by the new problem of consistency across the agency's science. Donn Viviani, head of the Regulations Analysis branch, wrote to all other chiefs of the OPPE: "Intermedia transfer, conflicting and/or duplicative assessments, different regulatory control levels, inconsistent health numbers, lack of coordination, sound familiar? They should, every one of these "integration" issues has been previously identified and studied, recommendations presented and, in at least several cases, some 'fix' implemented."

For him, the problems were emerging again because program offices had no incentive and interest to share information and work, and were getting no credit for integration. The memo spoke of program offices as "fiefdoms." [19]

To stop the conflict, a small group of four people from all concerned offices was set up, becoming the so-called ADI policy group in the RAF. Agency toxicologists worked through all existing ADIs to analyze the source of the many differences and to think about ways of resolving them. The authors of RAFG had already seen the problem: There were a lot of choices made in the course of calculating an ADI, and any of these choices could be manipulated for regulatory offices to attain the kind of result they wanted—not necessarily in terms of setting a level of risk, but in terms of justifying the amendment of adopted rules, or justifying not doing so (NRC 1983). If a decision-maker thought that it was better not to change an ADI, even though a new study indicated a new lowest adverse effect level (LOAEL; adverse effects of the substance appeared at a lower dose, so presumably more frequently), she or he would compute the difference between the LOAEL and the dose found in humans, called the *margin of safety,* or *safety factor.* The argument would be that, somehow, it was sufficient, even though the computed margin differed from the one that had to be applied conventionally to the LOAEL to determine the science-based ADI.

RAFG, in fact, provided the key to analyze this problem. It was not that regulators always directed the science, or that science was unable to respond to regulators' needs. It was that, in critical cases, any member of the relevant regulatory office (scientist, lawyer, or decision-maker) could play with any of the parameters of a risk assessment—the intermediary world of "science policies." That generic notion helped to depict a constituent problem for the whole agency and legitimized a standardizing effort, led by this newly formed, cross-agency ADI policy group. The group moved to delete the language of safety factors and of ADI. Safety and acceptability were judgments, so the terms empowered regulators to formulate these judgments without consideration for the conventional methods defined by scientists. It standardized a set of "uncertainty factors" instead, to apply in order to create a scientific "reference dose." A variety of uncertainty factors could be chosen, depending on the level of available data.

The concept of *uncertainty factor* had an enormous advantage over that of *safety factor*: it was an explicitly judgmental and qualitative notion, not a realistic one like that of *safety factor* (which tried to account for true biological differences between species). It was also more contextual because levels of uncertainty could vary from one risk assessment to another. It was a way to get "a better approximation of how a particular chemical should be considered,"[20] unencumbered by the shadow of the truth of what the chemical really, precisely does in the body of an animal or human. The group also decided on the creation of an electronic registry of reference doses for hundreds of chemicals that various program offices should take as a basis for their decisions, the Integrated Risk Information System (IRIS).[21] The group did not reduce the range of possible choices in scientific assessment of risks, but it explicated and standardized this range, helping to establish a more regulative design to routinely agree on chemical values, for IRIS.

So, alongside the guidelines—which were an essential instrument for the agency to be auditable by the courts, but did not strictly standardize the practices across all program and regional offices[22]—the RAF, as an institution to organize a cross-agency or external review of the assessments performed in the agency, worked as an institution that generated choices for decision-makers. It did so in a way that was accountable, or at least readable from the outside, thanks to the commonly accepted risk assessment/science policy/risk management grid. It was soon hailed by many as a very astute and important new institution that helped to avert many controversies. Dorothy Patton, who administered the RAF for several years starting in 1985, found that offices knew they had an interest in referring to the RAF issues on which they lacked expertise and on which they felt other offices would have competence or could offer opportunities for publicly challenging the agency: "Basically, the offices that had the responsibility to issue a standard, and were under pressure and lobbying to do so, on arsenic for instance, were looking for stronger arguments, reassurances, better science, by coming to the forum."[23] The overall opinion on this committee was that it did become a sort of neutral space or no man's land between offices, and that many issues were sorted out there, thus averting and avoiding conflicts among offices, disciplinary cultures, and science policies. At the same time, the ability to converse among offices was at the cost of a kind of cultural homogenization, with the main platform for discussions there being toxicology and the hazard identification–hazard characterization part of the risk assessment orthodoxy, at least in the first few years of operations.

Managing the Revision of the 1976 Guidelines

The utility for the RAF to develop risk knowledge for the whole of the agency, and not for one office, was also tested through the process of revising the 1976 guidelines—and probably even more radically than for the establishment of noncancer guidelines, given the political sensitiveness of the cancer issue. These guidelines were interim, and had mostly been applied by the CAG to develop health assessments or criteria documents on behalf of regulatory program offices for water and air. In the meantime, the CAG had expanded inside the OHEA and was replicated in several other groups (exposure assessment group, mutagenicity assessment group, etc.), that had also drafted guidance. But many of these documents had simply not been finalized and officially applied in the agency. The guideline-making process was reignited, with a view to definitely put them to use across the agency to ensure consistency.

The difference with the earlier efforts to establish guidelines was that the OPPE was directly involved in developing them, which meant a closer involvement by the heads of the agency, and Alm specifically. Russell pointed the way forward:[24] More guidelines should be developed to implement a different approach of quantitative risk assessment, particularly of exposure assessment. Guidelines would be of enormous benefit, to improve coordination across the agency (without having to resort to a vain centralization of risk assessment activities in one office, such as the ORD), and to limit discretion at lower levels and clarify the agency's "science policy." This new bureaucratic theme of accuracy and explicitness in risk assessment, combined with economists' favorite idea of what constituted "efficiency" and "consistency," corresponded to the CAG's ambition to expand the use of quantitative risk assessment across the agency. Betty Anderson had always been clear that the 1976 interim guidelines, and other pieces of guidance developed thereafter, would need to be regularly updated as science progressed. In the meantime, RAFG had convincingly made the case for developing more guidelines, in particular to explicate experts' uncertainties and judgments.

Starting in July 1983, Elizabeth Anderson took the lead for that project, and leaders for the development of every new guideline were nominated. Each of them would manage a group with representatives of the various offices. The project was designed to be a short, but intense six-month effort, helped by the accumulation of experience in the OHEA. The process took longer than expected. The twelve full-time equivalents and $600,000 of extra funds requested by the OHEA to perform the work were never obtained. Because the OHEA was unable to divert resources from the routine and time-consuming tasks of developing health assessment documents or exposure and risk assessment documents for the Air and Water offices, the work was perceptibly delayed. The participation of other offices in the process was just sufficient, though patchy. Key offices were absent from some of the meetings, or they were slow to respond to the request for comments on drafts. The OPTS, an important player in the agency concerning risk assessment, and one of the main targets of the planned extension of the application of guidelines, did not attend the meetings or submit written documents. Only after the draft was completed did the office raise concerns in a lengthy document. On several occasions, the leaders of the workgroup asked the assistant administrators and Al Alm to weigh in and stress the importance of cooperation and hard work. While the redevelopment of guidelines was in no way conflictual, the inclusion of a larger set of people from across the agency amplified the "differences of opinion on several important issues."[25] The standard methods and criteria for carcinogen risk assessment stabilized by the CAG were discussed on four levels: terminology, presentation of exposure calculations, weight of evidence, and, once more, extrapolation to low doses.

Terminology: Deploying a Standard Knowledge Representation

One of the first issues to come up was terminology. An official from the OPTS maintained that in the first meetings, "it was apparent people were using terms in different ways"—something that the confusing definitions of exposure, health, or risk assessment in existing documents could only confirm. He soon developed a "Regulatory Decision-Making Nosology" memo to "help demystify matters."[26] It straightforwardly transcribed the definitions laid out in RAFG, showing how the existing scheme used in and around the CAG (two qualitative questions, followed by a quantitative risk assessment) would profitably be replaced by a reference to RAFG's generic scheme for risk assessment, comprising four components: "hazard identification, dose-response assessment, exposure assessment, and risk characterization. The first of these, hazard identification, is a qualitative risk assessment. The three remaining components comprise quantitative risk assessment."[27] His assistant administrator endorsed the adoption of this vocabulary in the drafting of the guidelines.

One such problem emerged in the distinction between the concept of dose-response assessment and exposure assessment. The ambiguity was that there is no agreed-upon definition of the point on or in the body where exposure takes place. In this context, *exposure* means contact with the chemical. But does it mean contact of the visible external envelope of the body as a whole, or exposure of a particular organ, or of the cell? Toxicologists extended their jurisdiction as far as they could, to the whole body, claiming that they could analyze, quantitatively, the effects of a chemical on the physiology and biology of the organism taken as a whole. This scale defined their territory of expertise. The exercise of exposure assessment concerned what happens outside this boundary. It works with the applied, or external doses of the chemical, while the internal doses was the purview of toxicologists. The separation between applied and internal doses is the line that defines the respective domains of exposure people and dose-response people in the agency. It could be seen as a bureaucratic schematization of the body, which allowed the articulating the various expertise to be applied in a standard decision-making process. Exposure calculations came after dose-response assessment.

It was thus, at that very time, in 1983, that the agency was reorganized by a set of categories that henceforth defined the discrete competences and roles of its scientists, fitting them into a broader organizational scheme to turn out decisions. Just like RAFG had subsumed the various scientific analysis of the hazard under a generic notion of "risk assessment," so too the memo described previously and the guidelines put the emphasis on dose-response assessment—and, implicitly, quantitative, probabilistic risk assessment—to embody the agency's expertise. While it might seem natural to move in that direction, given the precision and order in the integrated scheme, it was by no means an innocent move. As the audit of risk assessment practices would later reveal, along with a survey of risk assessment in regional offices, the quantitative aspect of risk assessment was not the easiest part of the job, and certainly not the most practiced one. Most offices were doing what was now called *hazard identification,* demoted to an initial step in a broader scheme, whereby decisions were to derive from a contained quantification of the risk. This change was facilitated by the authority that this knowledge representation had in addressing controversies. Promoting this formalism was not just a prescription for embracing quantitative risk assessment among reluctant staff; it was also a shift toward another method of treating uncertainty—one that was more bureaucratic, in as much as it was applicable without deep scientific knowledge, and thus applicable across the entire agency. RAFG's anatomy of risk assessment, replicated in the agency in the "nosology" memo mentioned above,[28] carved out a number of recognizable, commodified knowledge items that officials across the agency, however closely involved in the particulars of a risk assessment, or however knowledgeable in the science, could extract from a risk assessment document.

One of the striking aspects of the risk assessment guidelines eventually published in 1986, was that they included detailed, suggested formats for the presentation of scientific information, accumulating each of these elements of information. Uncertainties were an explicit component of these knowledge schemes. Hence, the big change, from 1976 to 1986, was that risk assessment documents would include explication of choices made in the scientific analysis (e.g., on exposure assessment, as discussed later in this chapter), rather than justifying preferences for one or the other option in terms of best professional judgment. Itemizing knowledge in this way was an effective method for a bureaucracy because, if well delineated, these items would work to define categories of people and groups, organizing them in a broader organizational mechanism for making decisions. And indeed, the guideline revision efforts implied going one step further from RAFG, by specifying the boundaries between these elements of knowledge and the associations among them in order to ensure that no more confusing organizational knowledge politics were created in the agency.

Manufacturing Risk Presentations

The ways in which exposure assessment was generally handled in the agency's health and environmental assessment documents constituted the second problem that the guideline working group had to address. First, the behavior of a substance in the environment was the source of multiple variations and uncertainties: the nature of a chemical substance changes as it is transported, degraded, and so on. To account for these dynamics, exposure assessors are forced to rely on models. For instance, as arsenic was emitted into the air, the EPA used dispersion models to be able to calculate, with an unavoidably high degree of uncertainty or variations, the dose at which populations in the environment of the point-source would be exposed. Second, exposure assessment would seek information about the lifestyles, location, and consumption practices of various subpopulations at a level of detail that no existing database could match. To compensate for the patchy monitoring data, assessors modeled qualitatively the situation of three fictitious individuals: the exposure of a "typical person" (undefined), that of a person living close to the source of the hazard, and the "maximally exposed individual"—a person exposed to the hazard 24/7, for an entire lifespan.

In practice, risk assessors tended to be selective in their presentations of the estimate, reporting only the one that was based on the notion of the maximally exposed individual. This reduced the ability of anyone to subsequently participate in the choice of the most useful calculation for addressing the regulatory question. The selective presentation of exposures invisibilized an operational choice, and indeed even a moral dilemma that the EPA as a whole was to face in its decisions: either taking measures to protect the small number of individuals who suffered most (e.g., by removing a plant emitting extremely high doses of a dangerous chemical at a given site) or reducing average emissions over the country to reduce average levels.[29] Ruckelshaus was personally concerned about the equity behind these choices, and that translated into Alm and Russel's request, in the process of revising guidelines, to eliminate the practice of exclusively advancing data on the maximally exposed individual to present real-life monitoring data or presentations of ranges of exposures and uncertainties instead.[30] The suggestion to risk assessors not to use only worst-case exposure, but also "realistic exposure scenarios," had in fact been the object of a dedicated memo of Alm to all assistant administrators.[31]

The revised text did not address the issue at length and made no clear statement about the relative value of data and models. It emphasized the uncertainties and sophistication needed to treat the problem of exposure assessment, and it referred readers to a separate guideline on exposure assessment. The resulting guidelines (EPA 1986b) handled the problem by shifting from a black-boxed expert judgment to a different, algorithmic model. The guidelines stipulate to work with monitoring data, and then construct real-life monitoring profiles, shedding light on more vulnerable people; where insufficient, to model exposure in a probabilistic way; to perform a sensitivity analysis of the model; and to present all uncertainties involved in the modeling as part of the results of the exposure assessment.

Having articulated that sequential reasoning, the authors of the guideline expounded it through a flowchart—a formal visual technology of uncertainty reduction that minimized the reliance on trust in the judgment, competence, and experience of remote risk assessors in another office (see figure 7.1).

Governing Judgment

The methodological concept of WOE was the third aspect in line, and it was discussed far more extensively. There too, Alm and other policy analysis staff in the agency showed an appetite for these arcane concepts that were supposed to structure the judgment of risk assessors. The interest and availability of high-level administrators for the science used in their agency varied widely, but Ruckelshaus and Alm were deeply interested and involved. Alm, in particular, was a voracious reader, including of scientific documents.[32] According to another senior scientist of NCEA, he was respectful, asking many questions and trusting the competence of his staff instead of imposing his directives. In those days, he requested and received dozens of memos clarifying the science of risk assessment and the dilemmas involved from his most senior and experienced scientific staff. Betty Anderson had produced one such memo outlining what a WOE was. The topic had become a bone of contention between the CAG and the OPPE that was reflected in an important amendment of the 1976 guidelines that derived directly from the controversial case of perchloroethylene (PCE), also known as *tetrachloroethylene* or *perc*.

PCE was a popular solvent used in the dry-cleaning industry, the regulation of which was under consideration by the Water Office, the Air Office, and the Office of Toxic Substances simultaneously. OHEA was working on a risk assessment of the substance on behalf of these offices. In April 1978, its CAG had published a preliminary health risk assessment report for the substance. In October 1980, another group of scientists, also from OHEA, worked with the Water Office to set an ambient water quality criterion. In 1983, the process for regulating the substance accelerated, and the CAG revised its health assessments. Back in 1976, the CAG had used no other way of classifying the level of hazardousness than a simple characterization

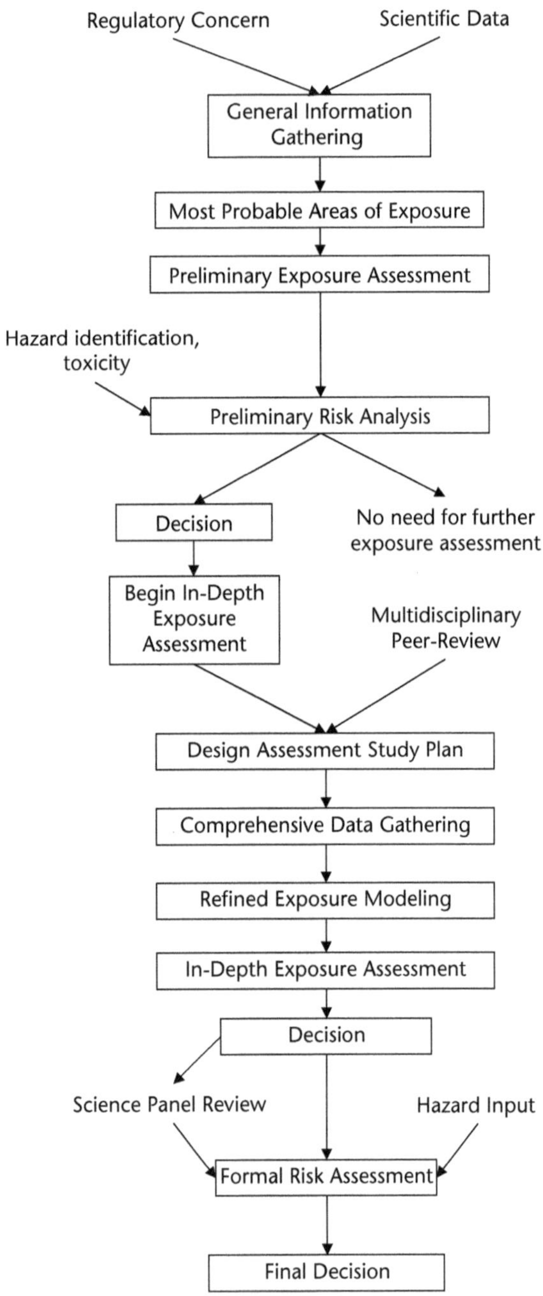

Figure 7.1
Decision path for exposure assessment (adapted from EPA 1986b).

as either substantial or suggestive. By 1982, it had evolved to consider that the scheme used by the International Agency for Research on Cancer (IARC) was the only one with substantial institutional use and international acceptance (EPA 1982). The scheme devised by CAG on the basis of the IARC scheme, a class-based aid to judgment, was as follows:

- A substance for which there is "sufficient evidence" to support a causal association between the exposure and cancer will fall into group 1: carcinogenic.
- Where evidence is "almost sufficient" or "suggestive,"[33] the chemical will be put in group 2: probably carcinogenic to humans. Group 2 is divided into higher or lower degrees of evidence: 2A and 2B, respectively.
- Where evidence is "inadequate" or "limited," the substance should be considered "possibly carcinogenic to humans": group 3. [34]

CAG's conservatism was reflected, specifically, in the way that it defined and used group 3. To IARC, a group 3 chemical is a substance that "cannot be classified as to its carcinogenicity to humans" because of inadequate or limited evidence. The CAG, in contrast, argued that there was a continuum between "limited" and "almost sufficient" evidence, so it argued that a labeling as "possible carcinogen" was more appropriate than as "cannot be classified." CAG's revised group 3, in essence, seemed to recognize uncertainties and the possibility of false positives a little less straightforwardly than IARC's group 3. And it certainly contributed to make the entire scheme more regulation-forcing: If a chemical is a possible carcinogen, it could possibly be regulated.[35]

In spite of the limited evidence, CAG placed PCE in group 3. This was not acceptable for the Water Office, which was in charge of actually deciding on regulatory measures for this chemical and dealing with the concerned industries. Paul Milvy wrote to Al Jennings, the director of the Office of Standards and Regulations in the OPPE, to complain that the CAG had produced three different estimates of the risk of the substance between November 1983 and March 1984, finally concluding, on the basis of WOE, that the most conservative scientific view was to consider the substance a probable human carcinogen. There was too much variation there to be able to establish a credible risk standard. More important perhaps, the Air Office disagreed with CAG's traditional conservatism, complaining that CAG distorted the classification scheme of IARC to fit with its conservative policy.

CAG placed PCE in group 3 of IARC "because the available evidence corresponds to the conservative scientific view that PCE is probably carcinogenic in humans." For Milvy, this original determination by CAG was "totally wrong." For the Water Office, CAG was distorting a tool for objectifying judgment, and thus forced the regulation of PCE. The fact that CAG modified its initial assessment only days later made things worse. This lack of continuity, in the context of conflictual relations with offices applying different policies, could only cause controversy. As Milvy noted, "because a simple, honest and straightforward approach was not adopted [by CAG], the resulting blatant contradiction fueled the controversy."[36]

This was precisely where PCE was a close call, as Russell noted in his memorandum to all offices contemplating a regulation of this substance. Russell asked to have a meeting on this as soon as possible.[37] The chemical was already being reported on in the press, and the EPA was unable to determine what action to take with regard to it. The Air Office interpreted the "limited evidence" classification as indicating that the substance could not be considered a carcinogen, so it should not be regulated. On the other hand, the Water Office, the hazardous waste office, and the Superfund office considered that these chemicals for which there was "limited evidence," and so were "potential" carcinogens, should still be regulated.

The March 1983 memo from the Air Office firmly challenged the IARC scheme. Was it not forcing these kinds of mistakes by artificially placing chemical substances in one group, "when in fact nature has provided a continuum of potency"? At the same time, would a new consistency-promoting classification inside the agency not reduce the ability of individual offices to engage with audiences, whether scientific, industrial, or other, thereby finally increasing the potential for controversies? The memo hit the bull's-eye: This was a perfect example of the high potential of a scientific assessment to cause major confusion outside the agency and to stimulate criticism of its decisions, if not legal challenges, at the very time when the OPPE was pushing an all-out policy of increasing agencywide consistency. Representatives of the offices concerned (Air, Water, ORD, and OPPE) met a few days later for what turned out to be a difficult, heated meeting. They agreed that consistency was a necessity, but they continued to disagree on whether the IARC scheme was the best classification and whether picking a classification was a scientific or a managerial task. Reporting the conclusions of this meeting to Alm and Ruckelshaus, Russell questioned the IARC classification: the agency never formally adopted the scheme, but the OHEA was using it as a matter of habit. The "policy implications" and "regulation-forcing dimension" of the scheme needed to be thought through, otherwise "programmatic conflicts will result and Agency policy will be shaped by default. Perhaps worse, the Agency will look like it doesn't know what it is doing and in the long run, lose credibility."[38]

During this episode, the OPPE and its particular synoptic view of the EPA, as well as pressure to streamline the science to make it conducive to shared decisions, seemed to dominate the ORD. Russell asked Elizabeth Anderson, the chief of CAG, for a clarification of the reasons of its use of the IARC scheme. In her response, Anderson aligned on the risk assessment–risk management framework, showing how the WOE method underpinning the classification of carcinogens fit into the risk assessment–risk management framework. She clarified that the WOE approach did not produce a decision by itself, nor was it regulation-forcing; in her words, it could not provide cutoff criteria to decide whether to initiate decision-making. This was a matter of risk management, and risk management needed to rely on other considerations on top of whether the chemical had the property of being carcinogenic (such as magnitude of exposure and size of the health risk). Along with these other pieces of information, "the weight of evidence that the chemical is a carcinogen supports or tempers the decision to initiate risk management or not." In her view, one could not black-box a risk assessment based on WOE in order to decide whether and how to manage the risk. No standard, general risk management strategy could be devised.[39]

Anderson essentially gave in to OPPE's attempt to redefine the WOE bureaucratic technology. Eventually, Russell suggested using a set of criteria other than the IARC because, in practice, they resulted in only two possible regulatory options: to regulate or not to regulate. More flexibility was needed, as were more precise criteria and more sophisticated weighing of the evidence. Alm concurred, as did the lawyers of the Office of the General Counsel of the agency, which put pressure on the OHEA to incorporate into the guidelines some elements of a WOE approach that would replace the IARC criteria. Eventually, the method was refined to be more sensitive to a variety of regulatory options. Five groups were defined:[40]

- A—Known human carcinogen (sufficient evidence from epidemiological studies or other human studies)

- B—Probable human carcinogen (sufficient evidence in animals and limited or inadequate evidence in humans)
- C—Possible human carcinogen (limited evidence of carcinogenicity in animals in the absence of human data)
- D—Not classifiable (inadequate or no animal evidence of carcinogenicity)
- E—Evidence of noncarcinogenicity for humans (no evidence of carcinogenicity in at least two adequate animal tests in different species or in adequate epidemiological and animal studies)

Taking Control of Extrapolation

The OPPE weighed in, finally, to discuss dose-response assessment and the CAG precautionary linear model of extrapolation—the Linearized Multistage (LMS) model—and the upper confidence limit. Alm requested a memo from Goldstein to explain, didactically, the scientific issues underpinning carcinogen risk assessment. Goldstein, who had Roy Albert, the former chair of CAG (see chapter 2), as a professor when he was an undergraduate, summarized the key debates in the scientific community (as concerned threshold, the variety of cancers, genotoxicity, and other topics) to conclude and give support to the use of more conservative ways of assessing these risks: "The traditional public health approach has been that in the absence of a reasonable degree of certainty, it is proper to accept the more conservative assumption of a linear no threshold relationship."[41] In most of its quantitative cancer risk assessments, the CAG was supplementing the use of the multistage model, borrowed from biostatistician Kenny Crump, with another convention of extrapolation: the so-called upper bound limit of 95%, a statistical confidence limit used to "force the multistage model to provide a linear term." The alternative method consisted in calculating a maximum likelihood estimate. The two methods produced variable results, the upper-bound confidence limit being the one that produced the most precautionary results. Both were consistent with the CAG's conservatism because the upper-bound limit was by definition higher than the most likely estimate, and that maximum likelihood estimate would be higher than an average likelihood estimate.

The OPPE questioned both, or at least aimed to qualify their use, notably because these implicit policies were sometimes considered inadequate for particular chemicals. On various occasions in the preceding years, the reviews of CAG health assessments by the agency's SAB led to acrimonious debates about the quality of the assessment and the opportunity of applying the usual linear no-threshold model to substances that could very well be interpreted as displaying a threshold. For Terry Yosie, SAB's staff director in these years, "the board was very strong on its recommendations that the EPA continue to diversify its expressions of risk probability" and stop relying on one model only, increasingly allowing nonlinear responses.[42] In 1980, PCE posed such a problem, as did formaldehyde. In December 1984, in a steering committee meeting on the substance, the OPPE representative underlined that there was a 500-fold difference in the estimated risk of cancer from exposure to formaldehyde, depending on whether one used the upper 95% confidence limit or the maximum likelihood point estimate of risk at the relevant doses. He required to present both risk estimates and to state explicitly which one would be followed, and for what policy reason. Arsenic was perhaps the most contentious of all, as the Water Office had decided to prove that this carcinogenic substance was a threshold substance, and thus to set a precedent of departing from the CAG's default linear model (Powell 1999).

Given that the OHEA applied the technique most of the time, it turned out to be an implicit policy that needed to be clarified in a new framework of thinking that distinguished more strictly between risk assessment, science policies (assumptions applied in the risk assessment), and risk management. As in the case of the WOE approach and IARC classification scheme, the CAG resorted to arguments that had a wide scientific consensus: the plausible upper bound "is generally recognized in the scientific community as a reasonable approach."[43] This kind of argument, now that a framework was in place that explicated the knowledge base used in the agency and established rules for coordinating judgment, was no longer valid. Anderson could not simply mention scientists' convention now that there was a public expectation, under the framework, to see how one consideration potentially leads to policy, decision-making, and which decisions are made.

These documents sketched out important changes in the practices of risk assessment in the agency, as well as an emerging new policy: Against the reliance on the LMS model only, the OPPE defended the idea that a variety of models may be used to estimate the risk of cancer at low doses and that, where possible, the most likely estimates, as well as upper- and lower-confidence limits, should be provided. Only a lack of data can justify the exclusive use of the upper-limit estimate of risk. Here, in essence, Alm and the OPPE were getting closely involved in the fashioning of what RAFG had called a "risk assessment policy," reformulated in the agency as *science policy*. For the first time in the young history of the EPA, a functional policy office, staffed by economists and policy analysts implementing the administrators' policy orientations, was getting involved in the details of science, redesigning it into an instrument of decision-making in the name of the controversy that it risked creating otherwise.

After extensive discussions, the CAG agreed to use the different methodologies side by side in a given assessment, and to compare their results. The 1986 guideline was also slightly less assertive than the 1976 one so far as the linear no-threshold model is concerned: "No single mathematical procedure is recognized as the most appropriate for low dose extrapolation in carcinogenesis" (EPA 1986a, 12), and could be applied to assess the suitability of models case by case, including a rationale justifying the use of one in particular: "in the absence of adequate information to the contrary, the linearized multistage procedure will be employed" (ibid., 13). No standard exposure assessment method was preferred, but the guideline indicated that a "cumulative dose received over a lifetime, expressed as average daily exposure prorated over a lifetime, is recommended as an appropriate measure of exposure to a carcinogen" (ibid., 13).

The notable change from the 1976 guideline, overall, was that these guidelines were structured according to the "more explicit terminology" (ibid., 2) of RAFG, rather than going by the two-question structure of 1976 or the qualitative versus quantitative assessment distinction made since. The 1986 guidelines structured informational bricks, components of a mechanically produced decision. They did so for unknowns, or nonknowledge, giving them a place in the scheme as *science policies*, a term that was absent from previous documents. In November 1984, Ruckelshaus wrote to program managers to indicate that the guidelines were soon to be published in the Federal Register, to "increase consistency in use of risk assessment and underline importance of scientific analysis itself as a factor in regulatory decision-making"[44]—a formula that was drawn from the 1985 guideline for risk assessment of the OSTP, showing how CAG lost autonomy not only to the OPPE, but also to external design actors trying to bear on the agency's way of making rules.[45] The CAG's early risk assessment guidelines had finally been translated into an agencywide position stressing the need for quality, accuracy, and consistency[46] and acknowledging that the judgments of experts constituted science policies. They had come to embody the new coherence between the agency's external, official mission and the value of reducing risks to health; its internal functioning and overall knowledge of risks, based on experimental toxicology and medicine; and a growing interest in epidemiology and exposure data, as well as recognition of the existence of uncertainties and of the possibility of reducing them through expert judgments and estimates.

Conclusion

This chapter has listed the initiatives concerning the reorganization and management of risk assessment in the mid-1980s, involving the leadership of the agency. During that period of time, the leaders of the agency have gotten involved in the formalization of what one may term *bureaucratic technologies* (the elaboration of guidelines, processes of cross-agency consultation, and distribution of responsibilities) so that the agency to be able to more surely turn out unified—and hence defensible and ultimately credible—computations of risk.

To be sure, these technologies were already in the making, and their importance was already sensed in some part of the EPA. Guidelines for the assessment of hazards and risks had existed for nearly ten years. The need for coordination between ORD scientists and regulatory offices was also recognized long ago. But this time around, the risk assessment–risk management assemblage provided a common point of reference to reorganize the work of scientists from the top down. More than a mere reorganization, the policies implemented here aimed to characterize the work of scientists in a completely different way, creating a clear classification of tasks and elements of knowledge needed to articulate a risk assessment that could be accepted across the agency and by top decision-makers. The risk design deposited a new set of identities in the organization in such a way that the various kinds of scientists became the components of an organizationwide system to guide the production of scientific information and interpretation of risks and uncertainties in order to make shared decisions. The science of risk became an object of this reordering of the organization. It became part of an agencywide architecture designed to limit internal dissensus and to increase the EPA's capacity to present uncontestable statements of risk to external audiences.

The particularity of that period is that the top managers of the agency, supported by the economists and decision scientists of the agency, orchestrated the work. They led the designation of the generic knowledge and practices defining the agency as a whole, giving it its identity as a knowledge-based policymaking organization. They could do so legitimately only because the political circumstances—the loss of credibility that the agency had suffered due to the improper use of science during the years of Gorsuch and inconsistencies among the multiple assessments of the same chemical—gave these people the necessary legitimacy to intervene deep into and across the organization to redefine the work of each entity involved.

8 The Rise and Fall of Comparative Risk Assessment

The leaders of the EPA may draw on a variety of possible designs to represent what the agency knows and what it does effectively. There are different ways of formalizing its generic object, the kinds of knowledge it uses to make its decisions, and how it manages to forge decisions on highly disputed issues. These designs vary according to the political configuration in which the agency is caught—the networks of supporters or adversaries that form around environmental issues and its action on these uncertain issues, and the inevitable controversies that ensue. The ambition to systematically measure the risks, costs, and benefits associated with decision projects lasted for most of the 1980s, despite some doubts as to the importance that William Ruckelshaus's successor, Lee Thomas, would grant to this technology, particularly after the departure of Alvin Alm, the deputy administrator who championed cost-benefit analysis and instilled the motivation in the agency to use that kind of information.

At the end of the 1980s, in a new configuration marked by renewed controversies over the EPA's priorities—stemming from its treatment of the discovery of supposed widespread risks from exposure to the gas radon and the pesticide alar, pressures on its budget in an aggressive Republican administration, and a changing national environmental agenda—the commensurative design assumed greater importance. During the term of Thomas (1985–1989), and even more so during the stint of Bill Reilly (1989–1993), efforts were made to create new knowledge representations and technologies to link risk assessors of various program or regional offices, so as to extinguish the uncertainty caused by these offices' nebulous and variegated ways of deciding which risk matters, and closing subsequent controversies concerning the EPA's inability to focus on the right subject. This mainly took the form of comparative risk assessment, a design that was concretely, albeit fugaciously, instituted in the agency between 1990 and 1992 thanks to Administrator Bill Reilly and his policy chief, Terry Davies, a member of the team that had designed the agency in the early 1970s and a long-time critic of its structure in regulatory silos. The mandate of Carol Browner, the first EPA administrator to be appointed by a Democratic president, in January 1993 marked a major shift, with a clear renegotiation of the role of risk assessment and cost-benefit analysis in EPA policies.

Bringing in Comparative Risk Assessment

In his job as assistant administrator for solid waste and Superfund between 1983 and 1985, Lee Thomas had frequently been in conflict with OPPE about his proposed decisions. When he succeeded Ruckelshaus in March 1985,[1] he felt that assistant administrators should regain some autonomy and be able to prepare and finalize rules without systematically referring them to the deputy administrator or to the administrator, as under Alm's cross-agency decision-making system (see chapter 6). Milton Russell and other managers of the OPPE, therefore, feared that under the new administrator, they would lose the transversal, policy development role that Ruckelshaus and Alm had instituted.

Russell devoted a good part of his energy in 1985 to convincing Thomas that the OPPE could play a very beneficial, supportive role in program offices. He particularly insisted that giving too much weight to program offices again could create problems: "In general, centrifugal forces in this Agency are powerful and engrained in its structure. Checks and balances are resisted. It will require constant attention to keep on track."[2] Russell maintained that the OPPE was there to make sure that all regulatory options were considered when the offices proposed a rule, and that cross-media problems were taken into consideration, even though program offices developed their decisions independently. In his view, the OPPE really should be considered a consultant to the administrator, helping to improve and evaluate the quality of decision options, and warning of potential future issues.

As the administrator, Thomas carried on using the same risk assessment–risk management language as the public reflection of the EPA's mode of action and goals as Ruckelshaus. In an address to the petrochemical industry, for instance, he reiterated that the EPA aimed to ensure continuous protection of the environment in the context of preserving economic growth, and that it did so "by carefully assessing the risks we face as an industrial society, and managing those risks effectively."[3] He took care to define the two processes of assessment and management, as though the creed needed to be further inculcated: "To assess the risk at hand, we gather as many facts as possible about the problem. This is a scientific process in which experts thoroughly review the extent of our knowledge and carefully design and conduct experiments to expand that knowledge. This scientific process gives us a basis for understanding the risk we face. It tells us what the risk is, what we know about it, and who is exposed. Then comes the hard part—risk management—deciding what to do about a problem once we are sure there is one.... The options before us include such things as new regulations, additional reporting requirements, new outreach programs or some combination of these and other approaches." Thomas soon endorsed the OPPE in its role, both as participant in the development of regulation by program offices and as monitor of the quality of these regulations. But he also pushed Russell to abandon an overly confrontational posture and to work as a consultant to offices for the development of their risk, regulatory, or cost-benefit analyses. In short, the idea was that the OPPE should be there to "push risk assessment/risk management,"[4] but not to do it instead of program offices, or to counter what program offices did.

The OPPE soon found a specific role for itself, which had not been attempted under Ruckelshaus: that of ranking and prioritizing issues for regulatory attention for the whole agency. Ranking risks, first, had the advantage of corresponding to Thomas's intention to rebalance the agency's agenda overall. A concern emerged in the mid-1980s already, that the EPA was dedicating too much attention, at least at the administrator level, to issues of health. At the time of writing of the flagship EPA report *Risk Assessment and Management: A Framework for Decision-Making,* several assistant administrators were concerned that the agency's leadership was placing far too much emphasis on health issues, when many program or regional offices were in fact struggling daily with pollution and environmental quality issues. Thomas, then assistant administrator for waste, was among them. Now the administrator, he was less inclined than Douglas Costle or Ruckelshaus to put emphasis on health protection as the global agenda for the agency. Between 1983 and 1985, he had managed the solid waste and Superfund programs, where health issues were not as predominant as in the

Air, Water, or Toxic Substances offices. Moreover, in the second half of the 1980s, the public's concerns seemed to be shifting back toward ecological threats as much as the health effects of chemicals—whether carcinogenic or other. Acid rain, protection of the ozone layer, or more localized cases of vast beach pollution were as frequently in the news as were chemicals. At the senior conference of May 1985, before his assistant and regional administrators, Thomas put forward four priorities, among which was the preservation of a strong scientific and technical capability, particularly from a cross-media perspective.

Second, there were a lot of ideas and competences in the OPPE for performing risk comparisons. This was not a new concept. Comparative risk assessment was already well established in the field of risk analysis (see chapter 2), and Dan Beardsley had developed a scheme to this end in 1977, in the context of toxics integration work. His Environmental Integrated Assessment Division was, overall, the comparative risk assessment shop in the agency. Beardsley had already advised Russell in 1984 to capitalize on the comparison among office priorities in order to prove the value of the OPPE's participation in interoffice deliberations. He argued that risk analysis, as a method for comprehensively reviewing issues and putting them on the same plane to orient broad decisions, was more valuable for the agency and the rationality of its policies than individual, in-depth risk assessment of substances. Rational and justifiable priorities for the agency would emerge from synoptic consideration of all issues, not from concentration and investigation of one after the other. Consistency mattered more than accuracy. Or, in Beardsley's words: "It is the continuing, consistent use of risk analysis that we believe is desirable."[5] The implication for the OPPE, which did not have the same level of access to the scientific data that program offices had, was to avail of its own data on environmental issues and to quantify and rank them. What the OPPE had was information about budgets for each office and program. Beardsley suggested to Russell that he compare the broad levels of risk for each environmental program across the agency, with the budget levels and priorities established by statute.

The OPPE inaugurated formal comparisons of risk in 1986. That year, Robert Wolcott, an economist in the office of policy analysis, undertook a comparison of the risks, costs, and benefits of a set of agricultural chemicals. The motivation was political: Thomas wanted to initiate more special reviews of pesticides than had been the case in the past.[6] In October 1986,

Thomas imposed an emergency ban on the pesticide dinoseb, and a definitive ban was concluded with the concerned companies in the middle of 1988. At the time, alar, and the hazards of its breakdown products to children and young people through consumption of apples, was prominent in the national news. The work of Wolcott would help in selecting products for special review and give managers resources to detect potentially concerning pesticides before they made it into the press and started to rock the agency's work program and priorities. The *New York Times* reported on the results of the comparative risk assessment of 1986 with an article headlined "Pesticides Finally Top the Problem List at E.P.A.," putting forward the words of the new head of the pesticides division, Steven Schatzow, that "[t]he pesticide problem is worse than ever" and that "virtually everyone" was exposed to that risk (Shabecoff 1986, B12). The EPA had rarely, if ever, singled out a class of risk in such a way in its young history.

In 1986, also at the request of the administrator, the director of the Office of Policy Analysis, Richard Morgenstern, began outlining a study to compare the risks addressed by the agency's major program elements and activities. The expansion of risk assessment techniques, along with increasing standardization due to guidelines, offered a chance to compare the issues that the various offices were dealing with. But the exercise was innovative: It was an attempt to introduce risk analysis in ways that were useful for planning, budgeting, and strategy, rather than only for individual programs and for quantification of health effects.

The methodology that was crafted was ecumenical. First, there was no exclusive emphasis on quantifying all risks on a common scale. The study would not be one of these "hard version" comparisons of risk (Hornstein 1992), like the one attempted by Chauncey Starr in 1969, based on a quantitative scale of benefits, measured in dollars, and risks, measured in terms of mortality. It involved a mix of quantitative assessment and qualitative judgment, by officials of the various program offices, of the agency's level of investment in each risk. Second, the exercise went beyond health and cancer.[7] Risks were defined as falling into four large types: cancer, noncancer, environment, and welfare. A workgroup was formed for each. The OPPE's balanced method for the comparative exercise reflected the rebalancing of the agency's overall agenda and a new strategy on ecological risks.[8]

The report came out in 1987 as *Unfinished Business—A Comparative Assessment of Environmental Priorities* (EPA 1987). In several ways, it was a surprise. First, the external reactions to the report were generally favorable. It was immediately picked up in congressional hearings, where the possibility to compare risks and modify priorities was seen by many as a very interesting methodological breakthrough. The document was, and still is, considered to be a landmark, groundbreaking study (e.g., Hammitt 1997). Second, the ranking of risks revealed by the report did not "correspond very well with EPA's current program priorities" (Morgenstern and Sessions 1988, 36). EPA's priorities were aligned on public perceptions. The authors indicated that indoor radon, indoor air pollution, ozone depletion, global warming, pesticide residues, and worker exposure, among other issues, were areas of relatively high risk and low EPA effort. The report gave force to observations that had emerged both in various parts of the agency and outside it, concerning its focus and action. It argued that the EPA's action was characterized by fragmentation, whereas the state of the environment should be approached in a holistic, integrated fashion. The cost of attacking risks one by one was much too high, and the effectiveness of individual risk strategies severely limited. That piecemeal strategy, moreover, would not be effective at addressing the problems that were topping the national agenda at the time, such as global warming or ozone layer depletion. The EPA could not continue to let its action be directed by public scares and the "chemical of the month" syndrome.

Thomas soon publicly endorsed the report and decided to make it publicly available. In all respects, it supported both his choice to continue using "Risk management" as the homogenizing motto of the agency, and his ambition to steer the agency's priorities back toward ecological and pollution problems. The actual consequences on the agency's operations were limited. Some attempts were made to redirect the budget toward the new priorities, such as global warming and depletion of the ozone layer. However, as the mobility of EPA funds was strictly limited by negotiations with the OMB and Congress, as well as the various tasks programmed by the legislation passed by Congress, these budgetary changes remained marginal. The report motivated Thomas to give greater support and resources to the Integrated Environment Management Division, inaugurated in the first years of the Toxics Integration initiative (see chapter 3). The Unfinished Business experiment also triggered a series of comparable comparative risk analyses at State level, by regional offices (EPA 1989).

Picking the Wrong Fight?

The EPA was not done with controversies surrounding its use of science and its manipulation of data to fit preferred decisions. In two cases at least in those years, the agency was accused of being not only conservative, but also alarmist, and of having pushed highly precautionary policies before clarifying the science. Those judgments on the EPA were also legacies of recent affairs that escalated and blew up in the space of just a few months, creating a public experiment for real-time observation of the peculiar modes of agenda-setting at the EPA.

Alar was the first case in point. The risks associated with this pesticide were the subject of an intense, short-lived national drama in 1989. On February 26, 1989, CBS aired a report on alar in a newsmagazine, including an advance look at the results of an analysis of twenty-three pesticides, including alar, by the Natural Resources Defense Council (given in a report entitled *Intolerable Risk*). The next day, the council released its report through press conferences in twelve cities, targeting the hazard of pesticide residues in apples and setting off a real panic among consumers of apples and apple juice. The action of the environmental group took the EPA aback. It found itself under a deluge of pressing questions from members of the public and the media, all in need of immediate, precise confirmation of the risks involved. The EPA had been considering the risks of that substance for more than ten years, but the deregistration of the chemical, once contemplated for its carcinogenicity, had been abandoned in confusing exchanges between the CAG, the Office of Pesticides, and the Science Advisory Panel, characterized by disagreement on the product's carcinogenicity.[9] The comparative risk studies of 1986 and 1987 had shown, however, that pesticide risk should top the agency's list of priority issues. Once the crisis had begun, the EPA did its utmost to reframe the controversy.

John Moore, former assistant administrator for OPTS and then acting administrator of the agency, issued a press release arguing that the Natural Resources Defense Council was misleading the public, but that the hazard, in case of long-term exposure, was real. The crisis continued unabated, spurred by a public campaign orchestrated by a public relations company under contract with the council. The actress Meryl Streep, leading a movement of "Mothers and Others," famously testified before Congress and on television shows about alar's dangers. The consumption of apples plummeted for months. The end of the story came with the deliberate withdrawal of alar by its manufacturer in June 1989. The EPA ended up being hard hit in its reputation and credibility for having admitted under pressure from the Natural Resources Defense Council that alar posed a risk and that farmers should stop using it (even though it had hesitated for years to actually make such a definitive pronouncement on the basis of animal cancer studies in its possession).

At the other extreme was the radon affair. Until 1984, the potentially high background level of this colorless, odorless, cancer-causing gas fell in the interstices of the EPA regulatory programs. The Superfund program could have legitimately tackled the problem of radon, which had been known since at least 1975 with the discovery of high levels of radiation in Pennsylvania. In 1981, the director of the program proposed a radon plan, but it was blocked by the OMB. Then, in December 1984, Stanley Watras, an engineer working on the construction of a nuclear plant, discovered extremely high doses of radon in his home due to infiltrations from a mine below.

This propelled the issue onto the EPA agenda and heavily damaged the agency's credibility. How could the agency overlook a substance potentially linked to 20,000 lung cancers per year—far more than any of the other substances that the agency considered to be high-profile and high-visibility issues—and affected up to 8 million homes? Even after the Watras discovery and recognition of the gravity of the issue, it remained very difficult to get a national radon policy off the ground. The EPA launched research to curb levels of exposure to radon, but was blocked by conflicts with the US Department of Energy and the OMB over the threshold under which the homes at risk of infiltration should be brought: 4 picocuries per liter, or more? Even inside the EPA, the director of the Office of Radiation, Richard Guimond, found it difficult to find support for the proposed policies. In 1988, at a key point in the radon controversy, EPA managers even suggested that Congress authorize an ORD budget with fewer monies for radon.

Eventually, in 1987's *Unfinished Business* report, radon was listed as one of the top carcinogens that the EPA should be concerned about. A radon action program was set up, involving surveys in schools, homes, and workplaces around the country, as well as investigation of high-radon-potential natural sites. A risk communication program was also established, with large public information campaigns. The EPA furthermore engaged in the testing and distribution of technologies to monitor radon levels.

Even though it remained obvious that radon was a hazard, this episode left a majority of the public with the impression that the EPA had overblown the issue, as no big cancer epidemic materialized (Edelstein and Makofske 1998). It harmed the agency's scientific credibility and spread the need, in the public, for an environmental policy rooted in "good science" (Abelson 1990; Brody 1991). Further complications emerged when the EPA proposed to regulate the levels of radon in drinking water, ensuring that they were in line with the levels that it had declared safe as part of its other action to reduce radon in indoor air. Congress imposed a moratorium on the regulation of radon under the Safe Drinking Water Act and requested a multimedia assessment of the gas in an attempt to put an end to the inconsistencies.

The public risk controversies of alar and radon had in common the fact that they called into question the reliance on animal studies in human health risk assessment research. These affairs showed that epidemiological studies and monitoring data were of utmost importance to characterizing the risks and, even more so, to defining appropriate, targeted measures. They were also of value to escape the unending interpretive controversies surrounding animal tests and the linear extrapolation methods.

The alar controversy played a particular role in the emergence of the problem of epidemiology, monitoring and data, because it caused Bruce Ames, a prominent biochemist and molecular biologist, to intervene publicly and minimize the risks of the chemical. Ames was renowned for his invention of a rapid test to identify carcinogenic chemicals (Ames 1979), which is now standard. While he was praised by the environmental movement for his decisive contribution to the detection of chemical risks, he gradually changed his mind concerning protection against chemicals, to the point of becoming an outspoken critic of environmental regulation (Proctor 1995). Starting in the 1980s, he developed a large database of chemicals, both natural and synthetic, and their cancer potency, showing that many synthetic chemicals were in fact safer than those that occurred naturally in food. He published an article in *Science* based on the results of his research, arguing that animal tests were not appropriate to calculate cancer risks (Ames et al. 1987) and taking on public regulators of chemicals for their science and choices.

Ames appeared in the second CBS program on alar, in May 1989, arguing that conventional extrapolation from high to low doses had enormously exaggerated the possible hazards compared to the risks linked to natural carcinogens generated by plants in defense against pest attacks. In a letter to the editor of *Science*, with the heading "Pesticides, Risk, and Applesauce" (Ames and Gold 1989), he ridiculed the focus on alar and reiterated his arguments about the unreliability of animal tests and EPA's misplaced focus on synthetic pesticides. The intervention by Ames mattered because his arguments motivated the manufacturers of a number of high-profile chemicals regulated as air pollutants (including EDB and PCE), to sue the EPA: Dow Chemical, Shell Oil, Accidental Chemical Corporation, together with the Chemical Manufacturers Association and others, launched their action in March 1991, drawing on Ames to demonstrate, like tobacco companies later on passive smoking (see chapter 9), that the EPA was acting in an arbitrary and capricious manner. The court did not heed their challenge.

The series of controversies were the sign of a particular climate surrounding the EPA, with various arguments compounding to downgrade the credibility of its research base. The agency was now appearing to choose the wrong targets and chasing ridiculous risks. It became common to hear in those days, including in Congress, that the EPA was wasting "vast amounts of resources chasing down infinitesimally small amounts of high-profile, media-sensitive, low-risk substances, resources that might have been more effectively used to boost jobs and American competitiveness or applied to other environmental problems or opportunities," as declared by the Republican senator Don Ritter (US Congress 1995, 155). In *Fortune*, a journalist ran a popular story called "The Big Clean-up Gets It Wrong" (Main 1991). After all, ten years after the major scandal, were people not relocating in Love Canal, that territory contaminated by Monsanto's polychlorinated biphenyl (PCB), and the case behind the adoption of the major Superfund decontamination program? In those days, journalists filled bookcases with such titles as *Toxic Terror* (Whelan 1993) and *The Asbestos Racket* (Bennett 1991), spreading the word that animal testing to spot cancer risks came down to a "mouse terrorism" (Whelan 1994b).

Reilly and "Relative Risk"

Bill Reilly became the new EPA administrator in January 1988, just before the alar controversy but after the publication of *Unfinished Business*. Reilly was one of the few Republican environmental leaders, and he had been chosen by President George H. W. Bush to show that he would deliver on the pro-environmental commitments made during his electoral campaign. Reilly's environmental credentials came from his membership in the Council for Environmental Quality in the 1970s and his presidency of the Conservation Foundation (Landy et al. 1994, 279). After being appointed, Reilly worked with Terry Davies, who was also at the Conservation Foundation, and Dan Beardsley to develop strategic priorities for his administration.[10] Both of these men had a strong inclination toward risk analysis, understood as the comparative analysis of policy priorities and decisions with regard to environmental hazards. Davies had been among the people who had pushed the EPA's design along functional lines, with a central role in opening an analytics office under President Richard Nixon. He was a member of the RAC, which proposed this reading of the administration of risk as structured by two processes of risk assessment and risk management, the latter including regulatory analysis. Beardsley was a policy analyst with the OPPE at the EPA until 1987. He was the leader of the Toxics Integration initiative under Costle and a key part of the enterprise of developing risk assessment and risk analysis during Ruckelshaus's second term as EPA administrator (see chapters 3 and 6).

This small group of three confirmed the strategic priority given to new global ecological issues such as global warming and ozone layer depletion, as well as a moderation of the toxics and health agenda. The group also forged Reilly's positive appreciation of *Unfinished Business,* and the virtues of comparative risk analysis as a tool for setting priorities in the area of environmental action, or as a way to take control of its priorities, over its principals in Congress or in the White House. Beardsley held the view that the main risk manager in the country was Congress, in that the laws it voted on and budgets it appropriated were de facto defining the priorities to which the agency was according resources and attention (Beardsley 1987). By engaging in comparative risk assessment, the EPA would gain a resource to demonstrate to Congress where, rationally rather than politically, the resource may be allocated.

Unfinished Business was still only an internal EPA report, even though it had been widely publicized and positively reviewed in the press. Reilly reasoned that having it peer-reviewed, like other kinds of scientific productions of the agency, would give it further authority and political clout. He and his top aides decided to use the SAB for this. It had already ventured into the area of relative risk in 1988. Its chairman had taken the initiative to form a research strategy committee to advise the EPA on how to orient its research efforts. The SAB defined "risk reduction" as the main overarching goal of the agency, above and beyond the goals defined by statutory programs, and noted that "EPA's basic mission is to reduce the level of risk to health and to the environment posed by wastes, residues and contaminants" (EPA 1988, 4).

In this report, the lessons of the difficulty of actually regulating the risks posed by widely diffused chemicals (carcinogenic or otherwise) were transparent. It argued, first, that instead of trying to regulate risks ex post facto by removing chemicals, the EPA should embrace a preventive strategy: act so that chemicals and pollutions are not generated in the first place. Second, the EPA should focus on those risks that it has the greatest chance of actually being able to reduce ("relative risk reduction"). Reilly asked the SAB to pursue this effort on relative risk through a review of *Unfinished Business* that would help him draw conclusions from this report for the strategic management of the agency's orientations.

In response to Reilly's request, the SAB formed a special committee, the Relative Risk Reduction Strategies Committee. The conclusions of this committee, laid out in a report called *Reducing Risk,* were that the EPA should rebalance its priorities, giving greater importance to ecological issues. It called for attention to be given to neglected high-risk issues, such as indoor air pollution and radon. *Reducing Risk* did not quantify all costs and benefits any more than did *Unfinished Business.* It was, to a large extent, based on expert evaluation of the potential for risk reduction and the gravity of environmental impacts of various sources. Interestingly, neither of these reports was criticized for being based on expert judgment or for overlooking massive uncertainties that necessarily moderate the reliability of a large comparative risk assessment exercise.

Reilly did not stop there—he asked a panel of high-level scientists to review the state of science in the agency in the context of limited resources for research, risk comparison, and priority-setting. In March 1992, the panel handed in its report, *Safeguarding Science: Credible Science, Credible Decisions,* to Administrator Reilly, concluding that the scientific basis of decisions was not sufficiently well organized. The report led to the creation of a council of scientific advisors, individually located in each programmatic and regional office. A new position of science advisor was established, under direct supervision of Reilly. For the time that was left before the end of Reilly's term, the science advisor experiment worked well. Guidelines for research quality started to be established and applied, and communication between the ORD and program offices was rapidly restored.

The *Risk Reduction* report supported the development of the Pollution Prevention Strategy and the launch of a program collecting ecological data (the Environmental Monitoring Assessment Program). It motivated Reilly to decide on an increase of around $7 million in the budget for the so-called geographic priorities, or holistic problems concerning such areas as the Great Lakes, Chesapeake Bay, and the Gulf of Mexico. Senator Daniel Patrick Moynihan, a Democrat from New York, tabled an Environmental Risk Reduction bill in 1992 and 1995, but it failed to pass each time. Only the bill H.R. 2910 became law,[11] applying the concept of risk reduction to specific issues such as mercury pollution (Susskind et al. 2001, 46). In the agency, the OPPE managed to institutionalize a procedure for program offices to report on the overall benefits of their programs. There was strong resistance there, especially by offices that ran expensive programs, such as Superfund. The OPPE forced offices to document the benefits of their programs only when they asked for supplementary funds (known as "budget add-ons") during the year.

Between 1990 and 1992, at the top of the agency, the image of risk assessment changed from a tool to inform risk-specific management measures to a tool to enable planning and prioritization; a tool for the administrator to steer the agency, as opposed to a tool for regulatory offices; a tool that manipulates information about safety, probabilities of health, and ecological effects, but also about the magnitude of a problem in terms of the costs of mitigating it, as well as the ease with which it could be brought down to acceptable levels. At the ORD conference on "Risk Assessment after Ten Years," Don Barnes, the staff director for the SAB, noted that the approach of RAFG provided no solution to rank risks and choose which one to tackle in priority (Barnes 1993).

Comparative risk assessment, or risk analysis, was the solution for the former and current leaders of an agency that had too often fallen victim to the unpredictable dynamics of public controversies about cancer-causing chemicals. Alm, who sat on the SAB committee that wrote *Reducing Risk*, perceived the EPA's environmental action as a "treadmill," with more offices, more personnel, more priorities piling up, and no ability "to get to all the problems" (US Congress 1994b, 197). The need for a mechanism to impose agency-level priorities was clear. Reilly was explicit about the need to move away from ungovernable controversies about chemicals, the inability to pin down levels of risk, and constant legal challenges against the agency.

This new take on risk analysis—as an agencywide planning tool rather than a problem-focused quantifying exercise—was a way to decide on greater numbers of environmental priorities, as opposed to being driven by episodic alarms such as Love Canal and oil spills, and constrained by laws passed by Congress. It was well noted in Washington and signaled a new kind of bureaucratic discipline. Terry Davies coordinated a project at the think tank Resources for the Future, on request from the OSTP, to assess how to generalize the kind of "programmatic Comparative Risk Assessment" that the EPA had pioneered. He proposed a similar framework for the administration of risk than the EPA was moving toward: "*When describing the risk-based disciplines, risk analysis is the most general term, which encompasses comparative risk analysis, risk analysis, risk assessment, risk management, and risk communication*" (Davies 1996, 5).

Another important event, attracting top EPAers, was a conference organized by Adam Finkel on the ranking of national environmental priorities (Finkel and Golding 1994). *Reducing Risk* was widely cited in congressional debates and hearings as an element in favor of the move toward greater expert-driven selection of environmental issues—in the terms that were then used, *prioritization* and *planning*, to be applied both to the agency's research programs and to the definition of its budgets. It concurred with what others were saying at the time. The White House's proposed budget for 1992 had mentioned "risk-based budgeting,"[12] soon followed by a prestigious Carnegie Commission on Science, Technology, and Government—including Alvin Alm, Justice Stephen Breyer, former FDA general counsel and RAC member Richard Merrill, as well as Gil Omenn—through its report *Risk and the Environment: Improving Regulatory Decision-Making* (Carnegie Corporation 1993).

Risk Analysis in a Democratic Administration

President Clinton's EPA was different—it placed less importance on proper risk and policy analysis, and in the calculation of the benefits of its policies, than before. Clinton's election put an end to three consecutive Republican administrations. He brought in Carol Browner, a former legislative director for Senator, now Vice President Al Gore, and formerly the head of the Department of the Environment of the state of Florida. Environmental policy changed more dramatically than during any of the last transitions, between Ruckelshaus and Thomas, or Thomas and Reilly. Browner's priorities, as she took the office, were pollution prevention, environmental justice, and innovation in control technologies (Fiorino 1995). The focus on risk clearly receded during this transition, with Browner's intention to shape a different EPA. Risk assessment and cost-benefit analysis stopped being presented in agency documents as generic policy tools, as they had been in the 1980s.

The OPPE in fact lost its prominent role of the 1983–1993 years—its "glory days," as one of the policy analysts of that period called them.[13] It no longer occupied the privileged advisory role to the administrator that it had acquired in 1983, and it was no longer managed by a political appointee: a signal of its decline in the internal hierarchy of offices. Because the Clinton administration was overall less adverse to environmental regulation, the agency got less pushback from the White House and the OMB during most of the 1990s, and the OPPE's most critical utility for the administrator was precisely in making proposed rules more robust with aspects of costs and benefits.[14]

The forays in comparative risk assessment of the early 1990s were not replicated. What remained of the comparative risk assessment drive of those years was a subcommittee within the SAB dedicated to residual risk, as well as other program-specific initiatives. One consisted of formalizing a system of technological evaluation inside the Office of Pollution Prevention and Toxics (OPPT), formerly the OPTS, which restored the importance of that approach by using the risk-ranking notion as a springboard. Effectively, engineers in various offices where pollution-reducing technologies were being discussed had always thought that risk assessment distracted from the achievement of actual reductions in pollution levels in industry processes. The Office of Pesticides as an office in precisely this situation, realized that the valuation perspective implicit in risk-ranking took advantage of the new agenda to try to push forward a new scheme for technology assessment. It legitimized the initiative by incorporating it into the framework of risk assessment, risk management, and risk reduction through the formal ranking and weighing of priorities (EPA 1994a). But no new rule or guideline emerged to make those processes more systematic in the agency. Besides local initiatives, the heads of the agency stopped supporting risk analysis across the agency, or did so only to ensure continuity of what had become essential processes in the agency, such as guideline development and coordination of risk assessment science through the Risk Assessment Forum.

More than a new internal rule, comparative risk assessment became a strategic bureaucratic screen against the various audiences demanding changes in the agency. Thus, comparative risk assessment continued to feature prominently in public expositions of the agency's way of making decisions and of its priorities. Risk analysis remained an important theme for representing the agency, materializing its rationality for audiences, and constructing the credibility of its choices. Such was the case during congressional hearings on the ORD. In the early 1990s, the financial situation of the ORD seemed sufficiently bad for a group of industry leaders and environmental activists to write to Congress, asking legislators to appropriate more money for the ORD than pledged by the White House (US Congress 1992).

The ORD budget had never really recovered from the cuts of the Ronald Reagan era. In constant dollar terms, it was even in decline, and it seemed to have an aging workforce as well. At the same time, the strength of its research in human health risk assessment was in question, as was its way of deciding on risks to investigate in short- and long-term research efforts. Several attempts at reforming the EPA's science and the ORD emerged from Congress and within the EPA. Congress requested a report from its Office of Technology Assessment. It then took various legislative initiatives, introducing a Risk Assessment Improvement Act, in terms of which the EPA administrator was to report to Congress on ten environmental issues correlated to the highest environmental risks, "with respect to which uncertainties could be significantly reduced through research."[15] The head of the ORD had to establish a strategic plan for his office and procedures for ensuring the quality of research.[16]

At another hearing to discuss two proposed bills about environmental quality and environmental justice, Lynn Goldman, assistant administrator for the OPPT brought in by Browner maintained that the paradigm of risk assessment and risk management "established" by the National Academy of Sciences and "adopted" by the US EPA, had now been enlarged to include new undertakings: it was now "helpful to distinguish between risk assessment, risk management, comparative risk analysis, and risk communication" (US Congress 1994a, 16). The statement that she put on the record of Congress included the graph shown in figure 8.1, enlarging the RAFG-based

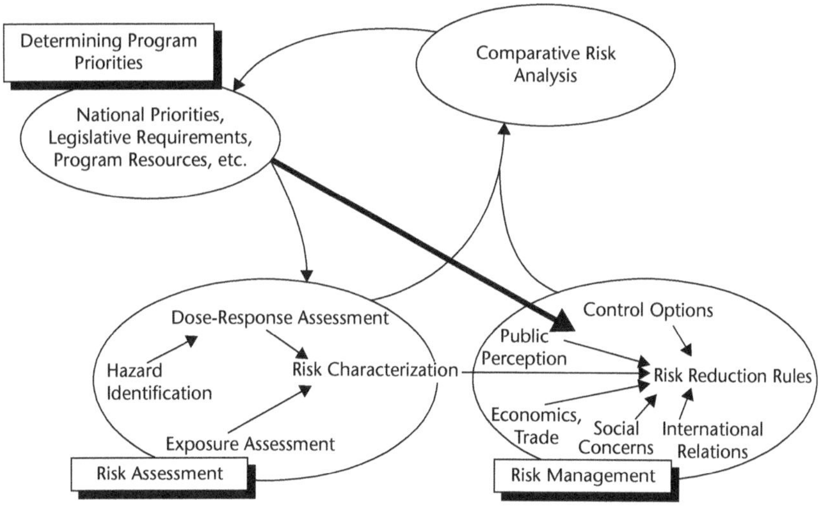

Figure 8.1
An integrated risk decision-making design incorporating comparative risk analysis (adapted from US Congress 1994a, 33).

scheme to incorporate a "comparative risk analysis" process for the agency. The bottom circles represent the standard modules of risk regulation that RAFG conceptualized and that were institutionalized in the agency in the 1980s. The processes are run in accordance with an a priori, upstream determination of national priorities. Risk assessment and risk management generate information that feed a new module of comparative risk assessment, which in turn is used to inform this broad agenda.

Comparative risk assessment was even more important as a design in the relation to the White House, which was working on the topic for itself, using comparative risk assessment and risk ranking as a political solution to demonstrate the balance existing in its environmental and economic policies. In October 1993, President Clinton issued Executive Order 12866, which superseded the Executive Order of 12291 issued in 1981. The EPA was still directed by this new Order to conduct benefit-cost analyses for significant regulatory actions and to select options which "maximize net benefits" to the extent permissible by law. But "net benefits" were now parenthetically qualified to include "potential economic, environmental, public health and safety, and other advantages; distributive impacts; and equity" (Federal Register 1993, 51735).

Clinton's executive order provided more flexibility in environmental policy because it explicitly recognized a role for nonmonetary considerations in policymaking (Goodstein 1995). The executive order materialized in the setting-up of a working group, led by the OMB's OIRA, alongside the OSTP, with the participation of a large set of federal regulatory agencies. The group came out with risk analysis principles that essentially perpetuated the rationality of policy analysis and options development—and was thus not to be confused with the more routine practice of health or environmental risk assessment: "'Risk assessment' is the term that is used for the process of describing or characterizing the nature or magnitude of a particular risk. Risk assessment is the foundation of risk analysis. The terms are sometimes used interchangeably, but assessment is a gathering, the assembling, and the arraying of information on the risk. Risk analysis includes not only risk assessment but also risk management, how you are going to respond to this risk, risk communication ... Risk analysis, taken as a whole, provides a means for organizing scientific, technical, social, and economic information" (Sally Katzen, quoted in US Congress 1994d, 10).

Conclusion

From the creation of the EPA until the beginning of the 1990s, the comparison of costs and benefits of various environmental policies has constantly gained importance in the agency. With administrator after administrator, and through the succession of deputy administrators and leaders of its policy office, this expertise has slowly institutionalized, to embody a part of what the agency does and how, for its audiences. There is a strong, symbolic aspect to this—especially when the rationale of analyzing and comparing risk emerges in the public discourses of the leaders of the agency (e.g., before Congress)—but also a material one. Guidelines for risk comparison and risk-ranking were developed; policy analysts received a formal, official function of performing agencywide, synoptic analyses of the risks that the agency was to tackle; risk analysis was concretely articulated with the work of scientific risk assessors; and even the agency's prominent SAB was consulted, and it contributed to manufacturing and applying these methodologies.

The notion of risk analysis that was formalized at the beginning of the 1990s to generalize risk comparison, risk-ranking, and environmental valuation may not appear to be that different from scientific risk assessment. Indeed, the risk assessment–risk management framework explicitly combined them. But behind seemingly arcane debates and subtle choices among equally generic terms lay a fundamental difference in the design that various leaders and experts of the agency pursued: a reductionist one, aiming for accuracy in the description and explanation of a risk taken individually, against one of comprehensive attention to sets of problems emerging in the environment as a whole and analyzed synoptically. The latter design emerged in the agency for different reasons and at a different moment than quantitative risk assessment. It appeared when the agency's mission and policy started to be questioned by a Republican Party that was itself becoming less environmentalist. The controversies that led the agency to design itself in this way typically concerned the priorities of the agency, its way of choosing among the environmental issues to pursue, the fact that it was seen to be "asking the wrong questions" (Landy et al. 1994).

Risk analysis was, then, the bureaucratic technology of the day—the tool to limit contestation, to reframe public debates about environmental issues, and to make the outputs of the agency more acceptable and effective. But this technology was also more short-lived: Its importance declined as soon as the political atmosphere evolved and as the agency got less contested in its choices—at least provisionally, when Clinton accessed power, and before the Republican political landslide that allowed the party to take total control of Congress in 1994. Risk analysis, as a technology for producing rational administrative decision, closely reflects the cycles of political support and contestation that the agency goes through. And the early 1990s mark the beginning of a cycle of politicization of the EPA's science and ways of making decisions. Risk analysis—its rise and fall—reflects it, but so will risk assessment, and soon. A period of intense and contentious definition and formalization of the science of the agency had indeed begun.

9 Scientization and the Reform of the Risk Assessment–Risk Management Framework

In the early 1990s, around ten years after William Ruckelshaus officially embraced the notions of risk assessment and risk management to demonstrate how the EPA administered environmental problems, and as that new discipline materialized in the way the agency made decisions, calls multiplied to go beyond what the authors of RAFG supposedly prescribed. Having become the reference for how the agency works, the risk assessment–risk management framework was now accused of instituting a too-strict separation of science and policy. What was seen as an improvement in the early 1980s—a clearer definition of the actual roles of toxicologists, biostatisticians, exposure scientists, but also economists, policy analysts, lawyers, and other administrative officials in the formation of a decision—was now becoming part of the problem. People had embraced the roles too ardently and were enjoying the comfort and security of these delimited responsibilities too much. There was apparently not enough communication between the two. Risk decisions were produced following fixed science policies or so-called default assumptions rather than adjusting to the specifics of the data, uncertainties, and objectives of risk managers, in a constant dialogue and characterization of the risk.

This criticism was in part misdirected because RAFG was not actually calling for a separation of science and policy and the risk assessment–risk management framework had not been applied in exactly this manner in the agency. Among other things, it helped invent new modes of exchange between the ORD's scientists and regulatory programs, for instance. As the previous chapters have shown, it legitimized more, not less involvement of the managers of program offices and of the agency in the rules of risk calculation.

What this criticism reflected was the emergence of a new series of attacks on the EPA's scientific capacities and conservative policies that it applied to chemical risks. The early 1990s was a moment in which opponents of the agency chose to depict risk assessment as being strictly about science and delegitimized the application of default assumptions and science policies to risk decision-making. Industry think tanks and lobbyists, the OMB, and an aggressive Republican vice presidency, as well as scientists pushing for the application of the latest uncertainty-reducing methods, coalesced to argue that the agency was applying rigid, conservative assumptions and not using the best possible, most advanced science. This campaign compounded with the difficulty of bringing legal procedures involving major chemicals to a close. Pressure from Congress on the agency did the rest. This context of the early 1990s is one in which a different design—a scientistic, predictive one—emerged, based on the belief that uncertainty can be reduced thanks to the progress of scientific knowledge, and that an agency should not be organized to make decisions unless this is the case. This dispute translated into a contentious redesign of the EPA, with various people inside the agency, as well as audiences outside, fiercely competing to revise risk assessment guidelines and define standard procedures for risk characterization. Science, once more, was the medium of the reconstruction of the material identity and expertise of a contested agency.

Iconic Risk Assessments and Emerging Agency Failure

The stint of Bill Reilly as EPA administrator witnessed an accumulation of complicated cases of chemical regulation. In at least three major cases affecting a variety of regulatory programs run by the agency—namely, controlling dioxin, asbestos, and arsenic—external challenges in court, led by industry or environmental groups in succession and compounded by internal discussions and inconsistencies, created nearly inextinguishable regulatory controversies and corroded the credibility of the EPA's apparent formal way of making integrated decisions.

Dioxinlike compounds form a wide range of substances, sometimes described as the most potent known carcinogens based on animal studies. Epidemiological studies are not convincing, given the difficulty of estimating the levels to which people are exposed. The EPA regulates dioxinlike compounds under various statutes as a pesticide, as a contaminant found at

Superfund sites, as an air pollutant, and as a water pollutant. Since 1983, the regulation of the levels of dioxin in effluents had been the object of several episodes of tension, notably in 1983, when it was established that political appointees in the agency were unwisely trying to save the substance from regulatory intervention. The CAG had produced the first health assessment for use across the agency in 1981. In 1982, the EPA promulgated effluent limitations and technology-based standards for the pulp, paper, and paperboard industry.

For at least the next five years, the dioxin issue was low on the agency's research agenda, going from assessment to reassessment as new experimental and monitoring studies were published, and as various industries, the OMB, and environmental groups maintained pressure on the EPA to publish more or less conservative estimates of its carcinogenic potential. Following a front-page article in the *New York Times* in 1987 about the presence of high levels of dioxin in paper products, the agency formed a group to produce an integrated assessment. Meanwhile, a group in the RAF was asked to revise the earlier estimate. Departing from its traditional approach of using the upper bound, the RAF group decided to choose a risk value near the midpoint of the dose-response curve in order to increase the acceptable dose slightly.[1] This estimate, of course, did not end the process.

The agency was challenged in court by environmental groups and was under pressure to agree with the NGOs, given revelations in the press that it had cooperated with the pulp and paper industry to produce monitoring data (Powell 1999). The consent decree signed with the Environmental Defense Fund and National Wildlife Fund (NWF) included the obligation for the agency to perform a comprehensive risk assessment of sludge, water effluents, and products made from pulp in the bleaching pulp mills, deliver a multimedia risk assessment by April 1990, and then to propose regulations to control pulp sludge disposal and address the discharge of dioxins and furans into surface waters by October 1993.

Opponents to the EPA's conservatism, for their part, put pressure on the agency via a carefully staged scientific conference. The so-called Banbury conference was organized in 1990 by industry groups and by Michael Gallo, a toxicologist who had consistently battled for the consideration of models of biological mechanisms in risk assessment (or what he called "biologically based dose response models" or BBDR modeling), and against the standard and protective mode of linear extrapolation followed by the EPA since the end of the 1970s. The aim of the conference was to discuss the possibility of a molecular analysis of the biological steps preceding the observed effects of dioxin. For Gallo, a conclusion of the conference was that "if we can't depart from the linear model in the case of dioxin, for which we know so much, then we can't do it for anything" (cited in Roberts 1991). Inside the conference, debates were intense, but by virtue of a vigorous public outreach program led by the Chlorine Institute, it appeared to the rest of the world to have resulted in a consensus that the risks of dioxin probably had been overestimated due to the conservatism of the EPA's assumptions (McGarity and Wagner 2008).

The Banbury conference left the EPA enthusiastic about the possibility of finding an actual scientific consensus to develop regulations for dioxin (Roberts 1991). In 1991, a National Institute for Occupational Safety and Health epidemiological study on 5,000 men over a 13-year period found no increase of cancer among workers exposed to low levels of dioxin. Both events pressured Administrator Bill Reilly to announce, in April 1991, a comprehensive reassessment of cancer and noncancer risks of exposure to 2,3,7,8-tetrachlorodibenzo-p-dioxin (TCDD) and other compounds. Scientists in the ORD and the SAB went to work on the issue once again. Meanwhile, Vernon Houk, a scientist in the Centers for Disease Control and Prevention (CDC), argued publicly that dioxin was a weak carcinogen, after all, and that the decision made in 1981 to buy land for $33 million and evacuate thousands in Times Beach, Missouri, was mistaken (Kelly 1995).

The reassessment inspired Reilly to say that dioxin was a "model risk assessment" for the agency. Two officials of the OHEA called it a "flagship risk assessment" (Preuss and Farland 1993) because the agency had made a concerted effort to put its risk assessment into discussions publicly with academic scientists, the regulated industry, and environmental groups. Yet nothing indicated that these new procedural engagements worked. Despite years of assessment and reassessment, accumulation of studies, and clarification of methods and judgments to be applied to such cases, the EPA was nowhere near closure of the case. If anything, dioxin was still a battleground (Bailey 1992). The years that had lapsed, the costs of reducing levels of dioxin in the rules that the EPA had proposed, and the millions spent studying dioxin and evaluating the resulting studies denied the exemplary nature of the reassessment process.

Unfortunately for the EPA, there were more high-profile controversies surrounding iconic chemicals in those years. Ozone was one. It had been the object of a high-profile court case back in 1981 [*American Petroleum Inst. v. Costle,* 665 F2d 1176, 1185 (D.C. Cir. 1981)], which confirmed the agency's 0.12 ppm standard. During the 1980s, however, the problem of longer exposures to ozone surfaced: The respiratory effects of ozone appeared to last more than an hour beyond the end of exposure. In 1987 the agency's scientific committee for air toxics, CASAC, asked the agency to include new data on the effects of longer exposures. After the relevant office, the Environmental Criteria Assessment Office (a part of OHEA), had drafted a supplement to the criteria document, the committee sent a closure letter to Reilly. The CASAC was split, but some of its influential members, such as Morton Lippmann and Bernie Goldstein, argued that they had studies showing that the effects last beyond one hour of exposure. In 1989, Bill Reilly discreetly dropped the idea of revising the ozone standard. Congress was in the process of reauthorizing the Clean Air Act. As the criteria of risk and rules for regulating air pollutants were likely to change in the very short term, it seemed a bad time to try and bring to a halt the debate on the health effects of ozone that had split the CASAC.[2] In 1991, the American Lung Association sued the EPA to force the agency to complete the revision. The court ruled in favor of the association, forcing the EPA to revise the standard by March 1993.

Perhaps the case of asbestos was even more symbolic of the sort of paralysis that risk assessment seemed unable to avert, or perhaps even accentuated. Asbestos had been listed by the agency as a hazardous air pollutant as early as in 1973. Later, the OPTS envisaged regulating the substance under the TSCA. After several iterative health assessments, an agencywide point of agreement seemed to have been found in 1986, following a multimedia assessment of asbestos orchestrated by the OPTS. The office proposed banning the product, having discerned a linear, no-threshold dose-response curve, despite claims from industry groups that too much uncertainty remained for a precise shape of the curve to be determined. The risk was deemed to be at the level of 1 in 1,000 for workers, and 1 in 1 million for the general public. A ban would thus prevent around 2,650 cases of cancer every year. The rule was finalized in 1989, but with a lowered estimation of the benefits in terms of cases of cancer avoided (down to a range of 160 to 200).

Administrator Bill Reilly and John Moore, the head of the OPTS, shared the view that substantial evidence—notably a class A carcinogen classification by the IARC—was available to back the decision, despite remaining uncertainty about differences of potency between different kinds of fiber. A symposium held at Harvard University's Energy and Environmental Policy Center, and follow-up papers in *Science*, nevertheless maintained the debate alive. It was, however, suggested that the risks linked to exposure to asbestos were lower than those related to exposure to indoor air (Powell 1999). In the ruling on *Corrosion-Proof Fittings vs. EPA* of 1990, the Fifth Circuit Court of Appeals did not uphold the ruling, arguing there was insufficient evidence to argue for these levels of risk, and that the cost-benefit analysis and cost per life saved were unreasonable. This ruling was a massive blow to the TSCA program and blocked any future attempt to ban substances under that statute (Stadler 1992; Powell 1999; Boullier 2016).

The inorganic arsenic dossier, finally, had been provisionally closed by the decision of the assistant administrator for the Pesticides and Toxic Offices—judged strange by some in the agency[3]—to allow the use of an uncertainty factor of 10, was reopened in 1989 by the combination of two events partly external to the agency (Powell 1999, 210). First, a coalition of environmental groups sued the EPA for not meeting a court-ordered deadline for promulgating this standard. Second, the SAB recommended the revision of the risk assessment performed by the RAF, in light of the likely existence of a mechanism of detoxification of the body, established by the recent methodology of physiologically based pharmaco-kinetics (PBPK)[4]—that the SAB was generally pushing the agency to embrace. An interagency working group was established to do so (concretely to revise the reference dose in the IRIS database), but it failed to agree on which uncertainty factor to apply to the experimental data. Hank Habicht, Reilly's deputy, decided to use a factor of 3 in order to arrive at an estimate that would preserve the maximum concentration limit that the Water Office had previously defined for the substance.

Although the institutional mechanisms for reviewing research and science policies had been activated on that dossier already—the SAB and the RAF notably, including the recently experimented external peer-review workshops—the issue motivated the ORD to work on the issue again separately. An ORD working group under the Health and Environmental Research Laboratory (HERL) looked at what research would solve the case, arguing that with a proper animal model, major uncertainties could be solved in just a few years. The Office of Drinking Water also asked another researcher in HERL to develop a research agenda on arsenic, particularly in the PBPK area. Other academic groups were getting involved too, such as the Society of Environmental Geochemistry and Health (Powell 1999).

The scientific battle between academic researchers, an industry-supported task force, and EPA scientists continued for years, without allowing any agreement to emerge on what research agenda to pursue, or even on whether further research would help reduce the uncertainty. The most concrete, immediate result of the process was that the EPA was unable to meet the deadline imposed by the court—determining a maximum-level concentration by the year 1995—as a result of a lawsuit initiated by the NGO Coalition.

The Postconservatism Campaign

What made those cases particularly difficult to handle was that the bureaucratic knowledge and standard designs used by the agency since the 1980s were now becoming objects of controversy. These cases were controversial because they were dealt with in an atmosphere of conflict concerning the EPA's normal method for deciding chemical risks. Terry Yosie, formerly the staff director for the SAB of the agency under Ruckelshaus, called this moment the era of "postconservatism."[5] In the postconservative era, decisionistic design, resting on the use of default assumptions to bridge risk data and risk management, started to be heavily criticized. Postconservatism is not a spontaneous movement, but rather the result of a new political configuration involving several political factors.

The first factor was the changing environmental policy platform of the George H. W. Bush administration. During the first half of his term (1988–1990), Bush conceded a few environmental policy initiatives, notably the reauthorization of the Clean Air Act.[6] But this "modest rebirth of federal regulatory programs ... receded during the administration's final two years" (McGarity 1998, 8). No other major legislative initiatives were undertaken until the end of the mandate. Vice President Dan Quayle, at the head of the newly formed Competitiveness Council, and Richard G. Darman, the new head of the OMB, became much more aggressive in their supervision of the EPA. For instance, the OMB reviewed many major rules, sometimes creating long delays and complicated negotiations with the EPA before the agency could issue these rules. Nearly 80 percent of the rules sent to the OMB for review were suspended there for longer periods of time than had been foreseen by Executive Order 12291 in 1981 (Fiorino 1995, 72–73).

The OMB also opened a debate on risk assessment methods, waging a strong attack on the agency's preference for conservative extrapolation. In

1990, it targeted risk assessment in an aggressive report decrying conservatism (OMB 1990). The report compiled a critique of all of the analytical choices made in a typical risk assessment by the EPA. It attacked not only the issue of extrapolation from high to low doses, but also the question of exposure, and indeed all of the areas in which an assumption or inference is needed to arrive at a single risk number. The OMB articulated a critique against risk assessment that became very popular in those years: the so-called compounded conservatism (Burmaster and Harris 1993; Bogen 1994; Cullen 1994) or cascading of assumptions.[7] Its "current regulatory issue" argument went as follows: "Suppose there are ten independent steps in a risk assessment and prudence dictates assumptions that in each instance result in risk estimates two times the expected value. Such a process would yield a summary risk estimate that is more than 1,000 times higher than the most likely risk estimate" (OMB 1990, 26). The OMB also argued that that when the EPA took precautionary stances simultaneously at the levels of the choice of the most sensitive species, of the maximum tolerated dose to test in the animal, of a linear model of extrapolation to establish the risk at low doses of the chemical, of body weight versus surface area conversion, to extrapolate doses from rats to humans, of calculating excess risk based on the life-time cancer risk to the maximally exposed individual, and of a calculation of the excess risk based on an upper-bound estimate instead of a maximum likelihood estimate, it ended up producing a synthetic estimate of the risk that could be "a million times higher" than the "real" risk. This, in turn, supposedly led the EPA to focus on "trivial carcinogenic risks while failing to address substantial threats to life and health" (ibid., 14). The OMB illustrates this point by comparing the EPA's decision to ban EDB (see chapter 5) to its decision not to regulate aflatoxin, a naturally occurring substance found in peanut butter that is a million times more risky.[8]

The OMB did not have scientists at its disposal, but it did not shy away from concretely discussing the scientific foundation for each of these choices. On the more critical discussion of extrapolation, it compared the results obtained from five models to show that they could vary by two orders of magnitude. That movement was given a massive push, and the EPA was further shaken, because of the engagement in the controversy of the prominent biologist Bruce Ames, in the context of the debate on the regulation of the pesticide alar (see chapter 8). It was also accentuated when one of *Science*'s deputy editors, Philip Abelson, recorded all these facts in an aggressive editorial. For Abelson, the case was clear: "Considerable evidence is already available that the standard EPA approach is outdated and more will be forthcoming as detailed studies of metabolic and physiological processes are made. Bruce Ames and his colleagues have produced substantial evidence that results of effects of huge doses of chemicals in rodents are often misleading.... The EPA still sets guidelines on carcinogenic risks based on the limited information available during the 1970s. The agency needs to update its regulations as new facts are discovered" (Abelson 1990, 1497).[9]

The attack of the OMB resonated with the actions of the chemical industry in Congress. The revision of the Clean Air Act gave the industry the excuse to join in the battle against the EPA's risk assessment standard methods. The 1990 amendments to the legislation presented a complicated package of variegated legal developments resulting from political compromises in Congress. The amendments introduced several new provisions, inter alia: to classify areas where air pollution levels persistently exceeded the NAAQS (nonattainment areas) and to develop plans to deal with each area; to tighten automobile and other mobile source emission standards; to require reformulated and alternative fuels in the most polluted areas; and to establish an acid rain control program with a marketable allowance scheme. But it also redesigned the EPA's core method, risk assessment, through Title III of the 1990 Amendments on hazardous air pollutants from stationary and urban area sources.[10] The discussions on this title centered on the fact that the EPA simply had been too slow in regulating carcinogens, or perhaps was deliberately avoiding developing standards, given the impossible conditions that it had to fulfill, by law, to regulate air pollutants (Graham 1985). Given the short deadlines and the recurrence of intractable risk debates on each substance, the EPA had succeeded in regulating just seven substances in around ten years (Sexton 1995).

Congress moved forward to change this, reducing the role of risk assessment in the formation of standards. According to the complicated "Technology First, Then Risk" approach defined in the Clean Air Act of 1990 (EPA 2000b), the Air Office now had to identify and apply the technologies that could reduce the emission of such pollutants from key sources ("maximum available control technology," or MACT), and then perform a risk assessment for this fraction of the risk that was not eliminated by these technologies by an ample margin of safety (set at the level of 1 in 1 million risk of developing cancer). Congress helped make a manifest leap forward for environmental protection, compared to the many frustrated attempts in the 1980s to increase action against air toxics (Oren 1991). But in so doing, it spawned a "monster of complexity" (Quarles and Lewis 1991, v) and strongly restricted the autonomy of the agency in crafting how it would use science, and which science it would use to make decisions. Congress, in effect, acted as the primary risk manager rather than the leaders of the agency doing so (Goldstein and Carruth 2003).

In those years, the AIHC was still very actively lobbying Congress on the subject of risk assessment. Nine years after having pushed Congress to launch a critical study of regulatory agencies' handling of science through the NRC, it repeated the trick in a favorable, Republican-dominated Congress: it found a senator to push an amendment to the revised Clean Air Act, for Congress to provide funds to the NRC to review the EPA's risk assessment methodology for air pollutants. This manifested itself in section 112 of the amended Act, on hazardous air pollutants, in the form of a requirement for the administrator to "enter into appropriate arrangements with the NAS to conduct a review of (A) risk assessment methodology used by the Environmental Protection Agency to determine the carcinogenic risk associated with exposure to hazardous air pollutants ... and (B) improvements in such methodology."[11] The section went into the details of what should be discussed: the method for assessing the carcinogenic potency of pollutants (dose-response); and the methods for calculating exposure, "for hypothetical and actual maximally exposed individuals as well as other exposed individuals."[12] The Act also required the EPA administrator to revise the guidelines for carcinogenic risk assessment, explaining whether and how the separate recommendations of the NRC and the SAB were taken into account in the revision.

The Red Book Problem

Using the NRC framework helped both to create a public representation of what was going on inside the agency—claiming a separation of risk assessment and risk management, or science and policy as in Ruckelshaus's public discourses (see chapter 5)—and to organize the interactions between scientists, regulatory analysts, lawyers, and political appointees. In the terms adopted for this study, the framework effectively functioned as a bureaucratic screen, but the screen was no longer effective. A set of controversies had revealed new inconsistencies. The risk assessment–risk management framework no longer functioned, and Ruckelshaus's public separation of science and policy was no longer convincing, in light of these issues. The early 1990s sounded the demise of the RAFG, as calls rose to replace its basic catechism.

Opponents to protective environmental regulatory interventions argued that risk assessment and risk management were too separate, and in so doing, they advanced their own vision of risk assessors as irresponsible environmentalists who clad their preferred policies in the language of science. Richard Wilson (see chapter 2) and William Clark, two proponents of cost-benefit analysis and longtime opponents of the EPA's conservatism,[13] argued that risk managers contented themselves with a single-number estimate of risk, without searching for the uncertainties and assumptions incorporated in it, while risk assessors failed to pay any attention to the purpose of their assessment. They claimed that the separation of risk assessment and risk management created mutual irresponsibility: "[S]cientists have used the separation as an excuse to go off into a corner and do irrelevant research" and "Risk managers have used the separation as an excuse not to understand the science and to insist on a 'simple' statement of risk as a single number" (Wilson and Clark 1991, 191). The OMB, in its 1990 report on risk assessment at the EPA, argued along similar lines. The biggest problem, the office maintained, was that science-policy choices were not assumed as such by risk assessors and never owned up to by risk managers (OMB 1990). The industry argued that with such a separation, regulatory decisions were not science-based any more. Scientists were ignored. Uncertainties were neglected by policy managers, who could advance their preferred policies no matter what risk assessors demonstrated.

Supporters of environmental measures did not agree. But they still concurred with the idea that separation had become too strong in the agency. In the context of the increasingly tangible failure of risk assessment, and dose-response calculation in particular, to produce definitive and protective decisions, NGOs, with environmental health scientist Ellen Silbergeld at the forefront, had started to doubt and articulate strong criticism of the standard risk assessment process. At the same time, they pointed to possible alternatives (Levenstein and Wooding 1997; O'Brien 2000). At a 1991 congressional hearing on the strengths and limitations of science in policymaking,

Joe Thornton, a policy analyst for Greenpeace, vehemently argued against risk assessment: "The major real-world use of risk assessment has been to approve pollution. Regulators use it to set acceptable levels of pollution and to issue permits for polluting facilities. Highly technical risk assessments have also been used by industries to intimidate citizens working to protect their communities from pollution" (US Congress 1991; see also Kuehn 1996; Thornton 2001).[14] A wave of criticism of risk assessment emerged, making it an inherently negative and adverse technology of decision-making for environmentalists.[15]

Silbergeld, then with the Environmental Defense Fund and one-time member of the SAB, wrote in 1991 that the EPA had gone too far in separating risk assessment and risk management, creating the conditions of an "uneasy divorce" (Silbergeld 1991). She argued that due to a lack of communication between risk assessors and risk managers, no commonly agreed way of dealing with uncertainties could be defined. The solutions found were designed in a piecemeal fashion by whoever found themselves faced with the uncertainty. The resulting inconsistencies were particularly striking where decisions were delegated to states, regional offices, or program offices. As risk assessment had become highly sophisticated and less mechanical, states and regional offices were having a hard time defining risks. They had trouble knowing which office to trust when different offices assessed the same substances.

Thus, the scientization of risk assessment created a new problem of inconsistency, and ultimately of ineffectiveness. Risk assessment could not deliver on the promise of revealing what the risks were, or of enabling the imposition of protective policies—an idea that would become stronger only later, among environmental health scientists, as well as environmental groups. Attentive to the sociological and political determinants of what counts as science and objectivity, and emphasizing the importance of deliberation in the construction of socially robust knowledge, Sheila Jasanoff, a professor of science and technology studies, concurred. In a paper published by the *EPA Journal* in 1993, she explained that risk assessment was suffused with arcane policy choices: "core elements of the risk assessor's work: how to select among competing models, how to balance conflicting scientific inputs, when to revise prior assumptions, how to register and represent uncertainty, and when to hold out for more scientific information … are firmly planted in the policy domain. Too rigid a separation between risk assessment and risk management seems in the light of this analysis to be both naive and misguided" (Jasanoff 1993b, 37).

Between these two positions, another point of view on risk assessment–risk management was expressed by Goldstein in a way that was more sympathetic to the dilemmas facing the EPA. He saw no reason to actually bring risk assessment closer to risk management or to revert to the decisions made in the aftermath of the EPA crisis of 1981–1983 and of RAFG. One simple reason for this, he argued, was that the separation never really existed,[16] and that risk assessors did what they did to be helpful, constructing as precise a view of the risks as possible given the presence of uncertainties. If there were one thing to fix, it would be to require that risk managers define more precisely what they expected from risk assessors. That was also the perspective of Donald Barnes, the staff director of the SAB. Looking back on ten years of thinking in risk assessment–risk management terms, he cautioned that the separation was not possible and advisable: "[S]ome degree of interaction between risk assessment and risk management is essential if the risk assessment answers are going to address the risk management questions" (Barnes 1993, 12).[17]

But whatever the real situation was inside the EPA, it meant that the risk assessment–risk management framework no longer worked as a bureaucratic screen, as if people had learned to decode what was really happening behind those notions. Tellingly, these new prescriptions came with more critical comments and a simplification of the content of RAFG. Most people now were arguing that the report prescribed the separation of risk assessment and risk management. Wilson and Clark (1991) said that RAFG and CORADM (wrongly described as having "considerable influence" on the EPA [ibid., 188]) articulated a paradigm of separation of risk assessment and risk management, which the EPA embraced, but with unfortunate results. Perhaps the single most direct caricature came from Sedman and Hadley, who spoke of the risk assessment–risk management schism being "consecrated" as a "universal" rule by RAFG (Sedman and Hadley 1992, 192).[18] But even Don Barnes, within the EPA, had a similar view.[19]

From this point onward, the nuances in the report were lost on most of the subsequent readers, who believed that RAFG prescribed separating science from policymaking, when the authors of the report actually were advancing a sophisticated, dichotomous definition of politics as "weighing policy alternatives" (NRC 1983, 3) (the politics in risk management) *and* as choosing inferences in risk assessment (the politics inside risk assessment). That reduction of the report to a supposed model happened at a time when the RAFG started to be called more systematically the "Red Book," and its subtle recommendations perhaps supplanted by a simplified slogan of necessary separation of risk assessment and risk management. This expression appeared around 1990 (Wilson 1990) and became more common from 1992 on.[20] At a time when the EPA's structures, forms, and operations were being modeled and simplified, history was made by drawing a line between "then" and "now," with RAFG embodying the "then."

The fact that there was too much separation of risk assessment and risk management, then, could mean several things. Within program offices, the various roles and expertise of health and environmental scientists, policy analysts, and lawyers or regulation writers were separated, placed in different internal compartments. The Water Office was in such a situation, having created the OSTP in 1991 to host its entire scientific staff instead of having scientists sit next to offices in charge of regulatory programs. It could also mean that there was less interaction between scientists in program offices and OPPE policy analysts, as OPPE was less politically supported and interoffice examination of proposed rules (and options) were less lively.[21] In addition, the interaction between ORD risk assessors and program offices seemed perfectible. The action of the so-called regulatory oversight group within the ORD, tasked with these interactions, did not have satisfactory results (Powell 1999). Finally, regional offices and states continued to have a hard time following the assessments produced by the various offices for substances that they were interested in knowing more about in order to establish depollution measures for local contaminated sites. The internal newsletter *Risk Assessment Review*—the publication of which stopped in 1994—was not sufficient to circulate information about new risk assessments, or it gave incomplete information.

The risk assessment–risk management framework, once a harmonious model mixing protective and commensurative designs, that helped assembling the EPA—creating articulations between natural scientists and economists, between ORD and program offices, between headquarters and EPA regions, and so on—suddenly became the essence of a dysfunctional agency. Various factors compounded to cast doubt on this design. One was the observed failure to produce accepted decisions about chemical risks. Risk assessment work was iterative, not linear, and often led to complicated, convoluted histories of assessments and reassessments, with many twists and turns toward low or high estimates. Industry groups could commission new studies to counter the risk assessment of the agency; research and scientists with various affiliations could also independently publish studies to push the EPA to go in one direction or another; and of course, industry or environmental groups challenged the risk assessments in court. Risk assessment, in short, was not conducive to accepted regulatory decisions.

Another factor was the development of new knowledge and techniques to assess the biological modes of action of chemicals that made the science policies used by the agency—the standard assumptions that it applied in the majority of its decisions—less credible. Under the pressure of more or less instrumental prescriptions to refine the science, analyze the uncertainties, explore variable levels of exposure, collect data and study plausible biological mechanisms before applying the no-threshold hypothesis, and to have risk assessments peer-reviewed before turning them into decisions, risk assessments seemed to have acquired a life of their own. Behind the formal question of the separation, the real problem was that RAFG had delegated the discussion of assumptions and sciences policies to scientists. As these were becoming more complex, less standard, it seemed important to discuss them more openly (Jasanoff 1993b).

To inscribe these new scientific developments into the EPA's standard design and way of making decisions, the chemical and petrochemical industries seized on the concept of risk characterization. Risk characterization was a part of the paradigm, a term coined in RAFG. But neither the concept nor the practice of it had really been developed since the early 1980s. Starting in 1989, though, the AIHC, an industry group that had led the campaign to deprive regulatory agencies of their scientific mission, felt that this aspect of "presentation of risk assessments," or of "risk communication to decision-makers," was a major weakness in the EPA's policy and practice (AIHC 1989). The AIHC took the lead in the organization of the first workshop in this area in 1989, with the financial support of the EPA and the Society for Risk Analysis. In the meeting, the EPA risk managers pooled their needs in terms of risk assessment, and on the basis of this, they drew up a list of very basic recommendations. These included that a risk assessment presentation had to be comprehensive and understandable; its applicability and usefulness had to be clearly stated, credible, and defensible; it had to address "contentious issues" and explicate the reasons for choosing "critical scientific assumptions"; and the relevance of the risk assessment for the regulatory framework of interest had to be explicit (AIHC 1989; see Bier 2001 also on this).

While none of these recommendations was really innovative (many had been at least outlined in 1984 in the OPPE's guidelines), they were now much more detailed, and they singled out risk characterization as a specific step in the process, a new generic competence in which the agency had to demonstrate its proficiency.

Another workshop, organized two years later, resulted in a publication called *Improving Risk Characterization*,[22] which stressed the need to provide full descriptions of qualitative and quantitative elements and to include a "candid description of uncertainties" associated with each component of the assessment (AIHC 1992), in parallel with a report and article by researchers of the Harvard Center for Risk Analysis, including John Graham (Gray 1993; Gray et al. 1993). The AIHC eventually proposed its own guidelines (AIHC 1995).

While the documents emphasized the broad consensus on risk characterization during those years, they also advocated a particular version of risk characterization—one in which uncertainty was presented as irreducible and as an argument against regulatory intervention; one in which, then, risk characterization evolved to become a kind of gatekeeping mechanism so that an assessment of risk did not automatically translate into a regulatory measure. Beyond the more general and innocent language about the need to bridge science and decision-making, the AIHC guidance actually advanced a far more radical proposal. A slight change in vocabulary was promoted at the level of whom risk characterization was for. The AIHC spoke of "an iterative process designed to be interactive with end-users," of which there was a "diversity" (AIHC 1995, 216). The risk assessment process was no longer a pipeline spewing out estimations for an EPA regulation writer; instead, risk assessment was conceived of as producing transparent risk estimations available to stakeholders, including the industry. The risk characterization memorandum also deconstructed the mechanism of risk assessment: the practice of assembling different kinds of studies and evidence about the hazard, the dose, and the exposure in order to produce a risk estimate. Echoing the "corpuscular approach" promoted by the Supreme Court in the case of *Daubert v. Merrell Dow Pharmaceuticals* (McGarity 2003), the AIHC memo argued that each of the components of a risk assessment should be separately summarized and made transparent. Most important, the AIHC argued that "any quantitative description of risk, exposure, potency, or other risk elements should be expressed as a range" (ibid., 217).

For the AIHC, much like the researchers of the Harvard Center for Risk Analysis, single point estimates of risk were unrealistic and false. Uncertainty and variability could not, and should not, be neglected and truncated by the imposition of a policy threshold that defines levels of intervention. The dispersion of results should be expressed as much as possible to avoid false positives and unnecessary interventions. The demand to include within risk characterization a comparison between the risk being studied and other comparable risks was part of the same rationale.

The kind of risk characterization that the industry pushed, therefore, aimed to go beyond the agency's mechanical risk assessment–risk management framework of the early 1980s, putting uncertainty to the fore and installing a design that would not generate regulatory measures as fluidly as the RAFG scheme seemed to do. Renewed political opposition to the EPA in Congress helped further formulate and advance this alternative bureaucratic design.

Congressional and Industry Pressure on the Passive Smoking Issue

The issue of passive smoking, or environmental tobacco smoke (ETS), provided the opportunity for the EPA's adversaries to launch intense, concentrated attacks on the agency and, indirectly, to reinvent its typical decision-making processes.

Passive smoking was one of the few prominent issues that a newly established Office for Indoor Air (inside the Air Office) was taking care of. Indeed, the reauthorization of the Superfund legislation in 1986 had mandated the EPA to look into problems of indoor air risks. And though no regulatory powers were granted to the EPA on this particular topic, it still had a role: to inform the public and deliver advice about the private management of these risks. Along with radon (discussed in chapter 8), passive smoking was an issue that the new office championed. The case seemed clear and ripe for intervention against so-called secondhand smoking: Evidence of the link between cancer and nonsmokers' exposure to the smoke generated by smokers existed.

In 1989, the EPA published a fact sheet about ETS, followed by a compendium of technical information, data, and studies to back the claim made in the fact sheet to the effect that passive smoking was responsible for approximately 3,000 lung cancer deaths annually in nonsmoking adults.

Building on previous reports by the NAS and the Surgeon General, and based on its own 1986 risk assessment guidelines, the EPA proposed that ETS be categorized as a class A carcinogen. In June 1990, the office published a policy guide indicating possible options for reducing the risks of cancer from passive smoking—namely, ways of instituting smoking bans in the workplace. The EPA started to be accused of launching an irrational crusade on secondhand smoking, but it was simultaneously accused of cozying up with the tobacco industry on this issue (Anonymous 1990). The latter acted first, strengthening the first representation of the action of the agency publicly, through the Congress.

The EPA's actions, of course, were important to the tobacco industry, but they also piqued the interest of Thomas J. Bliley, Jr., a Republican congressman from Virginia, first elected in 1980, who was the ranking minority member of the Subcommittee on Oversight and Investigations of the Committee on Energy and Commerce. Bliley contacted the tobacco company Philip Morris as early as 1983 to coordinate their action against the so-called war on tobacco policy (Muggli et al. 2004). Philip Morris extensively supported Bliley's campaign in Congress against the EPA's actions. Their first coordinated acts were to prevent the draft risk assessment of the Air Office from going through the SAB, and to compel some scientists in the dedicated SAB group to soften their position, to the benefit of the tobacco industry. This did not prevent the report from going to the SAB, and the SAB to advise Administrator Reilly to classify passive smoking as a class A carcinogen.

The first hearing took place in April 1991 before the Subcommittee on Health and Environment of the Committee on Energy and Commerce (the "Waxman hearing" on indoor air pollution), with Representatives Henry Waxman and Bliley submitting the deputy administrator of the EPA, Hank Habicht, to intense questioning. Waxman, a Democrat, queried Habicht on the true desire of the agency to take on the problem, when its public information tools seemed to have little effect, as well as the fact that the instrument of the TSCA had hardly been used in the past decade. Bliley, in turn, attacked the credibility of the agency's risk assessment, pointing to criticisms from the SAB and suggesting that a member of the SAB panel had deliberately withheld a study from consideration—one that would have decisively changed the face of the final meta-analysis. He also demanded to know from Habicht why the technical compendium had not been reviewed by the SAB, a question that Habicht ducked.

Bliley continued to pressure the EPA, with the help of Philip Morris, sending Reilly no fewer than eleven letters arguing that the agency had made various fundamental procedural and substantial mistakes (e.g., delegating risk assessment to a contractor) and was in effect biased in its science. The letters criticized the EPA for not including all relevant epidemiological studies in its review, particularly negative ones. Bliley made a connection between the case of ETS and the rules governing WOE in the agency to demonstrate a more general wrongdoing. Further, he criticized the fact that the agency and the SAB had applied a 90 percent statistical confidence threshold on epidemiological studies, thus endorsing studies that would have de facto been excluded from consideration had it applied the 95 percent limit that the 1986 guidelines imposed—or so he argued.[23]

Bliley's action was coordinated with other tactical moves by the tobacco industry. The first was the sound science campaign (Samet and Burke 2001; Baba et al. 2005), a wide public relations and scientific lobbying operation devised to attack the credibility of a range of studies demonstrating the link between cancer and smoking. The campaign consisted of installing in the public mind a high standard of "sound science," against which nearly any study could be put in doubt, which effectively paralyzed agencies that wanted to use science in their decision-making (Wagner 1995). It resonated with the chemical industry's push to revise risk assessment methods and to embrace the science of modes of action, the biologically based dose response models, pharmacokinetic models, over supposedly unscientific defaults.

On the judicial front, Philip Morris and RJR Nabisco (the makers of a number of cigarettes, including Camels and Pall Malls) sued the agency, arguing that the SAB report was unscientific, arbitrary, and capricious.[24] Their argument was not only that the studies included in the meta-analysis had too dissimilar designs to be subjected to a numerical comparison of their results, but also that the SAB had applied a 90 percent statistical level of confidence in its review, as opposed to the EPA risk assessment guidelines and the 95 percent confidence level standard. The case would be the forum for an unusual, detailed attack on the substantive, scientific grounds of the EPA's decision to classify ETS as a carcinogen (McGarity 2003a). In its final ruling, the court went beyond what constituted the hard-look doctrine and the normal deference to agency science, and found that EPA disregarded information and made findings on selective information; did not disseminate significant epidemiologic information; deviated from its Risk

Assessment Guidelines; failed to disclose important findings and reasoning; and left significant questions without answers.[25]

The Deputy Administrator's Initiative on Risk Characterization

At the time of the ETS affair and the OMB's controversial initiatives against the EPA's risk assessment methods, Habicht was also leading the federal Interagency Working Group on Risk Assessment, under the aegis of President Bush's science advisor, David Allan Bromley. The group—yet another interagency initiative after the IRLG (see chapter 2) and Ruckelshaus's IRMC (chapter 5), this time under greater supervision from the White House—was supposed to examine opportunities for collaboration on methods, research, and other issues of interest to agencies engaged in risk assessment. Habicht did not seem to push for strong changes at the interagency level. He defined no calendar and no outcome for the group. But he defended his work at the EPA, outlining the novelty and benefits of what was being experimented with in the agency: increasing peer involvement, for the reassessment of dioxin; emphasis on risk characterization, to ensure that risk assessments of the same substance in air and water (dioxin, but also arsenic) were concordant; revision of various guidelines; and more research on uncertainty analysis, pharmacokinetic models, and dose-response models.

At a June 1991 open meeting, in the presence of industry representatives, Habicht claimed that "science should not anticipate the ultimate policy decision," that "exposure assessment is a key to good risk assessment" and that uncertainty should be reduced "as much as possible" (NAS 1991). He advanced these ideas to counter the impression that the EPA had given in the ETS affair—to wit, that risk assessment was overridden by a predetermined antismoking policy and exposure assessment had been poorly executed. Habicht's presentation at the meeting showed, at least, that he was addressing risk assessment practices in such a way as to counter criticism that the agency was facing at that time on this particular issue.

Against Habicht's apparent reluctance to put out concrete new standards and guidelines for application across the federal government, the White House decided to move forward with its agenda. The OMB, with support from Vice President Quayle, started developing a new executive order to impose principles and standards for risk assessment across all agencies, particularly targeted at the EPA as in the early 1980s under President Ronald

Reagan. The Competitiveness Council was building on the efforts of a lobbyist who was well known to the president and to the OMB in particular: OIRA's former chief, Jim Tozzi. Tozzi had left the federal government in 1983 to become a business lobbyist, helping the tobacco industry and other companies fend off environmental and health regulations. Over the years, he created a number of organizational vessels—officially, think tanks—such as Federal Focus Inc. and the Institute for Regulatory Policy, and he organized a series of forum meetings to discuss the principles of risk assessment and the use of science in policy. The latter move was spurred by the tobacco industry's crusade against the regulation of ETS by the EPA. The meetings resulted in the publication of a thick report to the Competitiveness Council, *Toward Common Measures: Recommendations for a Presidential Executive Order on Environmental Risk Assessment and Risk Management Policy,* which Vice President Quayle embraced enthusiastically (Federal Focus 1991). It suggested the adoption of a new presidential order that would outline four principles: (1) regulatory agencies must develop a list of "risk priorities"; (2) science used in risk evaluations must be separated from management decisions; (3) risk management decisions must be cost justified; and (4) the magnitude of risks must be communicated in relation to familiar risks (Borelli 1992). As Reilly's mandate was coming to an end, Habicht, grilled in Congress on the passive smoking case and feeling the pressure from the imminent White House intervention on risk assessment standards, went forward with an initiative on risk characterization.

One main reason pushed Habicht to consider this concept as a legitimate response to the controversies of the day. The risk assessment–risk management conceptual tandem had lost its effectiveness in influencing the agency's decisions and representing this order to the outside world. Inconsistency in the decisions of the agency had increased, stimulating discourse about the EPA's inability to know what decisions needed to be made. Around 1990, the agency had attempted a cluster approach to bring representatives of programs together and align their risk assessments, but that project did not go very far.[26] At about the same time, the high-level panel of experts that was reviewing the state of science in the agency (the forthcoming *Safeguarding the Future* report) had come to the conclusion, from its thirty interviews across the EPA, that the agency was bad at communicating to Congress, decision-makers, and the media about the uncertainties residing in science and how they were (or were not) handled in the resulting decisions. The experts had made a recommendation that the science advisor be responsible for making sure that "full, documented discussions" of the scientific pros, cons, and uncertainties, as decisions and policies are developed (EPA 1992e).

Another major internal problem of agency integration came from the increasing sophistication of the models and the supposed precision with which cancer risks were computed. These numbers hardly left any choice to the decision-maker but to regulate. Bill Reilly, the EPA administrator, had expressed the feeling that he sometimes was suppressed by a "tyranny of numbers"[27]—and confusing numbers at that. One particular problem was that the 10–6 threshold was not always interpreted in exactly in the same way across the EPA: some were assessed in terms of population risk, others in terms of individual risk.[28] In other words, as risk assessment was becoming more scientific—meaning, infused with more biology, on the modes of action of substances, and more computational work, with statistical models—risk managers had more difficulty making sense of the numbers. Ten years earlier, it had seemed sufficient that risk assessors clarified what their assumptions were, so that risk managers could interpret and work with their numbers. The assessment process was now much less mechanical. It involved long lists of choices of model parameters, case-by-case considerations of the biological modes of action of chemicals, and other complicated steps. In short, risk assessment was getting scientized, employing more sophisticated biological models and modes of calculations that regulatory officials were less well equipped to question.

All in all, risk characterization crystallized the controversies of the day surrounding how the agency should decide about risks. The outside pressure for the notion of risk characterization—notably the initiative of the AIHC on this concept—meant that the agency had to forge its own approach of the concept in order to be able to draw benefits from it. Habicht asked Peter Preuss, a scientist and experienced risk assessor, head of the National Center for Environmental Assessment in ORD (formerly OHEA), and a policy analyst to write the memo. It was circulated to the entire agency on February 26, 1992, articulating three main principles that needed to be implemented in all offices: "1) risk assessment information must be clearly presented, separate from any non-scientific risk management considerations; 2) key scientific information on data and methods and their uncertainties should be identified in the risk characterization and a statement of confidence should be included that identifies all major uncertainties along with comment on

Table 9.1
Categories of participants in the risk assessment–risk management process (adapted from EPA 1992d)

Groups generating risk assessments	Groups using risk assessments	Risk managers
• Scientists and statisticians in the ORD • OPTS and other program offices • Carcinogen Risk Assessment Verification Endeavor (CRAVE) • RfD/RfC (Reference Dose/ Reference Concentration) workgroups	• Program offices • Regional offices To develop regulations through existing databases (e.g., IRIS, ORD health assessment documents, CRAVE and RfD/RfC workgroup documents).	• Agency managers • Decision-makers

their influence on the assessment; 3) it is Agency policy to present information on the range of exposures derived from exposure scenarios and on the use of multiple risk descriptors" (EPA 1992d, 5).

The first principle read like an implicit procedural defense against accusations that the EPA was dictating to scientific risk assessors what evaluations they needed to turn out for EPA policies to go forward. Again, this was the industry's main point in the ETS affair: Risk assessment had been used to find evidence to support a long-determined policy crusade against tobacco. The report defined the EPA in the terms of the risk assessment–risk management framework, distinguishing three groups in the agency, as described in table 9.1.

The goal of this reassertion of the virtual architecture of risk analysis was meant to define responsibilities for the credible articulation of science with policy: Everyone in the agency had to be aware of the need not to mix scientific and nonscientific considerations when dealing with a risk assessment.

The memo discreetly instituted a major shift. Until then, and since the second term of Ruckelshaus as EPA administrator, it had been acknowledged that risk assessment was not a scientific exercise, but a mix of science (e.g., establishment of a dose-response relationship based on the administration of various doses of a chemical to a test animal) and science policies (assuming that carcinogenic chemicals are toxic at any dose and applying a linear extrapolation model from the actually tested doses to infer toxicity at low doses). It argued, in contrast, that risk assessment could be cleansed of "non-scientific considerations" (EPA 1992d, 4), as if it were an exercise that could be separated from the administrative, regulatory actions of the agency. The memo was reflecting the ongoing change in the meaning of risk assessment outside the agency: its scientization, under the influence of substantial knowledge of biological mechanisms of risk causation (physiologically based pharmacokinetic modeling, especially biologically based dose response modeling). It seemed to imply that the determination of science policies had to be brought under the control of risk managers. They would no longer be set in guidelines and would not be a form of extension or codification of a risk assessor's preferred professional standpoint. They needed to become debatable scientific choices, to be made explicit to the managers of the agency.

The memo leveraged two other concepts to bring to life a new representation of the EPA's science and decision-making mode. One was uncertainty analysis. The former guidelines for risk assessment were not strong on uncertainty analysis, or even on uncertainty characterization. The carcinogen risk assessment guidelines of 1986, for example, discussed the inferences made by risk assessors but did not venture into the formalization of uncertainty analysis in any depth. They exuded a kind of professional confidence in the idea that science policies took into account the major uncertainties, and that risk assessors would be able to make their choices expertly. The guidelines noted that the results of quantitative risk estimation were uncertain, specifying that the WOE should be presented to cover this potentially important uncertainty. These guidelines did not say much more than that "major assumptions, scientific judgments, and, to the extent possible, estimates of uncertainty embodied in the assessments are presented [in the risk characterization section]" (EPA 1986a, 2).

The 1986 guidelines, in that respect, reflected a limited practical engagement with uncertainty analysis in the agency,[29] which appeared more and more unacceptable with regard to the fast developments of biological and environmental sciences on this front. In the early 1990s, uncertainty analysis in the area of environmental risk developed quite rapidly. Adam Finkel, at Resources for the Future (Finkel and Evans 1987, 1990), and Morgan et al. (1992) clarified the terms of both kinds of uncertainty to expect to see and characterize in environmental risk assessment: parameter and model uncertainties. Suter (1990) and Barnthouse (1992) made great strides exploring these methods in the context of complex ecological risk assessments. In the early 1990s, uncertainty or sensitivity analysis of biological, pharmacokinetic models appeared as well, under the impetus of biostatistician Frédéric

Bois, who showed how probabilistic methods such as Monte Carlo analysis may be used to assess and correct uncertainty in parameters (McKone and Bogen 1991; Spear et al. 1991; Woodruff et al. 1992; Woodruff and Bois 1993). John Graham's Harvard Center for Risk Analysis also addressed the topic, indicting the use of single point estimates of risk by regulatory agencies (Graham et al. 1991) and testing the elicitation of judgmental probabilities in experts to document the inevitable spread in quantitative risk estimations due to uncertainty (Evans et al. 1992).

Therefore, the Habicht memo updated the agency on these developments. It recorded the recent insistence in the scientific networks surrounding the agency on the discussion of uncertainties: "A discussion of uncertainty requires comment on such issues as the quality and quantity of available data, gaps in the data base for specific chemicals, incomplete understanding of general biological phenomena, and scientific judgments or science policy positions that were employed to bridge information gaps" (EPA 1992d, 10). Uncertainty, so the memo argued, came in the form of "measurement uncertainty"—variations in measured exposures for instance—or "data gaps." Uncertainty analysis provided qualitative characterizations of each of these two kinds of uncertainty. It went into some detail: Key scientific information about methods (e.g., choice of models and species) had to be highlighted; a statement of confidence in the assessment that identified all major uncertainties had to be provided; and information had to be presented on the range of exposures derived from exposure scenarios and on the use of multiple risk descriptors (population risk, individual risk, risk for most exposed individuals and so on). More important, it enlarged the notion of uncertainty to include the ambiguity residing in the qualitative judgments of risk assessors. In stressing the qualitative, narrative nature of risk characterization, the memo was rehearsing an old point made in RAFG[30] but often forgotten in the practice of computing numerical estimates of risk.

The second (and probably more important) policy development stemming from the Habicht guidelines was the requirement that risk assessors calculate exposure in different ways: "EPA risk assessments will be expected to address or provide descriptions of (1) individual risk to include the central tendency and high end portions of the risk distribution, (2) important subgroups of the population such as highly exposed or highly susceptible groups or individuals, if known, and (3) population risk" (ibid., 21). The guidelines did not endorse one risk assessment policy, but it translated the outside controversy on the best way to capture risk as outlining the range of possibilities and asking its assessors to use them all, to the farthest possible extent, and justify their choices. Again, while exposure to chemicals and risks was in no way a new concept, the Habicht memo was the first occurrence of an agency-level exposure policy—at least since Alvin Alm's initiative on the issue in 1983–1984 (see chapter 7).

Of the three domains of knowledge comprising risk assessment (hazard identification, dose-response assessment, and exposure assessment), exposure assessment was the least well developed "limb" of the risk assessment process.[31] An effort was made in the ORD to develop this area (see chapter 7). A 1986 guideline on exposure assessment detailed these methods over no less than eleven pages. A subsequent document on exposure-related measurements (1988) discussed the issue even more extensively. In parallel, the Superfund guidelines, published in 1989, described in detail the range of estimates that should be presented as part of the characterization of risk, particularly in the exposure part. But guidelines did not by themselves promote exposure assessment expertise, and knowledge of the exposure of vulnerable populations continued to appear insufficient, particularly as the question of exposure of children to pesticides gained salience in the public debate at the turn of the 1990s. In 1988, while the reauthorization of FIFRA stalled in Congress, the EPA decided to bring in the NAS again, requesting a study of the particular risks of dietary pesticides to children and infants. The report came out a number of years later, in 1993.[32] *Pesticides in the Diet of Infants and Children* (NRC 1993a) "really changed the paradigm" in the Office of Pesticides.[33] It forced and guided the office to revamp its methods and sources of data for exposure work, specifically concerning the characterization of the exposure of children. It veered toward combining the distribution of food consumption and the distribution of residues, and the report led it to think about the topic of "cumulative risks" (i.e., the risks from exposure to several pesticides simultaneously) as well.[34]

Habicht had the full support of Administrator Bill Reilly in his endeavor.[35] The risk characterization initiative was a highly political one, given the pressure of the OMB and Competitiveness Council, as well as the constant oversight of Congress, not least during the ETS affair, but also the initiative of the industry on risk characterization itself. As mentioned previously, the AIHC was using this notion to argue that uncertainty should prevent the agency from deciding and that risk characterization should be the moment of the decision-making process where assumptions of risk assessors are disclosed to and discussed with stakeholders, such as regulated businesses.

The risk characterization version of the industry was a tool to advance a very conservative agenda: minimizing regulatory intervention. Liberals and public health advocates understood the notion quite differently. For them (e.g., Finkel 1991), the whole point of risk characterization was not to produce an ever-more-refined range of estimations, which would only paralyze intervention for a longer time. Rather, it should be about determining the level of uncertainty incorporated into the calculations, in order then to make a decision either to perform further research to reduce or eliminate the uncertainty or to start making a decision and intervening to reduce the risk, the uncertainty being minimal or, on the contrary, being too large and complicated to deal with.

Habicht did not take sides for either the AIHC-Harvard version of risk characterization (as an incentive for ever-more-refined uncertainty analysis and more precise point estimates of risk) or the decisionistic version (risk characterization as a way to define levels of uncertainty and truncate scientific analysis of risks to move toward decisions). But it walked a fine line between these two versions in order to close the external debate and regain autonomy over internal reform of the agency. Like Ruckelshaus's June 1983 speech (see introduction and chapter 5), it was a moment when the internal workings of the agency were defined in a way that was presentable to a series of audiences that had conflicting views of what the agency should do. In fact, there was a strong communication initiative attached to the memo. Its immediate application in the agency was announced through a press conference, even before people inside the agency were warned of it. Reilly boasted to Congress about the initiative in 1992 hearings on the ORD's budget appropriation. The document was picked up almost immediately by various people in Congress (US Congress 1992). The memo achieved its objective—namely, discouraging Quayle from imposing any new risk assessment standards on agencies. In July 1992, the White House announced that this project had been shelved, officially because of the incompatibility of the initiative with the presidential reelection campaign.[36]

The initiative was less well designed to make an impact internally. This may have been due to its style: The content was often repetitive, and it hardly had the standardized process-and-outcome format that effective guidelines typically employ. Finkel, though very supportive (the memo picked up on his own ideas about uncertainty), described the document as "vague" and lacking the kind of "standardization that we need to move forward" (US Congress 1992, 125). No implementation plan was attached to the memo. People felt, in any case, that they were *already* doing what the memo described (Lynch 1996, Powell 1999).[37] The Office of Solid Waste rapidly published a document on the implementation of risk characterization guidance, showing that they did not have to actually change anything to comply with the supposedly new principles (EPA 1992c).

For the time being, the risk characterization initiative helped paralyze the various prescriptions stemming from the industry and the OMB. The risk characterization concept helped to show that there was effectively a problem at the level of the interaction between scientists and decision-makers in the agency. However, at the same time, it buffered the various pressures coming from the outside in order to help the agency follow its course.

Congress and the NRC Panels as a Battlefield

Risk characterization was an end-of-mandate initiative. By the presidential election of November 1992 and the change in EPA leadership, it had become clear that it had not solved all the problems of legitimate use of science in bureaucratic action, let alone ended the controversies raging outside, among the various audiences of the agency, about the right way of linking risk assessment to risk management. In 1992 and 1993, intense discussions unfolded in Congress, during multiple hearings held at the behest of Republican representatives and senators, about the EPA's science and research and its risk assessment.[38] Assistant administrators for research and development, for toxic substances, and often the EPA administrator directly (or his or her deputy) were called to testify alongside academics standing on one side or the other, from Finkel to Graham, from environmentalists to industry representatives, to engage with the kind of science that the EPA should be producing and using, and how it should be doing so. Issues of uncertainty and exposure knowledge were recurrent in testimonies and questions raised by insistent congressmen, notably James Scheuer, Don Ritter, George Brown, Henry Waxman, and Daniel Patrick Moynihan.

Bernie Goldstein, frequently called to testify, underlined on several occasions the need to increase efforts on exposure, one of the core issues of the

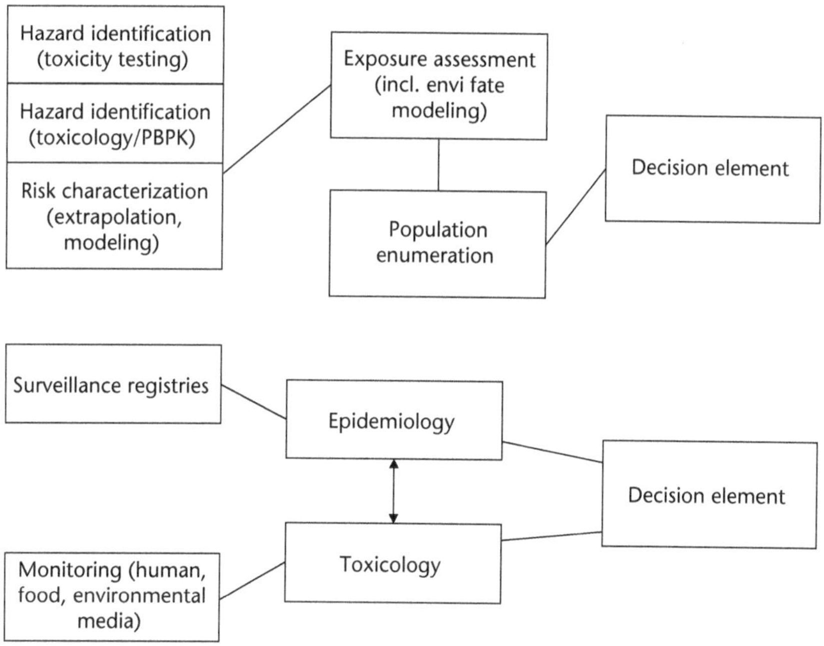

Figure 9.1
The "trinitarian" Red Book approach (top) versus a monitoring/surveillance-driven risk assessment approach (bottom) (adapted from Silbergeld 1994, 18).

risk characterization initiative. Ellen Silbergeld, then a scientist with the Environmental Defense Fund, also emphasized in Congress that epidemiological data, more than experimentally based predictions, was what now mattered more to improve the outcomes of risk assessment research, which was having such great difficulty actually delivering accuracy and protection against environmental threats. Investing in further refinements of dose-response extrapolation, through pharmacokinetics and understanding of biological mechanisms, seemed a dead end, as the case of dioxin illustrated (Silbergeld 1993). She outlined an alternative vision of what this would result in, as a kind of risk assessment process. Instead of incorporating the preference of toxicological testing and experimentally based predictions in the design (the top diagram in figure 9.1), two knowledge accumulation processes should run in parallel, both rooted in data collection and observations and united at the point of decision (bottom diagram).

Instead of simply expanding the dose-response characterization to a measure of risk for the whole population, with basic data about the amount of substance produced and simple estimates of the doses ingested by humans, Silbergeld pleaded for constructing two separate views of the risk—one experimental, the other observational. Epidemiology, in other words, should provide independent points of reference for a policy, not simply be abducted by toxicology as exposure assessment.

These hearings did not result in any strong initiative of Congress until 1994. But in the meantime, the debate about the agency's right design for the use of science in decisions had moved to the NRC once more. In 1993, the AIHC project of the late 1970s to have the NAS review the work of regulatory agencies had resurfaced. After the publication of RAFG, managers of the NAS opposed a bill, proposed by Republican senators Don Ritter and Dave Martin, to institute a new committee inside the NAS for the purpose of reviewing the risk assessments done by agencies. However, the idea of creating a committee to work on agencies' risk assessment methodologies and guidelines persisted. The NAS ultimately refused to institutionalize a standing committee for this mission.[39] The idea nevertheless appealed to the NRC's standing expert committee that was most affected by this proposal—the Board of Toxicology.[40] The board continued to discuss it throughout the 1980s, with the resulting decision to create a dedicated committee to review the EPA's science, the Committee on Risk Assessment Methodology (CRAM).

The committee chair was given to Bernie Goldstein, the former chair of the EPA's Clean Air Scientific Advisory Committee and former assistant administrator of the ORD (1983–1986). The committee included specialists in environmental health and occupational health, as well as biostatisticians, epidemiologists, toxicologists, and ecologists. Several members of the 1981–1983 RAC sat on it, including decision scientist Warner North; Franklin Mirer, a toxicologist and industrial hygienist from United Auto Workers; and Kenny Crump, the biostatistician who had developed the mathematical model endorsed by the EPA as a basis for the agency's preferred model of extrapolation, the LMS model.

The range of expertise and viewpoints about risk assessment was wide, with supporters of the default utilization of the linear model as well as clear opponents, such as the toxicologist Michael Gallo. Three industry-affiliated scientists from Mobil Oil, Dow, and Exxon were members of the group as well. CRAM's mission was to recommend changes to risk assessment across the federal government. Its federal liaison board included representatives of the EPA, as well as the FDA and OSHA, but its agenda was clearly focused on EPA problematics. The eventual CRAM report reinforced the emergent ideas in the fields of chemistry and biology research—namely, that risk assessment, instead of working with hypothetical statistical models to define the dose-response relationship, should make greater use of advances in biological knowledge of the modes of action of chemicals (i.e., of how substances concretely generate effects in the body at the cellular level). In other words, the report supported ongoing developments, already under consideration at the EPA, of controversial applications (see the case of dioxin discussed previously) centered around the use of pharmacokinetics and "biologically-based dose-response models," championed by Gallo. Scores of workshops and conferences took place during those years, often to present case studies revisiting the EPA's iconic regulated chemicals, to show how differently they could have been treated thanks to mechanistic knowledge (e.g., Conolly and Andersen 1993; Leung and Paustenbach 1995; Conolly 2002).[41]

The 1990 Clean Air Act amendments and the industry furthered the controversy around the risk assessment of air chemicals by the NAS. On the initiative of the industry, the amendments requested the EPA to get a report from the NRC on its risk assessment methods. Contrary to RAFG, this report had no institutional objective. The charge (which the EPA had no possibility to look into and amend)[42] requested a review of the methodologies used by the EPA, as well as of possible improvements, specifically in the areas of the assessment of carcinogenic potency and models for estimating exposure assessment. The charge explicitly pointed to the controversial "maximally exposed individual" assumption that the EPA had consistently applied since the 1970s in its risk assessments. Here, the NAS was required, directly by Congress, to evaluate the ways in which the EPA performed risk assessment.

The twenty-five-strong panel set up by the NRC in response to Congress included several biostatisticians, toxicologists, environmental and occupational health specialists, and environmental and chemical engineers. Only two lawyers were on the committee, one of whom left early in the process to join the Bill Clinton administration. A third person, Sheila Jasanoff, was also a lawyer originally, but more broadly, she was a social scientist with special knowledge in the field of science and technology studies, of which she is a leading figure. She was one of the two women sitting on the panel. There were few people there who had experience from within regulatory agencies, except a couple of scientists who were at several points members of the SAB, and the chair of one of its three scientific advisory committees, the Clean Air Scientific Advisory Committee: Roger McClellan (who was with the Chemical Industry Institute of Toxicology when the committee convened). As is usually the case, there was no explicit trace in the report of any policy preference or explicit bias of one or several individuals on the committee. But as is usually the case as well, the NRC made sure to pick people who would represent a variety of viewpoints in order to increase the chance of forging constructive recommendations supplanting existing conflicts between any two points of view. In this case, Finkel could be singled out as the one who would carry a public health perspective to the committee, while Roger McClellan would bring into the discussion a viewpoint that reflected the chemical industry's stance on the scientific issues that were discussed. Those two scientists, along with John Bailar (a biostatistician from McGill University), John Brauman (a chemist from Stanford University), and Lincoln Moses (a biostatistician and decision scientist, also from Stanford) were said to be the most vocal people of the whole group.[43]

The charge given to the NRC via the EPA manifestly built on the grid that RAFG offered and that the EPA had since incorporated into its administrative processes and guidelines. Accordingly, one of the key members of the panel recollects that the group overall considered the framework to be sensible. It did not try to reinvent it and did not spend time on it, but on the contrary used it as a scheme to review and evaluate the EPA's activities: *Science and Judgment* (NRC 1994) showed that in the first decade of thinking in terms of risk assessment and risk management, too much separation between the two was instituted. It announced several themes that would grow in importance and would find their utmost expression in subsequent reports, including the ideas that research at the EPA should be strengthened; and that risk assessment was not an end in itself, and that its relevance for risk managers was more important than acquiring truth. A couple of years after the deputy administrator of the EPA disseminated his guidelines on risk characterization, *Science and Judgment* (NRC 1994) also emphasized the importance of that practice.

The focus of the panel's discussions was on the evolution in techniques that the EPA could develop to address the most frequent criticism leveled at it (presented on page 6 of the executive summary and further detailed in the introduction): the use of default assumptions, the data needs, the assessment of multiple chemical exposures, and analysis and characterization of uncertainty and variability. There were, therefore, many topics on the table: a range of aspects of risk assessment that were structured according to the terms of RAFG and of EPA's existing, formal, bureaucratic knowledge. The design framed the controversy about the agency.

One topic, however, was recurrently discussed in committee meetings and prevailed in many of the panel's discussions without leading to any sort of consensus: the issue of defaults. The 1976 and 1986 guidelines of the EPA spoke of "estimations," "expert judgment," or, more positively still, of "science-policies" (EPA 1976, 1986a). By the early 1990s, however, *defaults* was the new term to designate the assumptions that the analysts of the EPA were using to calculate risks (such as the assumption that carcinogenic chemicals are toxic at any dose, and that a linear extrapolation model should be used to infer toxicity at low doses) (EPA 1990b).

Originally, the notion was used in engineering models. It then migrated to the area of exposure assessment, which was quite natural, given that this assessment did use models of dispersion and behavior in substances in various milieus. In 1992, an EPA document about pharmacokinetics applied the notion of default assumptions to the area of dose-response assessment for the first time (Federal Register 1992). Soon enough, the notion became nominalized. By then, the EPA was publicly known and decried quite simply for its use of defaults, and everyone could legitimately wonder how and when on Earth the EPA would consider moving away from these defaults, or making full use of the available science to depart from them. The default assumptions notion of environmental engineering translated into a public critique of the EPA's lack of knowledge. Mechanistic research and the developments in uncertainty and variability analysis in the area of exposure were clearly breaching the guidelines. By offering knowledge on the behavior of one of the substances in the body, the former in particular afforded an opportunity for the EPA—indeed, created pressure on the agency—to abandon its guidelines and employ the available biological knowledge rather than the standard linear extrapolation.

The theme of departure from defaults, as well as the idea that "EPA should clarify its standards for how it decides it should replace an existing default assumption with an alternative, ... a point to which the entire committee agreed" (NRC 1994, 615), pervade the report. This was in spite of the fact that the committee did not succeed in formulating a standard for when and how to depart from default options. It was precisely the issue on which the group was split. As Finkel recollects: "We all agreed on the four steps, but the issue kept coming up of defaults, assumption selection, how to bridge the gap when you have no knowledge of the fundamental chemistry ... we said that the EPA had the right idea: One should have a set of assumptions, generic, reasonable, to be the starting place. For instance, positive health effects in mice are reasonable predictors. That was not controversial, including on the industry side. But in the context of the Clean Air Act, and air risk assessment, lot of questions emerged around conservatism and overregulation."[44]

These discussions turned into an open conflict which, to the great despair of NRC staff, the committee never resolved. Instead, it was decided to include two annexes in the report—one for each position in the debate for or against the use of defaults—or, as one could put it, for or against trusting the agency's expertise and reasonableness. Finkel aimed to save and support the EPA's conservatism and protective goal. He defended one position under the label of "plausible conservatism," according to which the EPA should depart from defaults whenever "there exists persuasive evidence, as reflected in a general consensus of knowledgeable scientists, that the alternative assumption (model) represents the conservative end of the spectrum of plausible assumptions (models)" (Finkel 1994, 615–616): essentially, a twofold criterion for being scientifically up to date, and consistent conservatism, both of which were designed to allow regular updates of defaults based on new validated scientific knowledge, in line with an overall policy of reducing the risks of underestimation of the risk.

The alternative view, by the decision scientist and consultant North and the toxicologist McClellan, was the "making full use of science" one. Under this approach, conservatism should not guide the departure from defaults: Abandoning a default for a more specific analysis or set of data should be motivated by the willingness to embrace the latest and most accurate science.[45] This approach was unambiguously based on the belief that uncertainty would be reduced, or even eliminated, by continuing scientific research. The hope was palpable that regulatory agencies could get rid of defaults and "unidentifiable biases" (McClellan 1994, 2003) altogether. Accordingly, they argued that Congress should do more "to encourage EPA, other federal agencies such as National Institute for Environmental Health Sciences, and private sector organizations to plan and carry out research to reduce important uncertainties on the health consequences of toxic air contaminants" (McClellan and North 1994, 638), because in the progress of research and science lies the possibility to improve risk assessments.

In Finkel's view, uncertainty could at best be analyzed and characterized but not eliminated or reduced. Conservatism, and value choices more generally, was as inevitable as uncertainty was. In contrast, McClellan and North explicated in their text that uncertainty could and should be reduced, and that as scientific research continued, risks would be definitely and precisely known. Another significant difference between the two positions was that Finkel admitted that defining a criterion for departing from defaults, as both texts discussed, was a matter of policy. Finkel, therefore, did pronounce his policy views. He was of the opinion that the US population was favorable to a cautious heath protection policy, and that "'plausible conservatism' reflects the public's preference between errors resulting in unnecessary health risks and those resulting in unnecessary economic expenditures" (NRC 1994, 606).[46] Thus, while Finkel wrote as a scientist, with a decision theory background, as the authors of the "best science" appendix did, he also asserted clear policy choices. McClellan and North, on the other hand, did not explicitly venture onto this policy terrain. Only in the very last lines of their text did the two authors mention the "costly regulations based on conservative assumptions" (McClellan and North 1994, 640), perhaps thereby exposing their concern for the burden that health regulation creates on industries.

So, much like the RAC of 1981–1982, this committee was clearly divided into two camps. Unlike the RAC, however, the two sides were not of equal weight. Finkel admitted that, had they taken a vote, he would have been in the minority.[47] But also unlike the RAC, and unfortunately for the impact of this particular effort of the NRC, the committee did not find a way out of the conflict. Combined with the fact that the panel "tried to do too much" and "got lost in details,"[48] the conflict prevented the report from carrying a unique, neat message.

At the NRC, the report was not put forward as a major hit—quite the contrary, in fact. NRC panels are usually designed to produce recommendations that the entire panel can defend and project outside. Moreover, the NRC in general does not approve of the use of footnotes, appendices, or indeed any reflection of the dissensus expressed during committee work. Whereas RAFG and subsequent reforms in the EPA evinced an organizational order and design that ended the controversy of the early 1980s about the agency's legitimacy as scientific bodies, *Science and Judgment* (NRC 1994) failed to have such an effect. The risk assessment–risk management framework conveyed a public image of the EPA's operations, of an agency that derives decisions from a use of science. Science-policies were an essential element of this mechanical, decision-production design. Once the agency stopped being trusted in its defaults, this mechanical design collapsed and stopped resolving controversies about the agency's legitimate action. The RAFG model was still there to provide answers. But in the context in which the fundamental expertise of the agency was put in doubt, the notion that it would now work toward ensuring a better interaction between risk assessors and risk managers to adjust assumptions case by case did not seem to reassure anyone.

Postconservatism Institutionalized

Risk characterization was the best possible design proposition that Carol Browner, EPA's administrator from January 1993 until January 2001, could use to counter attacks on the agency. She picked up on it, with a decision to update the memo of deputy administrator Hank Habicht.

Browner had originally disregarded this topic. The transition from the previous team was difficult, if not conflictual, as she wanted to launch a whole new EPA and get a divorce from previous programs and priorities in a clear-cut manner.[49] Although she claimed that she wanted to put science at the heart of the regulatory work of the agency, following one of Clinton's campaign themes,[50] her actions did not show as much interest in this resource. It took her months to be able to fill the position of head of ORD, as the general counsel of the agency refused to relax the ethical rules preventing Bailus Walker from accepting the job because he had done research under contract with the EPA. She kept William Raub in the position of science advisor created by Reilly, but she left him out of the close circle of top aides on whom she relied. Raub soon left, and the position then remained vacant for several months (Stone 1994a, 1994b). Browner, nevertheless, did act on a recommendation from the report *Safeguarding the Future* (EPA 1992e). At the end of 1993, she expanded and revamped the Risk Management Council, turning it into the Science Policy Council, tasked with reviewing and deciding science policy issues that went beyond boundaries of regional or individual programs. It took up the function that the "oversight group"[51] led by Peter Preuss in the ORD, had, and which had been designed to coordinate the ORD's research and the science produced in program offices.[52]

Browner gradually turned to questions of science and risk analysis because she was soon faced with pressure from Republicans in Congress and their aggressive "regulatory reform" agenda deployed after the 1994 midterm elections in Congress, and the large victory of Republicans under the banner of Newt Gingrich's "Contract with America." The conservatives championed regulatory reform again and tabled a bill to mandate the agency to perform risk assessment and cost-benefit analysis for all rules.[53] Multiple proposals emerged in those years in the House or in the Senate, many of which imposed new requirements for risk analysis and cost-benefit analysis by regulatory agencies, presumably, on the side of Republicans at least, to limit the autonomy of agencies and create more occasions for review of their decisions (Graham and Sadowitz 1993; Anderson et al. 2000). In accordance with the White House initiative on risk analysis principles, Browner then resurrected risk analysis to counter regulatory reformers (Browner 1995, cited by McGarity 2001). In order not to leave this ground to Republicans, Browner decided to forge new models of operations for the agency. She relaunched, at the highest level in the agency, the interest in risk characterization. She resurrected risk characterization because the NRC argued for the need to improve this practice in *Science and Judgment* (NRC 1994). A new report, stemming from another division of the NRC, *Understanding Risk: Informing Decisions in a Democratic Society* (NRC 1996), also defended the importance of this sequence of decision-making. It was presented as a way not only to characterize uncertainties better, but also to adopt more deliberative modes of government. With the imprimatur of the Academies, the concept could help deflect and control external pressures to perform more analyses. A final reason was that Habicht's memo of virtually had no effect on the agency.[54]

Browner and her deputy set in motion a Risk Characterization Policy implementation team, largely attuned to what was in the memo of 1992. The assumption of the team was that people inside program offices were indeed separated by a "brick wall"[55] that needed to be brought down. Against the Harvard Center for Risk Analysis–AIHC coalition that pushed the subject of risk characterization between 1989 and 1992, risk characterization was reoriented to stress the need to explicate and discuss decision rubrics of risk, cost, benefit, and the values injected into the decision much more transparently. The whole point of the new memorandum developed by the working group in direct connection with Browner and with support from Senator Barbara Boxer (a Democrat from California),[56] was less to refine the ranges of probability estimates, as conservatives would have liked to see, than to construct a procedure enabling the expression of values injected into the decision and the crystallization of the latter. The team worked to bring program and regional offices together around a table. The idea was to develop projects in each of these offices and to organize agencywide workshops to get their feedback, pushing with each of them the agenda of clarity and consistency in risk management, which stakeholders were so eager to see implemented across the agency. Once the memo was out in 1995, Browner triggered a new process of formalization of risk characterization, based less on formalisms than on principles, and aimed at changing the culture of program offices (EPA 2000a).

Next, Browner accelerated the revision of the iconic cancer risk assessment guidelines. *Science and Judgment* (NRC 1994) had a provocative effect there.[57] The revision had long been in the making and was indeed announced several times (and postponed just as often) in several workshops (EPA 1994b, 1994c), resulting in the industry and associated scientists growing weary of waiting (Anonymous 1994). In January 1994, she requested the Science Policy Council of the agency to prepare an analysis and response to the NRC, in cooperation with the ad hoc committee that the EPA had set up to review *Science and Judgment*. The council stuck to the agency's traditional RAFG-inspired design and bureaucratic screen: It considered that *Science and Judgment* gave support to the continued use of defaults by the EPA, and only needed to make improvements, clarifying their scientific and policy basis, the criteria for departure from them (NRC 1994).

Interestingly the proposed guidelines first turned to RAFG to clarify what a default—or "inference option," in the terminology adopted in that earlier report—actually was, and to defend its use. The reference to RAFG caused the EPA to reiterate: "Since there is no instance in which a set of data on an agent or exposure is complete, all risk assessments must use general knowledge and policy guidance to bridge data gaps" (EPA 1996a, 15), and to add that "some gaps in knowledge and data will doubtless continue to be encountered in assessment of even data-rich cases" (ibid., 19). Perhaps what these lines indicated most clearly was that the EPA did not really believe in

McClellan and North's scientistic opinion that science alone would reduce uncertainty as time passed and knowledge progressed. The reference to RAFG also supported the claim that "[t]he choice of an inference, as the report observed, comes from more than scientific thinking alone" (ibid., 20) and varies enormously from case to case.

In other words, the proposal implicitly argued, *Science and Judgment* was wrong to associate the EPA with a systematic conservatism. The proposal noted, furthermore, that criticism of the EPA varied (accusing the agency of being too resistant to changing its defaults or questioning the basis on which to justify its actual departures from defaults), as did the proposed criteria for when to depart from defaults. The proposal recounted the disagreement expressed in the NRC panel and in the appendices of the final report between Finkel and North and McClellan. Quite clearly, the conflict among experts on the issue legitimized the EPA's decision not to formalize any criteria to depart from defaults ["No uniform checklist will fit all cases … a checklist would likely become more a source of rote discussion than of enlightenment about the process" (ibid., 17)]—a stark illustration of why and how dissensus in reports failed the NRC. The proposal contained a framework whereby "[i]f data support a plausible alternative to the default, but no more strongly than they support the default, both the default and its alternative are carried through the assessment and characterized for the risk manager. If data support an alternative to the default as the more reasonable judgment, the data are used" (ibid., 17–18).

The main building blocks of the new approach were agreed upon in a meeting of the RAF in August 1994 (ibid.), and then a workshop with twenty-five outside scientists organized a month later to foster discussion. First, most of the scientists agreed with the change to a narrative form of hazard identification. The agency would label them as "Known/Likely," "Cannot Be Determined," or "Not Likely," and the disappearance of strict letter classes would eliminate the definitive, regulation-inducing decision of whether a substance was a carcinogen.[58] Second, the use of defaults was to be minimized and greater consideration given to knowledge of modes of action, thresholds,[59] biological models, and the like. It recommended against the use of the maximum tolerated dose. Most scientists, including industry consultants, hailed the shift. Finally, the stress on epidemiological data led the press to argue that the EPA was moving away from reliance on animal testing and tumor identification in rats.

The 1996 guidelines directly reflect the trends embedded in other guidelines—on exposure assessment notably, as well as on risk characterization. Exposure of sensitive populations was given greater importance in this text, which also included a lengthy description of the appropriate practices for risk characterization. What it did not do, however, to the displeasure of environmental groups, was to incorporate a method of looking jointly at chemicals that have cumulative effects. That new approach and concept was reserved for subsequent guidelines.

The EPA encountered stronger opposition to its choice to relax the criteria of statistical significance for exposure studies. The new guidelines increased and specified the criteria of the 1986 guidelines,[60] but they dropped the "unlikely to be due to chance" formulation. The deletion of this language gave the agency greater leeway to consider studies that applied other thresholds of confidence, such as a 90 percent confidence level, as it had attempted to do in its policy on electromagnetic fields and on ETS (Gough and Milloy 1996). In doing so, the EPA was possibly trying to restore a balance: on the one hand, it gestured toward considering the possibility of not applying its defaults, particularly as concerned the nonthresholded nature of chemicals. But it relaxed the statistical confidence criterion to include more epidemiological studies and identify carcinogens that would have been overlooked by the reintroduction of a threshold approach. The same compromise was found in Congress to revise the Food Quality Protection Act. With intense involvement by Browner, the Delaney amendment was repealed, but an extra uncertainty factor of 10 was inserted for cases of evidence of fetal and postnatal developmental toxicity, as well as when data from toxicity testing relative to children were incomplete (NRC 1993a).

As Preuss recollects, the revision was an episode in a "twenty-year battle" between the cadre of agency risk assessors and a host of people (both inside and outside the agency) who aimed to influence their preferred approach: "Way back, in 1983, we said draw a straight line. In 1986, we said draw a straight line. In the 1990s, we said, draw a straight line. Our position never changed, because we said we have no idea what happens at low dose, what the shape of the curve is at low dose. So default would be straight line. But if you had evidence, you could change that. So there was a lot of fighting, a lot of pressure to reverse that thinking."[61]

The agency broadly maintained its preferred approach in that revision, but it allowed some flexibility. Since the first version of the 1980s, the guidelines had grown more adaptable and detailed over time. The flexibility that was introduced in the decision-making process, through the practice of considering, case by case, whether there was enough biological knowledge to not apply the chosen, default assumption, created more uncertainty for the EPA concerning the results of reviews initiated by the courts or elsewhere. It created the need to get more control over what scientists in the agency did, and risk characterization accomplished just that.

Conclusion

Throughout the 1980s, defining an uncontestable core of risk knowledge, demarcating the remaining uncertainties, and formalizing the policies applying in these zones of uncertainty have worked as a way to establish a legitimate functioning in the EPA. It did work effectively, to the point of becoming the foundation of a public identity for the agency, framing the perception of how it would make decisions for both external and internal constituencies. Charting the various forms of knowledge that are necessary to form a generically legitimate decision is the kind of design practice that was at the source of the risk assessment–risk management framework and its installation into the heart of the EPA during the 1980s. It is within this framework that the EPA could claim to apply something called *science* and depict itself as a science-based decision-making agency.

When this assemblage and the formalized bureaucratic knowledge and technologies it rests on are in turn contested or become the very object of the controversies that they were designed to avert, a new cycle of design sets in. This is what the period that is covered in this chapter illustrates. In the early 1990s, and more and more intensively since the technologies of science-based decision-making on which the agency modeled itself, the new design was gradually seen to fail. Risk assessment guidelines, with their set of knowledge and mechanisms to shape a decision, appeared insufficient in the light of the advances of scientific research on chemicals and environmental disorders. To be sure, many of these insufficiencies were manufactured. They were instrumental for the actors that wanted to cast doubt on the way that the EPA was making decisions, as well as on its supposedly conservative policies. Faced with violent criticism of its "default" models and the rise of new, sophisticated concepts of chemical modes of action, quantitative uncertainty analysis, PBPK, and the like—an overall scientization of the exercise of risk assessment—the EPA needed to revise the way in which it was performing its assessments, as well as the knowledge used in the process. More than that, it needed to find a way to restore the mechanical link between science and decision-making, in a context in which the methods and data for the scientific assessment of risks diversified and made this exercise less mechanical than it had been in the past.

Risk characterization—the notion that legitimate and credible decisions can be forged only if risk assessors and risk managers agree on choice of model's, statistical significance criterion, application of defaults, and so on—ended up providing the guide to this reform of the agency and of its formal knowledge and technologies. This did not happen without considerable debate, hesitation, and contention because, by now, the terms of the risk paradigm and risk-based decision-making had been appropriated by the multiple groups that want to gain influence over the way that EPA makes decisions—or over its decisions, anyway. And it certainly could not bring to a close the continuous scientization of risk assessment and the pressure to reduce uncertainties and predict hazards, which both risk scientists and adversaries of regulatory intervention exerted on the agency. More responses to these renewed controversies about the extent of uncertainties and the capacity of the EPA to solve them would be forged in time.

10 Beyond the Risk Paradigm

The EPA has been a risk-based agency during a large part of its nearly fifty years of existence. Around risk has emerged a whole set of formalisms—of the knowledge in use in the agency, and of the right way of producing decisions—that were instrumental to assemble the agency and construct its identity and legitimacy. These formalisms emerged from diverse disciplinary networks, with diverging views about what the source of the uncertainty was, and which knowledge should be used to forge credible decisions. A decisionistic bureaucratic design (carried by toxicologists, embodied in the technology of quantitative risk assessment and defaults) combined with a commensurative design (that of economists and policy analysts) to give form to EPA's standard way of governing problems. This EPA was defended by administrators and other political appointees, who enjoyed a reasonably high level of support from the key audiences of the agency, from courts to the OMB to the White House. The preceding chapters of this book have shown how the framework started to unravel—not all at once, but piece by piece, starting around ten years after the publication of RAFG. Comparative risk analysis lost momentum under the leadership of Carol Browner. Risk assessment was heavily influenced by scientists' and industry's efforts to fold this practice into a strict scientific definition, as if science could inform exact decisions. The reaction to all this was the need to recreate a design that allowed the explication of the limits of science and the definition of its boundaries with policy determinations. Those evolutions were captured and codified in a series of reflexive reports (NRC 1994, 1996; Presidential and Congressional Commission 1997) that accelerated the move from the risk assessment–risk management dualistic representation in order to promote risk characterization.

Moving forward another decade, these trends have only strengthened. Risk is nowadays much less present in the whole of agency discourse concerning the EPA's goals, science, and processes. It is more frequent to hear that the agency has risk-based programs, and that risk assessment is but one instrument among many, than to see the EPA administrator invoking risk reduction, safety, and health protection as the overall organizing policy principles. Thirty years after William Ruckelshaus's commitment to the idea of risk, which portrayed the EPA as a risk bureaucracy, other designs have emerged—namely, a new kind of bureaucratic technology termed "problem formulation, planning, and scoping." Like other concepts that previously gave form to the agency, these were in great part invented at the EPA, notably in those parts of the agency that were dealing with complex, systemic ecological issues. These concepts are now being pushed forward to show that the action of the EPA is not driven by what it is able to measure and calculate, but by what it decides to achieve. It is less acting on predefined objects of action—such as the risks of individual chemicals—than promoting indirect outcomes—sustainability of the environment—using the science that it needs to reach that end. That design helps the agency promote a new, autonomously defined agenda and keep control over the definition of the right science for policy. It serves to move away from the terrain of risk, one on which the agency has more and more difficulty actually imposing its policies, given the continued pressure that it is under to adopt more scientistic methods and always reduce scientific uncertainty.

Like previous episodes of formal redesign of the EPA, this involved its share of public controversy about its scientific capacity and its share of conflict to control the ground of science and to control the agency more generally. As in other episodes, the National Academies, the NRC in particular, is the site on which competing actors converge, and in which the controversy was reframed and new formalisms advanced. The 2009 report called *Science and Decisions* is, from this point of view, similar to *Risk Assessment in the Federal Government* from 1983: a turning point in the political life of the bureaucracy, a shift from dispute to design and an attempt for the agency to restore a capacity to govern the environment credibly through terms and technologies it owned.

A New Coherence: Sustainability

The shift from risk to sustainability as the overall framework of action and identity of the EPA was never as clear as during the first administration of President Barack Obama, when Lisa Jackson headed the agency. As Andrews

(2011, 247) recalls, "Jackson's stated goal has been to restore momentum to EPA's core programs—healthier air and water, and reduced risks from toxic substances—while also tackling emerging challenges such as climate change." Risk was not the way that the administrator defined EPA's approach and overall goals. Like the previous Democratic administrator, Browner, who stressed such objectives as environmental justice, Jackson placed climate change, environmental quality, and sustainability at the center of her agenda. Sustainability goals and adaptation became the new overarching imperatives for environmental action (Ostrom et al. 2002; Wennerston and Fidler 2007; Sexton and Linder 2014).

In those years, the representation of EPA's knowledge and of its typical, integrated way of making decisions also evolved. Paul Anastas, Yale University's prominent chemist and celebrated visionary behind "green chemistry," rationalized this change during his tenure as Jackson's chief of science (2009–2012). A couple of months after his nomination, he wrote to the entire staff of the ORD of the agency to announce a paradigmatic change. The goal of sustainability was now the "true north," for the agency in general, as well as for its science and research effort in particular. The path forward was to embrace technological innovation and to "couple our excellence in problem assessment with an equal excellence in solving problems" (Anastas 2010, 2012). According to his vision, risk would cease to be the methodological tool for the agency's science and decision-making: "[T]he issues of risk characterization, risk assessment, and risk management will be essential building blocks in all of the work that we do. But we're also going to ask ourselves: 'Is it possible to use our reductionist tools to characterize, assess and manage risk in ways that may be unsustainable?' I'd suggest the answer is yes" (Risk Policy Report 2010a). Anastas created transversal research programs, from which the various laboratories of the ORD get their funding. One of the eight research programs is dedicated to human health risk. At the same time as he changed the name and composition of the Science Policy Council (turning it into the Science and Technology Policy Council), Anastas got rid of several risk-related policy projects led by it.[1] The RAF ended up unanimated in those years and was essentially brought to a halt (Risk Policy Report 2010a).

The risk focus did not disappear, but it was demoted to just one kind of operation of the agency. Anastas commissioned a study from the NRC on the adoption of a new integrated sustainability perspective for the EPA. The study, dubbed the "Green Book" from the inception of the committee, explicitly aimed to replace the "Red Book" (RAFG) of 1983—not amending the paradigm, as previous NRC reports claimed to do, but replacing it entirely. The NRC committee was chaired by Bernie Goldstein, who saw sustainability as the new horizon for the agency.[2] The NRC 2011 sustainability report considers that risk assessment should be preserved as a key tool for decision-making. It should even be further developed, such as to address problems of cumulative risks. But it represents only one option or tool in the task of managing environmental problems.

In any regulatory process in the agency, risk assessment should now be preceded by a new exercise of "problem formulation, scoping, and planning": "At the early planning and scoping stage, project managers and analysts diagnose the issue or problem to be addressed. Upfront review of the nature of the problem, credibility of the science, and the decision and legal context helps in considering the nature of the assessment and decision process (Goldstein 1993; NRC 1996, 2007)" (NRC 2011). Only after this stage would analytical activities start, employing a variety of possible methods, including but not limited to risk assessment.[3] To naturalize this new framework of sustainability and help articulate these two sets of organizational reference points, the committee mapped the risk assessment–risk management framework onto their own sustainability flowchart (figure 10.1).

Each component of the sustainability framework, in practice, corresponds to a phase of the process of assessing and managing risk. For instance, the second phase of this process, in which risk assessment is planned and executed, maps onto this moment of the sustainability management process when other assessments addressing social, environmental, and/or economic dimensions would be performed alongside that of risk. In conclusion, risk assessment is not the fundamental, central process that RAFG presented, but its technical components are still the same, and they constitute it as a tool that should be employed to make decisions that enhance sustainability.

"Organizing Information" to Attack "Environmental Conditions"

The initiatives around sustainability resonated with an earlier, impactful study by the NRC published in 2009. *Science and Decisions: Advancing Risk Assessment* (referred to just as *Science and Decisions* hereafter) was written on request from the EPA directly [more specifically, by the National Center for Environmental Assessment (NCEA) inside the ORD] to review the risk

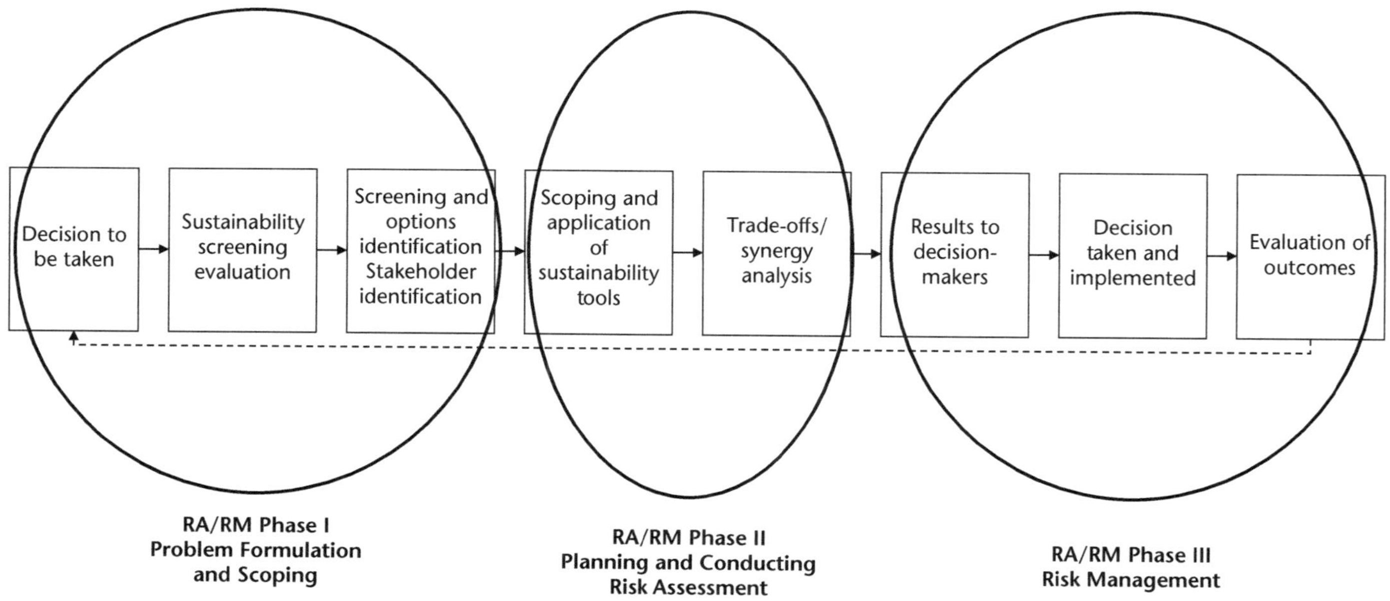

Figure 10.1
Correspondence between sustainability assessment, management elements, and the risk assessment–risk management (RA/RM) framework (adapted from NRC 2011).

assessment practices of the agency, evaluate progress achieved in the past several years, and point to directions for further evolution. The Committee on Improving the Risk Analysis Approaches Used by the US EPA, assigned to write this report, had the massive and ambitious task to "conduct a scientific and technical review of EPA's current risk analysis concepts and practices" (NRC 2009, 281), covering contaminants across all media, as well as health and ecological risk analysis. The panel had to formulate short- and long-term recommendations on probabilistic risk analysis, alternatives to default assumption choices, quantitative characterization of uncertainty, cumulative risk, risk variability, PBPK and biologically based dose-response (BBDR) modeling, and derivation of uncertainty factors. The cost of the study also gave a measure of the extension of risk assessment in the past thirty years and the ambitious charge of this new committee. RAFG had cost the NRC $500,000 in 1983 (around 1 million in 2009$), *Science and Judgment* $900,000 in 1994 (1.3 million in 2009$). This time, the budget reached $1.7 million.

The 403-page report advanced many new bold recommendations: developing a common approach to evaluate the risk of cancer and other effects based on the mode of action of the chemical of interest, being more precise and tailored in the analysis of uncertainty and variability, continuing to explicate default assumptions and defining standards for departing from them, and addressing simultaneously all sources of risk, in the spirit of cumulative risk assessment. All these technical recommendations were unified by a framework proposal: Risk assessment should technically be designed and tailored to the problem at hand. It should respond to the problems facing risk managers rather than being self-initiated and unguided. Taking stock of the practical limits experimented in the EPA with RAFG and the risk assessment–risk management framework was the easiest for the group. The limits were well identified by now: slowness of the one-substance-at-a-time approach, risk assessment as a central process leading to risk management, and lack of interaction between risk assessment and risk management. They were all the more easily recognized, as Joseph Rodricks, the former FDA toxicologist and chief of science, a member of the committee that wrote RAFG, and the supposed author of the fourfold scheme of risk assessment, was part of the committee. Rodricks could recall the genealogy of RAFG and its use at the EPA, as well as the particular context that motivated it—the political intrusions in the health assessments of benzene, formaldehyde, and bis(2-ethylhexyl) phthalate (DEHP) by political appointees in the first years of Ronald Reagan's administration (see chapter 5).

Like Anastas's sustainability initiative unfolding at the same time in the ORD, this panel lent weight to a new exercise of problem formulation, planning, and scoping. It pushed the integration of various streams of science away from partitioned disciplines of human health risk assessment and ecological risk assessment, or of cancer risk assessment and noncancer risk assessment. It pleaded against the use of "bright lines": defined doses, thresholds, and the like, below which risks are not supposed to materialize. Too much time is spent defining these limits for individual chemicals when instead the agency should be driven by the aim to improve overall environmental conditions and the effects of global exposure to chemicals in the environment. It suggests advancing further in the inclusion of considerations of variability of risks among populations. It also criticizes the use of uniform risk management measures, which leave the most exposed populations unprotected.

The report is called the "Silver Book," as if a sequel to RAFG. At the start of this work, James Reisa, a former EPA staff member and longtime director of the NRC's Board on Environmental Studies and Toxicology, who oversaw this panel, argued that the future report had the potential to become "the new Red Book"—a new gold standard for risk-based decision-making. The chair of the panel now recounts that "the joke is that we got silver!"[4] Indeed, the report advances a framework, and dutifully developed its own distinct graph, to contrast with the paradigm of yesterday (see figure 10.2).

The group capitalized on the formalisms developed at the EPA, particularly in the context of environmental assessments, to come up with a synthetic graph that leaves risk assessment intact at its heart, but flanks it with a planning stage, a risk management stage, and permanent stakeholder involvement. The framework tones down the importance of risk as the primary and overall object of the agency's action, choosing to promote a less specific action of decision-making to attack environmental conditions. Modules are defined to design useful risk assessment against the tendency to continually deepen scientific understanding of risk at the expense of timely intervention.

Compared to what was presented in the original definitions of risk assessment and risk management in the 1980s, the shift is clear. The process of governing a risk does not start with risk assessment. Options for decision-making should not derive from risk assessment alone. Risk assessors cannot

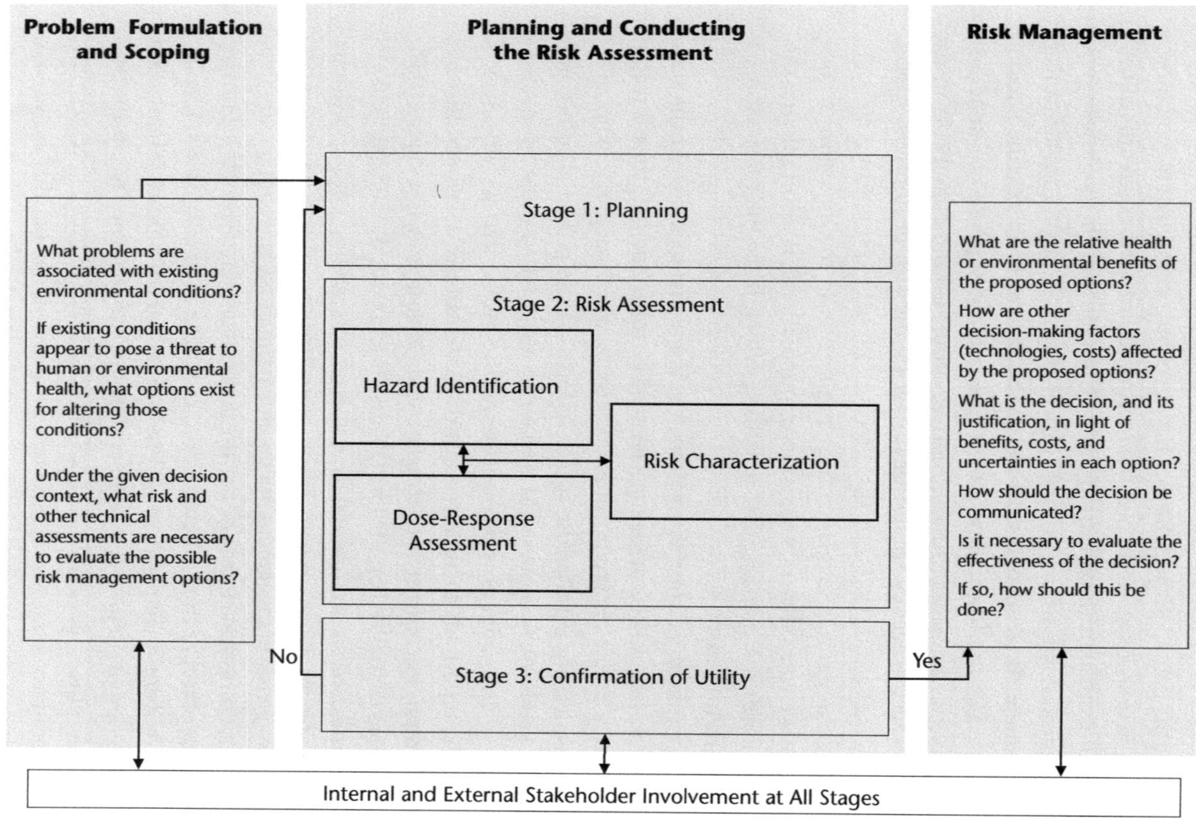

Figure 10.2

A framework for risk-based decision-making (adapted from NRC 2009).

use the assumptions, parameters, or models that they alone deem better. Rather, these elements should be chosen to improve the capacity to finally implement the decision. Also, risk assessment is defined more generically as the organization of information for decision-making.[5] The risk assessment–to–risk management algorithm is undone: Risk assessment does not so much forge decision points as it brings information to the decision-maker. By the same token, the framing of what the decision-maker does in terms of risk also recedes. The fact is: If one deletes just a couple of terms from the graph—mainly the words in the central box, like *risk, hazard,* or *exposure*—a framework is obtained that can be applied in literally any organization.[6] The image of the EPA as an agency that is concerned with risk and organized to deal with such an object is omitted from the design.

The agency soon responded to the NRC by holding a colloquium in 2010, with approximately 120 risk assessors and risk managers focusing on the lessons to extract from this and other concomitant NRC reports.[7] The colloquium was the occasion of a vast exercise of review and stock-taking of the agency's internal practices and approach of risk assessment: 116 agency risk assessors and risk managers were interviewed during the preparation of the event. A task force was then created under the Office of Science Advisor and the RAF. From *Science and Decisions*, it emphasized those recommendations that the agency was already actively working on (addressing the variability of exposure to risks, performing more uncertainty analysis, expanding the use of science to support or revise default assumptions, etc.), and the adoption of a "Framework for Risk Based Decision-Making," with elements on "Planning and Scoping" and "Problem Formulation," a focus on "Informing Decisions," and formal stakeholder involvement. The group thus favored a high-level response through new, generic designs and concepts applying to decision-making.

The working group offered a draft framework for human health risk assessment to inform decision-making (EPA 2012), with the overall objective "to improve the utility of risk assessment in the decision-making process." The document was not intended to supersede existing EPA guidance (which does not use such language as "utilization of risk assessment" in "decision-making"). It is meant instead as a supplementary reference for interpreting existing guidelines, helping agency staff assess the weight to give to the newly promoted concepts of "problem formulation, planning, and scoping" that were already mentioned in some of these guidelines. As

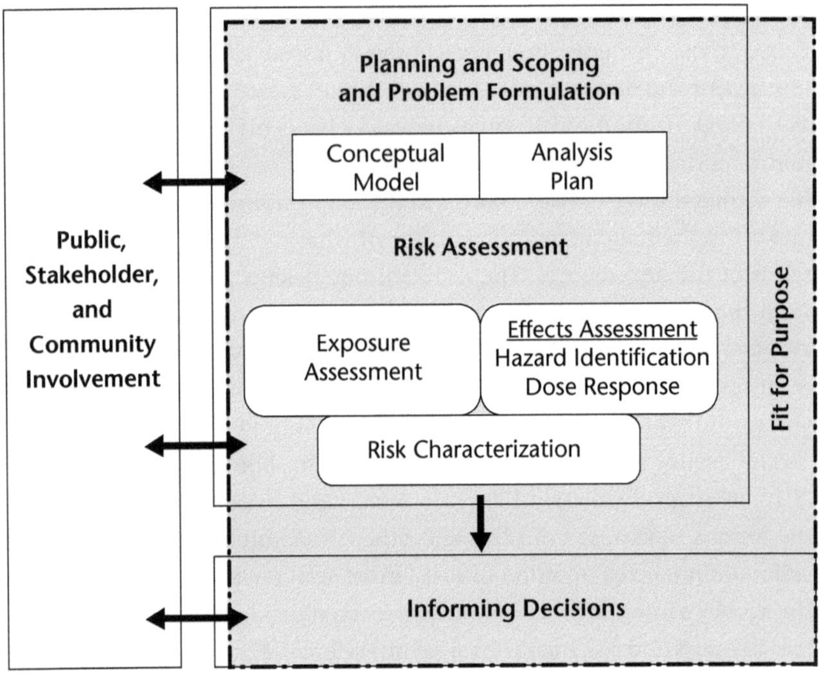

Figure 10.3
Framework for human health risk assessment to inform decision-making (adapted from EPA 2012).

before, these concepts are conveyed together by a graph (figure 10.3), in which the scientific process of assessing risk is incorporated into a broader process of production of knowledge to solve environmental problems.

From this date onward, the bureaucratic technology of *problem formulation* continued to be elevated as an agencywide discipline. The framework of 2012 was reedited in 2014. In that document, the 2012 flowchart was shoehorned into a wheel (figure 10.4) in an attempt to stress the permanence of a cyclical process in which one decision act feeds further reformulation of the problem, and to find a place for such principles that do not translate into linear bureaucratic operations, such as engagement with stakeholders.

The framework also accumulated a number of schematic lessons of the past[8]—the compatibility among assessments of health problems and ecological problems, and the need to produce a full risk characterization, qualifying the uncertainties and variabilities inherent in the final risk estimation. And in accordance with long-standing agency policy, it also emphasized

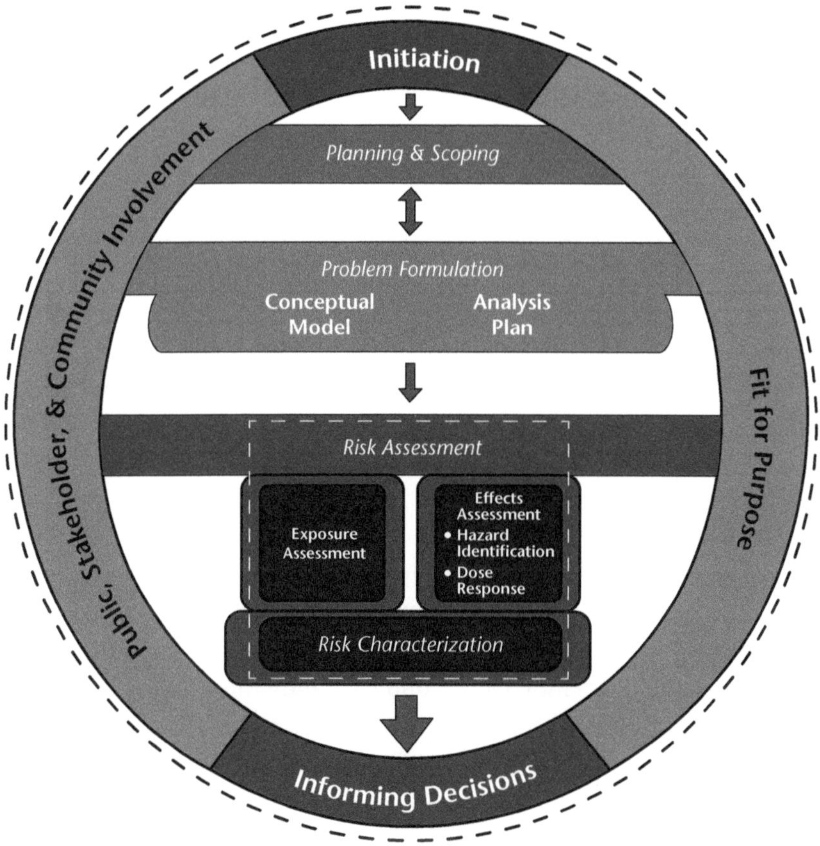

Figure 10.4
Framework for human health risk assessment to inform decision-making (adapted from EPA 2014).

the importance of scientific review and public, stakeholder, and community involvement (although it does not translate these principles into recognizable sequences of action).

Problem Formulation as a Bureaucratic Response to the Republican Assault on the EPA

This new design applying to the goals, processes, and objects of the agency did not emerge from the minds of detached academics. Neither were they imposed by the management of the agency. The various formalisms that were advanced here—*environmental conditions, problem formulation,* and *organization of information* as essential operations; *sustainability* and *environmental quality* as an overall rubric defining the objective of the organization—were selected in the context of a particular controversy about the EPA (notably its treatment of uncertainty and a continued accusation of being too conservative and not scientifically robust enough). They responded to the new "wave of attack" on the agency (McGarity 2012, 205) and its science in the early 2000s.

Despite evidence to the contrary during the campaign,[9] the election of George W. Bush as president in 2000 opened "another anti-regulatory era" (Andrews 2011, 247). The new president had chosen a relatively moderate Republican for the job of EPA administrator. Christine Todd Whitman, formerly the governor of New Jersey, was inclined to embrace the tradition of risk and cost-benefit analysis as other Republican administrators before her had done. She announced a shift back to "long-standing traditions" of "precaution, science-based risk analysis, and sound risk management, including consideration of benefit/cost" (Gibb 2001, 57), putting the battle between Congress and the EPA on risk analysis on hold. She resurrected the Policy Office (now the Office for Policy, Economics, and Innovations) and its role in reviewing regulatory proposals from other program offices, particularly those related to energy issues. But environmental policy in the new administration was made at the White House as much as at the agency. Whitman resigned after only two years, officially citing her desire to return to New Jersey but unofficially reacting to being frequently overridden by the White House and Vice President Dick Cheney in particular.[10]

Her successors, Michael Leavitt and Stephen Johnson, strengthened a number of environmental standards (e.g., standards for ozone and diesel emissions). But they also urged program offices to abandon the development of several important standards and policies and revived the kind of political interference in the work of EPA scientists that made headlines in the early 1980s. This time around, the EPA leadership edited scientific reports that alerted to the need to regulate industries to reduce global warming, as well as other technical reports on oil extraction or particulate matter pollution (US House of Representatives 2003; UCS 2008; see also Freeman and Vermeule 2007; Rest and Halpern 2007; Rich and Merrick 2007; Shulman 2008; Shapiro 2009).

The Bush administration also used the OMB to counter or impede environmental regulation and the science supporting it. George Bush appointed John D. Graham, a longtime critic of environmental policy and of the EPA, as head of the OIRA. This appointment was heavily criticized by environmental and public integrity groups (Hume 2001; Rampton 2001). At OIRA, Graham resurrected the kind of aggressive oversight and critical review of EPA practices that OMB had done during most of the administrations of President Reagan and the first President Bush, stalling many assessments or rules the agency proposed (Tomkin 2016). He won praise from his infamous predecessor, Jim Tozzi, who played this role under Reagan, before becoming an industry lobbyist (see chapter 9; Vogel 2012).

Graham was no newcomer to the field of risk regulation and risk analysis. Indeed, he had a whole career in this field, starting with his first job at the NRC, assisting Howard Raiffa when the statistician chaired the Committee on Risk and Decision-Making. He then got a PhD from the Heinz School at Carnegie-Mellon University, with a dissertation assessing automobile safety technologies, and did postdoctoral research at the Harvard School of Public Health. A specialist in risk-benefit and uncertainty analysis, he is a past president of the Society for Risk Analysis (1995–1996). Graham, as well as the colleagues whom he trained and mentored at the Harvard Center for Risk Analysis, have been remarkably consistent critics of the actions of such agencies as EPA and OSHA, demonstrating through numerous reassessments of agency decisions (many of which were funded by companies on the board of the center) that they were wrong in what they were doing, inflicting more costs on industries than creating health benefits for the society and the environment at large.

At the OIRA, Graham started by releasing a vast review of the costs and benefits of environmental and energy regulations (OMB 2003). As part of this exercise, the office requested comments on the practice of risk assessment to help the working group in its task. The survey was aimed less at assessing the value of whatever innovative risk assessment method and more at resetting the debate about the supposed precautionary approach that EPA perpetuates via its risk assessment practices—specifically via the default parameters and models that it applies during risk assessment when data are missing. The text of the notice builds on the themes that Graham typically defended while heading the Harvard Center for Risk Analysis: the precautionary nature of EPA's preferred conservative risk assessment methods; the unintended negative effects of risk reduction efforts, which often result in augmenting other risks that were not being targeted (the so-called risk-risk trade-offs); the need to reduce uncertainty as much as possible, and to calculate risks as precisely as possible, before making any decision.[11]

The agenda for the public consultation left no chance to the agencies: The notice called for comments about "how conservative defaults used by the EPA embed a precautionary approach" and "examples of approaches in assessment and management that appear unbalanced" (Federal Register 2003, 5499). The OMB made the numerous comments it received available to the EPA, specifically the more critical ones coming from the American Chemistry Council (formerly the Chemical Manufacturers Association). Most of them were critical of the EPA's work. They portrayed a decidedly conservative agency in its assumptions about risk (too inclined to be protective rather than to be right in its assessments), its persistent neglect of the uncertainties inherent in risk, and distortions of the science by its precautionary stances.

While the criticism was "not radical or particularly new" (EPA 2004, 8), the fact that it was channeled by the OIRA to lend support to its project of supervising the agency's risk assessments was sufficient to trigger a clear response from the EPA. Three political appointees of the agency decided to form a task force to produce an internal evaluation of risk assessment practices and "set the record straight."[12] The resulting staff paper on risk assessment was almost 200 pages long, mobilized seventy-four people across the agency, and was organized as a clear rebuttal of the criticism the agency received. The context was so important, and the substance so critical, that this staff paper, though not a guideline in any formal sense, had a major status inside the agency. It constituted a moment of coming together and of collective assertion of the policy of the agency in a moment of adversity. It was, thus, a key moment of codification and explication of its bureaucratic knowledge, recalling all that had been ingrained and practiced in the organization since its foundation, and perhaps even more so since its refoundation in 1983.

As in other places in the report,[13] the staff recaps the history of risk assessment practices and the numerous, evolving guidelines developed over the years for the exercise. Referring to the practice of "examining data before invoking defaults," of using modes of action, explaining defaults better, using uncertainty factors, clarifying choice of a "point of departure" from observed experimental data, and using environmental models, the panel demonstrated that at the EPA, "the derivation of risk estimates improves continually with the addition of newer techniques and relevant data" (EPA 2004, 141). This was essentially a way of saying that the OIRA was off the mark—many of the critical comments concerned practices that were no longer current in the EPA—and that it was making up controversy, so it was illegitimate to propose a refashioning of the agency.

The staff paper was produced right when the agency was finalizing its new carcinogen risk assessment guidelines—another fragment of bureaucratic knowledge that reflects the EPA's response to the controversy manufactured by Graham over the agency's supposed political distortion of science. Following the publication of the proposal for a new cancer guideline in 1996, a public debate emerged, and in response, an SAB review was performed, to which the agency fashioned a reply. A new draft was issued in 1999 and a final adoption occurred in March 2005. The process concluded the evolution that took shape in the 1980s and 1990s, such as definition of alternatives to the default linear approach; recognition of the need to analyze more data and use biological models; refined uncertainty analysis at the level of model choice or, within a given model, at the level of parameters; refinement of exposure calculation, particularly for sensitive populations (with a module for children exposure included, capitalizing on progress made separately on this issue); and revision of the WOE narrative classification. Overall, the evolution was linear, with less expert judgment allowed and more formalized criteria made available to choose among predefined analytical options. In short, and given the evolution of guidelines away from default conservatism, the EPA deflected the accusations of Graham and the OMB.

Graham's attacks continued, though. He created an Interagency Work Group on Risk Management, cochaired by the White House Council on Environmental Quality, using the need for consistency among agencies as an excuse to restrict the autonomy of the EPA—as Reagan had already done, in fact. The group prefigured a new procedure of reviewing the agency's assessments of chemical hazards, whereby the OMB would officially be allowed to comment on the risk estimates developed by the scientists of the EPA for the IRIS database on behalf of all other agencies and departments. It was designed to enable the US Department of Defense, the US Department of Energy, or the National Aeronautics and Space Administration (NASA) to intervene in the EPA's doings.

The process was implemented. It contributed to block the publication in the IRIS database of estimates for controversial substances,[14] some of which soon became the object of new federal-level policy drama—notably concerning perchlorate.[15] More crucially, Graham and Tozzi launched a two-pronged attack on EPA guidelines. Graham launched the *Risk Assessment Bulletin* (OMB 2006), which articulated principles and standards that the agency had to comply with in its risk assessment. The strategy was similar to what OMB, Vice President Quayle, and Jim Tozzi had attempted back in 1992: prescribing and framing how agencies calculate risks, address uncertainties, and develop regulatory measures on this basis. In that publication, he advanced the same agenda that he pursued while director of the Harvard Center for Risk Analysis, advocating analytical practices that mechanically minimize risks, such as providing central estimates of risk in addition to high-end estimates, requiring robust evidence of the adverse nature of the effects being targeted in risk assessments, and using ranges of risk obtained through probabilistic assessment instead of single risk estimates. The bulletin argues at length about the inherent, structural problem of uncertainty in health and environmental risk assessment and emphasizes the need to perform quantitative analysis of uncertainty, as well as sensitivity analysis of models, all to avoid finding risks where there were none—the so-called false positives.

Graham's launch of the bulletin was in all likelihood done in concert with Tozzi, who maneuvered to have the Information Quality Act implemented (Mooney 2004). That legislation created extra requirements for transparency of information in regulatory agencies, as well as justifications for industries to challenge agencies in court, after several failures of businesses to access large studies central to the EPA's decisions (on particulate matters and ETS) (McGarity 2003, 912–913). The *Risk Assessment Bulletin* was in essence using this imperative of transparency of regulatory information to justify imposing new analytical obligations on the agency.

A review of the bulletin—which Graham had required in the hope of giving a scientific imprimatur to it and propelling its recommendations into practice in the agency—was prepared by a panel of the NRC that included academics, but also several scientists with experience in risk assessment at the EPA or elsewhere.[16] Sally Katzen, the person responsible for the OMB's Principles for Risk Analysis of 1995, was a member, as was Joseph Rodricks, a member of the committee that had written RAFG.[17] The panel could not tell what problem the OMB was trying to fix, and it suspected that Graham was trying to rein in the EPA, particularly its actions on such issues as perchlorate and many other issues that, he thought, were being unduly considered for regulation based on indeterminate science.

The panel framed its review in less politicized but no less effective terms, fending off the new prescriptions by outlining the departure that they would create from the patient history of improvement of risk assessment methodologies in the agency (NRC 2007). The NRC panel pointed out that the bulletin insufficiently considered the state of risk assessment as performed by regulatory agencies, or the structure on which these risk assessments are based, including RAFG. The NRC review panel also stated that the suggested rules were unclear or confusing, which betrayed the fact that the OIRA was out of touch with the actual ways in which risk assessment was done in agencies. The conclusion of the NRC review report was a cruel assessment of the OIRA's effort: "The committee began with the working assumption that its role would be to recommend modifications [to what was being asked for], if necessary. After digging deeply into the bulletin and after extensive discussion, the committee reluctantly came to its conclusion that the bulletin could not be rescued" (ibid., 7).

The proposed bulletin is thus remembered as a "spectacular failure" (Shapiro 2007) of the OIRA/OMB, as well as a personal defeat inflicted on Graham, once known as the "regulatory czar" (Nakashima 2002, 35). The OIRA withdrew the bulletin, replacing it by a less controversial update of the 1995 risk analysis principles, developed under President Bill Clinton by Katzen (OMB 2007).

Science and Decisions as a Compromise

At about the same time as Graham left OIRA and launched the *Risk Assessment Bulletin*, his former colleague at the Harvard Center for Risk Analysis, George Gray, was in line to become the next chief of the ORD, as well as its science advisor.[18] It was Gray's turn to push uncertainty analysis and use the uncertainty argument to justify more extensive, repeated reviews of the EPA's assessment documents, but this time the efforts came from inside the agency.

As part of his candidacy as chief of the ORD, Gray had suggested creating a standing body, under the dual authority of the OSTP and OMB to review the EPA's scientific assessments. That proposal, which revived old projects of having a review board established within the NAS board, went under close scrutiny during Gray's confirmation hearings, and Gray had to backtrack under the pressure of Democratic senators (US Congress 2006). The EPA did not reject the new scheme entirely, though; it entered into talks with the NAS for setting up a standing panel (that risked overshadowing the agency's own SAB). The NAS was already playing a review role for the agency. In those years, more and more risk assessments were referred there: it dedicated panels to reassessing the EPA's work on dioxin, trichloroethylene (TCE), phthalates, tetrachloroethylene, and formaldehyde (NRC 2006a, 2006b, 2008, 2010, 2011).[19] With these reports, the NAS was now hosting controversies that originally had set the EPA against the OMB (not to mention other opponents of the agency's supposed conservatism). The cumulated impact of these reports was to push the EPA to think about revising the standard process for setting IRIS values, both procedurally and scientifically.

In nearly all these cases, uncertainty analysis was a bone of contention. On TCE, for instance, Gray in 2006 had asked the staff of the NCEA (formerly the OHEA)—an arm of the ORD and the main risk assessment shop in the agency—to perform a quantitative uncertainty analysis instead of a straightforward linear extrapolation. The NCEA staff and its director, Peter Preuss, resisted, arguing that the evidence did not support such a hypothesis, and that the exercise would lead in effect to relaxing the risk estimate. Producing a range of risk points, as Gray advocated, would not help risk managers, who needed clear decision points to determine whether to initiate the costly clean-ups of contaminated lands. For the risk assessors of the agency, running ten different models to get a shape of the curve at low doses was highly impractical. The choice of considering which model was supposedly right was purely arbitrary anyway in the absence of underlying biological experimental data. Using more models did not reduce the uncertainty that the risk manager had to face, making policy determinations. Rather, it was a theoretical exercise that only confused, and in practical terms, postponed, the decision (Freudenburg et al. 2008).

The staff of the NCEA resisted the calls to embrace uncertainty analysis because this scientific agenda was so closely associated with some areas of the academic field of risk assessment—namely, the Harvard Center for Risk Analysis—as well as with the OMB and with the Republican agenda on environmental policy. The refusal of the staff delayed the assessment and the planned review by the NAS. EPA eventually released its health advisory in 2008, proposing a smaller reference dose than in its last assessment ten years earlier. Although the assessment included a range of reference doses and made some strides toward uncertainty analysis, the NAS panel reviewing the document found that too many precisions were missing to allow for evaluating the quality of the application of the method.

The NRC, just as in the days of RAFG or Science and Judgment (NRC 1994, see chapter 9), became the new battlefield—the space in which the controversial redesign of the EPA's operations was fought. In the 2000s, the council had been assigned more and more frequently to review the EPA's risk assessments, gradually taking up the role of the SAB. The adversaries of the agency's decision-making autonomy frequently used the NRC to try and constrain it. In return, the EPA developed over the years a capacity to liaise with NRC panels, interpret reports, and negotiate the application of its recommendations,[20] to nearly turn the NAS into an ally. But with every new report, whether the NRC would eventually come out in support of the agency or of its adversaries was an open question.

In 2006, the Academies were approached concurrently by two different parts of the agency. Gray went to the IOM for a study of uncertainty analysis, hoping to get an authoritative request from that institute to the EPA to embrace uncertainty analysis much more. For their part, NCEA officials sought to enlist the support of the NRC to undertake a broad review of the risk assessment practices in the agency, particularly its evolution since the days of the RAFG. The initiative was meant to counter the controversial OMB bulletin and transform the climate that it had established. The initiative had another advantage: It would consolidate the conventional risk assessment approaches of the agency by bridging health risk assessment with the now fully developed ecological risk assessment approach. That the initiative for the report came from the "defending" part of the EPA—the one under pressure from OMB on IRIS—is confirmed by the fact that the funds for the study came from the NCEA itself. The expectation and hope was that the NRC review would bolster the credibility of risk assessment methods in the agency.

The IOM study on uncertainty was delayed by the EPA's refusal to appropriate funds for the study,[21] so Gray could not formally endorse it: He left the agency in 2009. The report requested by Preuss, of the NCEA, was released as *Science and Decisions: Advancing Risk Assessment* that same year. It seems that both Preuss and Gray closely followed the setup of the study. They had the chance to give precise input on the charge and to take a look at the preliminary selection of panel members. As ever, the panel composition balanced environmental and public health perspectives—contributions from scientists who were associated with NGOs or industry and had a great deal of experience in toxicology and risk assessment, as well as engineering and biostatistics. Two former colleagues of Gray and Graham in the Harvard Center for Risk Analysis were part of the group. Tom Burke, a medical doctor, epidemiologist, and former public health official in the state of New Jersey, and at the time a professor at John Hopkins University, chaired the group. People such as Bailus Walker, Joe Rodricks, and Lauren Zeise had much experience in the area of chemical risk assessment, including by regulatory agencies. Burke was someone who was very aware of the methods of risk assessment and the risk assessment–risk management framework. Back in 1983, when he was working for the state of New Jersey, he had been trained by Rodricks on the then-new approach at a national conference of state environmental and health officials.[22]

Given the issues on the agenda of the committee and its "eclectic"[23] composition, there was potential for controversy, and even failure. The committee that produced *Science and Judgment* in 1992–1994 had never managed to get everyone to agree on the same argument about defaults, leaving the NRC staff with a bitter taste in their mouths. And there was another potential disagreement this time, surrounding the extent to which agencies should pursue scientific sophistication in risk assessment instead of closing uncertainties and knowledge gaps by default science-policies, as it had always done. For some panel members, this quest for scientific precision, which the EPA had come to accept (embracing the paradigm of chemicals' "mode of action," BBDR modeling, but also attempting to apply more probabilistic risk analysis and Monte Carlo approaches to reducing uncertainty), was what prevented the EPA from actually delivering more and quicker assessments of substances to use to fill in the IRIS database. Others thought that the calculation of a reference dose, or bright line below which no risk can be expected, was outdated. It needed to be repealed in favor of new scientific methods, including quantitative uncertainty analysis. The latter seemed to be Gray's preferred perspective on the debate, as he came to ORD in October 2005 with an agenda on uncertainty analysis and more frequent recourse to nonlinear modeling. But with this issue, which "falls right on the US political divide,"[24] complicated discussions could be expected. The NRC staff wanted to avoid a repetition of the failure of *Science and Judgment*.

The various recommendations that came out of the group bear the mark of these oppositions, as well as of the interventions of the chair in order to come up with formulations that everyone could sign on to. The chapter on uncertainty and variability analysis clearly stated that there remained more to do for the EPA in this regard. But the chapter did not push formal, quantitative uncertainty analysis all the way. Rather, it defined different degrees of sophistication in uncertainty analysis and a tiered approach, recommending that the agency be clear as to how and why it chooses to analyze uncertainty qualitatively or quantitatively. It did not attack the EPA on risk-risk trade-offs, but it did argue that this risk-risk dimension and benefit-cost comparisons could be encouraged via dedicated guidelines.

Chapter 5, on the reference dose concept, and chapter 6, on the use of defaults, were most difficult for the panel to complete. Chapter 5 did not actually recommend getting rid of the reference dose idea, even though some in the group considered it a "lame duck."[25] But some methodological perspective on the improvement of the concept helped soften the language and saved the reputation of the concept, while still pushing the EPA to evolve. The chapter argued for a harmonization of cancer and noncancer assessment—and specifically for the need to compute an overall measure of the risk of noncancer effects (comparable to the "there is a 10–6 risk of developing a cancer from exposure to chemical *x*"), in such a way that the policy office could include noncancer effects in their analysis of economic impacts. Chapter 6 recommends the agency go further in the development of explicit risk management considerations underpinning the use of defaults, or the search for further quantitative evidence of risks. The next major chapter is chapter 8, which pushed concepts of "problem formulation, planning, and scoping" to denote the preliminary work pertaining to risk management and framing risk assessment. It was not controversial, though the committee had to decide whether to introduce other concepts, such as that of "solution-focused" risk assessment, which was tabled by Adam Finkel.[26] Under great pressure from the NRC and EPA officials monitoring the work of the group and reviewing the report, staunchly opposing any form of too radical organizational revision, the group did not go in that direction. Finkel did not try to have an appendix included in the report to outline the concept, as was in *Science and Judgment*.[27]

The final scheme bears no trace of the *solution* language advocated by him; it falls back on the more neutral language, already ingrained in the EPA, of *problem formulation*. In the text of the report, it is converted into a language that has been spoken for decades already, including in RAFG: that of "options." Risk management options should be on the table *before* a risk assessment is undertaken, so that this assessment focuses on informing these options. The context in which the panel worked—a request from the EPA for a report that would review, rationalize, and essentially give support to ongoing changes in the agency—contributed to polishing the recommendations to make them more agreeable in the agency: The risk assessors of the agency could argue that they already were doing problem formulation.

Bureaucratic Lineages: The Internal Origins of "Problem Formulation, Planning, and Scoping"

Overall, it seems, the uptake of these external suggestions was limited, or it was slow and incremental, subject to political impetuses at a later stage. Most suggestions were discussed in a specially established forum called the Alliance for Risk Assessment, put together with the not-for-profit risk assessment consulting company Toxicological Excellence in Risk Assessment (TERA), founded and managed by ex-EPAer Michael Dourson.[28] The dioxin assessment, for instance, was split in two parts (the cancer, and noncancer parts), preventing the use of unified dose-response methodologies for cancer and noncancer effects, and the calculation, by economists of the cost of the latter. Uncertainty analysis, though expanding, remains a limited practice. In combination with the NRC reassessment of the EPA's report for phthalates (NRC 2008), *Science and Decisions* contributed to accelerating the work on cumulative risk assessment, balancing the large investments in more sophisticated dose-response research, and following the new biological, mechanistic paradigm of "toxicity pathways" (NRC 2007).

There was great activity around the IRIS process after 2010. Several actions were done to relaunch the program, following the controversies of 2006–2010 concerning the credibility of the estimations and the delaying effects of the OMB's and other departments' comments on the agency estimates. The proposals for reform after 2010 used suggestions presented in *Science and Decisions*, notably those surrounding "problem formulation," a conceptual stage that now became more official in the new IRIS process developed after Obama's 2008 election. But the staff appeared to resist the idea of making this problem formulation stage an area for more explicit considerations of risk management options in order to improve the utility of IRIS assessments. The IRIS staff within NCEA also feared that turning problem formulation into an occasion for interaction with stakeholders, would offer the industry too much access to the assessment process. In short, the experts and bureaucratic specialists of human health risk assessment—once the flagship bureaucratic technology for an EPA in search of credibility—were using the same strategies than those that led, back in the 1980s, to promoting risk assessment. But this time, in a context in which risk assessment was definitely owned by specialists and other prescribers of methods that they no longer controlled, it searched in house again for those formal techniques that restored, or maintained, its capacity to act.

The fact is that, like risk assessment in 1983, "problem formulation, planning, and scoping" already pertained to the agency's bureaucratic knowledge and decision-making technologies. It was simply that until that time, these concepts were limited to certain parts of the agency. Only the political context—the new Republican assault on the agency beginning in the 2000s—explains why these concepts were used by the leadership of the agency to evolve a new organizational identity. The controlled response of the EPA to *Science and Decisions* in terms of "problem formulation, planning, and scoping," greater uncertainty analysis, and greater tailoring of risk assessment to decision-making needs is the result of a slow institutionalization of new practices and policy visions. The agency leveraged concepts that had been forged in the ORD throughout the 1990s for two main purposes: the reorganization of the agency's research programs and of its laboratories, and the development of methods to calculate ecological risks.

The design of the agency's research programs is a continuing problem. On several occasions at the end of the 1970s, during the second term of Ruckelshaus as EPA administrator (see chapter 7), the agency had to tackle this major, inborn problem of integration of its research and regulatory missions. In the early 1990s, a new series of reform started. The first episode was, as ever, the agency's internal design effort, materializing as the 1992 report *Safeguarding the Future: Credible Science, Credible Decisions*. The committee that produced this report had advanced a framework to enable the EPA to adjust its research plans to the characteristics of environmental problems in a broad sense. The problem of the agency's research soon deepened, becoming the subject of congressional hearings. The many years of budget decline, the loss of 400 employees since 1980, the rate of retirements, and the growing perception in the agency that ORD scientists were too detached from the realities of regulatory work (see Powell 1999 on this) all compounded to generate a feeling of crisis within the EPA's research.

Following a request from Congress to Browner, the MITRE consultancy did an assessment (MITRE 1994), reviewed by the Research Strategies Advisory Committee of the SAB (EPA 1994e). In parallel, a NRC report (NRC 1993c) and a Carnegie Commission report (Carnegie Commission 1992) came into the picture, advocating the creation of megalaboratories. An internal EPA steering committee then considered recommendations stemming from these reports and chose to follow the path recommended by the Carnegie Commission, with one key adaptation: The four laboratories should be based on the risk assessment–risk management paradigm because so much of the science of the agency was defined by the four broad categories of the framework, as illustrated by the thinking applied in health risk assessment and ecological risk assessment guidelines.

In a major reform in 1995, the head of ORD at the time, Robert Huggett, decided to align the structure of EPA research on the "blueprint" of RAFG (Patton and Huggett 2003, 1337): One laboratory focused on hazard identification, one on dose-response, a third on exposure research, and the last on risk reduction technologies under the rubric of risk management (EPA 1994d). Before this reform, the ORD was organized into five offices along disciplinary lines[29] (EPA 1990a). After the reform, all research activities were reframed as a building block of risk management. The 1996 *Strategic Plan for the Office of Research and Development* offers the diagram shown in figure 10.5 (EPA 1996b, 4), where "The risk assessment process is one component of the overall process of risk management" (2), a process initiated by the identification of the problem. Such was the basis of the "new risk-based organization" of the ORD (EPA 1996b, 51).[30]

In 1995, a couple of years after this major reorganization, the new Republican-dominated Congress reopened this issue. It pushed the agency to rethink its research programs. As part of the Departments of Veterans Affairs and Housing and Urban Development Appropriations Act of that year, it requested that the EPA obtain an independent review of the overall

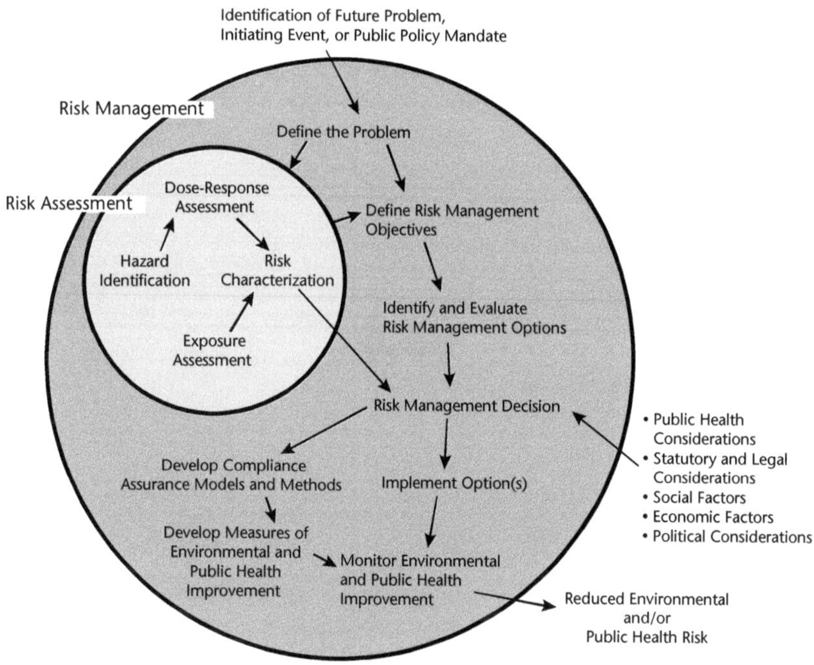

Figure 10.5
The scientific and technical contributions to risk management (EPA 1996b).

structure and management of its research program and evaluate the scientific peer-review procedures that it used. Once more, the NRC was called upon to perform this review, which resulted in a report called *Building a Foundation for Sound Environmental Decisions* (NRC 1997). It argued for the need for another organizing principle besides risk: "Not all environmental issues can be assessed and ranked within the risk paradigm. The more complex and global the problem, the more difficult the task of risk assessment will be. While the risk-based research strategy is sound, it must be augmented and adapted to encompass potential and emerging risks as well as current ones" (ibid., 11). The report puts forward the theme of ecological complexity. The utility of performing risk assessment is assessed in the context of this new horizon and overarching goal: knowing and addressing complex environmental issues. The report introduces a framework in which research and science developed by the agency is coupled with the regulatory process and "problem-driven research," but not conducive to it in a linear and mechanical fashion (see figure 10.6).

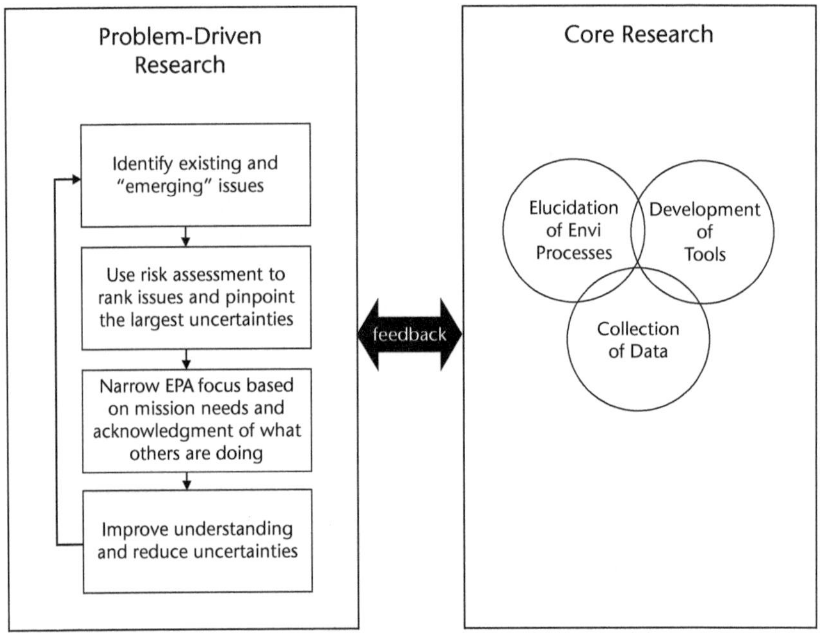

Figure 10.6
A framework for research at the EPA (adapted from NRC 1997).

The report defends the use of risk assessment as a priority-setting tool, not as an uncertainty-eliminating or predictive one: "[R]isk assessment currently is the most satisfactory approach for setting research priorities in the environmental arena. It is particularly valuable in identifying areas of uncertainty that need to be resolved in order to achieve more accurate assessments" (NRC 1997, 46). To reframe the EPA's science and go beyond the reductionist focus on risk, Paul Anastas picked up on some of the ideas introduced by an NRC report of 1997, which similarly argued that risk assessment was too much of a reductionist method to approach complex problems that the agency typically would have to face. Problem formulation was inherited from the work in the agency on ecological risk assessment. That concept was essential to the relatively marginal formal discipline of ecological risk assessment. In that discipline, the collection and analysis of data are preceded by a stage where the chemical of interest is considered to be diffuse in the environment, and a number of stress effects in this environment are created.

Before anything else, a plan of analysis must be laid out to define which stresses are actually relevant to consider, and with what data and modes of calculation they may be captured and measured. Such was the major difference that the specialists in the national Oak Ridge Reference Laboratory had emphasized when asked by the ORD back in 1983 to produce an ecological risk assessment of synthetic fuels on the basis of the RAFG.[31] Ecological issues then were not subjected to risk assessment methods. The Oak Ridge National Laboratory had extensive experience in nuclear risk analysis and associated methods of fault-tree or event-tree analysis. Glenn W. Suter II and his colleagues had to innovate because these methods that calculate the probabilities of a predefined event are reductive: "We found it was not really applicable to ecological systems. Because they are not affected in terms of a 'component' failing, causing failures in other components. You could force ecology into this formalism, but it was not the sort of formalism that we thought was natural and logical for ecological issues."[32] Their approach was a more relaxed formalism, in which complexity is embraced through a problem formulation stage, leading to employing a range of possible analyses, including the more quantitative, model-based risk assessment. The whole discipline, in their view, did not correspond to a rubric of risk analysis. They coined their own disciplinary category, "Technical support for decision making under uncertainty," and emulated the just-published RAFG by producing something generic, and framework-like,[33] as in figure 10.7.

The method was not taken up in the EPA immediately both because it came from outside the agency and because it was not in line with the emerging routine of how to link such an assessment process to decision-making, which the framework did not address at all—after all, Suter and his colleagues were not working in a regulatory agency. People in the agency already had their own ways of doing assessments, and they were not sure what they would gain from a new concept and terminology. So it took a while for the EPA to actually elevate the methodology of ecological risk assessment to a broader, cross-office status.

The early 1990s saw concrete advancements on the topic, with the RAF taking care of developing a framework for ecological risk assessment in which risk assessment disappeared in a broader analysis step, succeeding problem formulation (see figure 10.8). This concept was coined by the

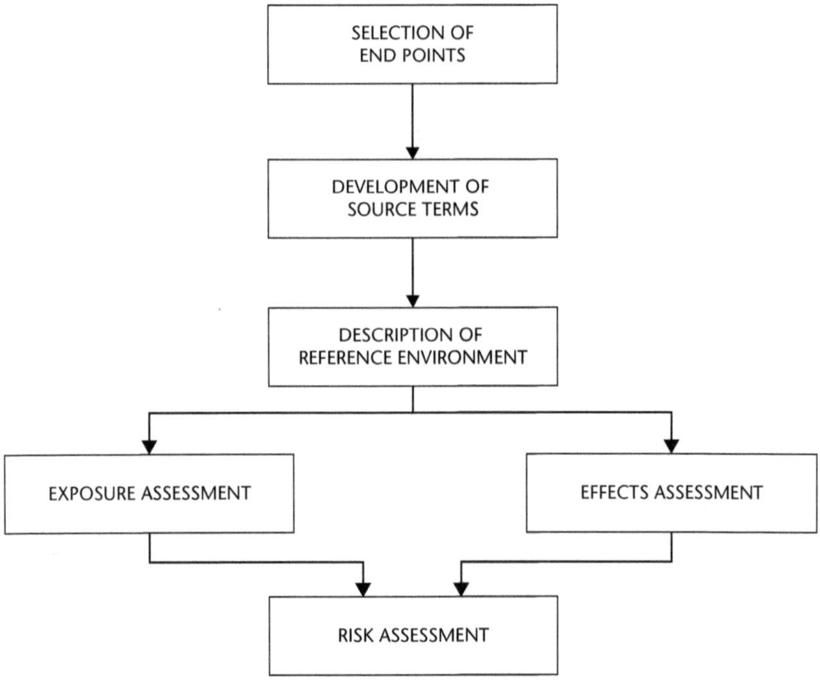

Figure 10.7
Flowchart for ecological risk assessment of toxic chemicals; developed by researchers at the Oak Ridge National Laboratory for the EPA (adapted from Barnthouse and Suter 1986).

dedicated group of the RAF, not by the Oak Ridge researchers, out of the need to organize the link between science and policy.

The agency adopted a formal guideline for ecological risk assessment in 1998 that incorporated the notion of problem formulation, replacing the reductive risk assessment process by a more integrated[34] exercise of analysis that assembled the necessary data and calculations to respond to the problem. It did so right after other prominent reports advocated the importance of "problem formulation"—the NRC's *Understanding Risk* (NRC 1996) notably, but also the report of the Presidential/Congressional Commission on Risk Assessment and Risk Management (Presidential and Congressional Commission 1997).[35] That commission had set risk management as the overall goal, context, and framework of environmental policy, within which analysis of risks was to be performed. In its report, it insisted that the challenge consisted of making risk assessment a substance-by-substance,

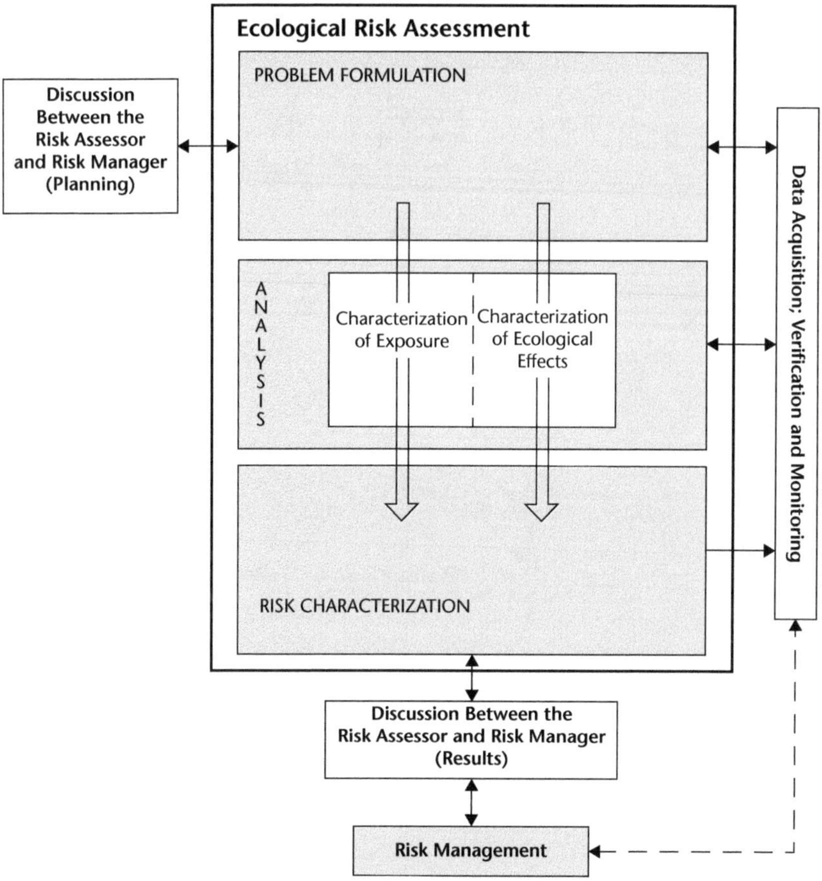

Figure 10.8
A framework for ecological risk assessment (adapted from EPA 1992a).

reductive approach to risk that was better suited to effective intervention global public health issues. Reasoning graphically in terms of a decisional wheel (at the initiative of the physicist and engineer of the panel)[36] helped select, generalize, and give coherence to the whole set of prescriptions (see figure 10.9).

A formulation of the problem and clarification of the context in which it is posed comes first. Before engaging in an analysis of the risks posed by individual agents or substances, the problem/context stage must clarify the multidimensional elements and complexity of the problem. Stakeholder engagement is a constant task throughout the cycle.[37]

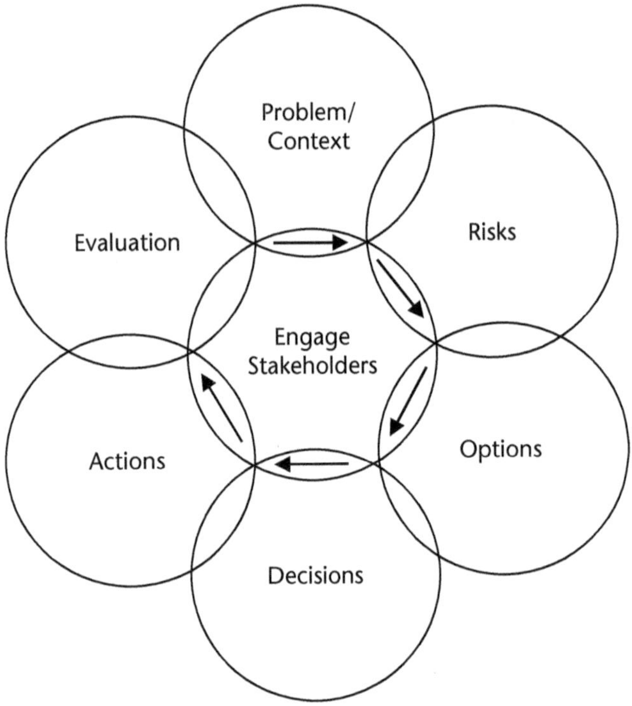

Figure 10.9
The circular design of the Presidential and Congressional Commission on Risk Assessment and Risk Management (adapted from Presidential and Congressional Commission 1997) decision wheel.

Conclusion

In the 2000s, the controversies surrounding the handling of risk by the EPA have shifted to its failure to actually regulate chemicals and reduce risks, and particularly the slow pace at which it concludes scientific assessments of chemicals and makes management decisions. As the introduction of *Science and Decisions* makes clear—as do many other NRC reports during this period—the knowledge and technologies applied to make decisions on these objects seem imperfect. This observation was both realistic and manufactured. It is realistic because the sum of the chemicals to regulate, and the time it takes to agree on the risks of any single one of them, make it impossible for the agency to simply envisage that it could finish the work someday. But it is also manufactured because, over the years, risk assessment has grown more complex, infused with propositions to always know the risks with greater precision and predict them more accurately before attempting to decide anything. This scientization, fueled by the scientific capacities of the regulated industry, causes the inability to decide, the continuous delaying of the moment of decision—paralysis by analysis. The failure to regulate chemicals individually is compounded by the more and more compelling observation, in several corners of environmental and health sciences, that health risks are not determined by exposure to one substance at a time. Risks would be reduced, and environmental health preserved, by analyzing and treating biological and ecological systems in their full complexity, taking into account the set of stressors and factors affecting them.

This means that the 2000s represented an entirely new design configuration—one in which the standard knowledge and technologies embedded in the agency lost credibility; one that evoked the ambition and competence of many actors inside and outside the agency in order to refashion its processes so that it appeared capable of addressing complex, long-term issues. Science, again, lay at the heart of these dynamics. It was both the object of the controversy and the source of a new design. At the heart of the controversy was the EPA's continuous reliance on default, protective assumptions. As explained several times in this book, defaults help the agency make decisions. They are the cornerstone of a particular way of designing itself, of formalizing a way to move from a necessarily uncertain and inconclusive science to environmental policy decisions.

This decisionistic, protective design, however, was discredited by demands to use ever-more-refined uncertainty analysis and discover true risks—a strategy of elimination of rather than compensation for uncertainty. "Problem formulation, planning, and scoping," in turn, embodies a more pragmatic and holistic design to defect the latter predictive, scientistic approach. What was essential to the progressive, pro-environment proponents of "problem formulation" was to avail themselves of an organizational mechanism by which ambitious protective goals are defined every time a new issue emerges, since science could no longer ensure the legitimacy of protective decisions by itself.

At this time, the outcome of this new configuration is unclear. A new design has emerged, which is driving the reform of the organization now and probably will do so for many years, though slowly and contentiously.

Science remains at the heart of a conflict between progressives, who argue that it should be used pragmatically to help achieve goals of environmental and health protection; and conservatives, who insist on moderate administrative intervention and, accordingly, maintain that we should wait to make any decision until we can predict the outcome with absolute certainty—a daunting criterion. Today, President Donald Trump's EPA is being redesigned in exactly this way.

Conclusion

The EPA has been and remains a major institutional site for the development of risk sciences and their use in regulation and policy. It is an organization through which risk assessment and risk management were conceived, tested, and gained credibility as a standard for the administration of uncertainty. The EPA, in return—its constitutive expertise, the material organization of the production of decisions, indeed its overall image and legitimacy—was shaped by risk decision-making. The history of risk and the EPA represents a complex case of mutual construction of one science and one agency, because risk decision-making is not a unified rationality. It is comprised of several rationales crystallizing over time in the midst of the many controversies engaging the EPA, or parts of it.

This book has examined multiple episodes connecting controversy to the shaping of the credibility of an administration. Back in the 1970s, the goal to cancel the registration of a range of pesticides linked to serious pollution problems and cancer risks, and the difficulty of obtaining such decisions in the courts, given the structural uncertainty surrounding what causes cancer and the merit of getting rid of any substance suspected of causing cancer in laboratory animals, led to the replacement of fixed judgment principles by a mixed decision system. This system rested on both calculation and judgment criteria, taking the form of a risk assessment guideline (chapter 3) that was essentially a procedure to force the explication of criteria of toxicity, and then of decision-making. That technology benefited from the nascent science of toxicological extrapolation and medical carcinogenesis, but also from the procedural competence embodied by Alvin Alm, the agency's premier policy analyst. A few years later, the enduring dispute concerning the EPA's standard use of a linear extrapolation method, combined with the outright, arbitrary intervention of deputy and assistant administrators under Ann Gorsuch to rescue benzene, formaldehyde, and DEHP, intensified the debates concerning the boundary to be drawn between scientific calculation of the risk and political judgment about the acceptability of these risks and decisions to reduce them. This was the context that led to this other technology of legitimate decision-making—the separation of risk assessment and risk management (chapter 4).

Once established, this design helped articulate provisional solutions for many of the complicated, apparently undecidable cases of chemical risks, where the legal authority of the agency and the credibility of regulation development process were insufficient to impose a decision on disagreeing parties. The ozone controversy—particularly the difficulty to set an ozone standard in the absence of consideration of costs—anchored the practice of risk management as a "weighing of policy options" (chapter 4). The controversy surrounding arsenic in air led to the Tacoma public hearing experiment and, indirectly, to the stabilization of risk communication processes and principles (chapter 5). The formalization of a comparative risk assessment methodology is inextricably linked to the revelation of widely diverging assessment of iconic chemical substances, such as formaldehyde, at the end of the 1970s, but even more so with the intense questioning of the agency's priorities in the radon and alar affairs (chapter 8). The deployment of a risk characterization discipline in the agency, starting at the end of 1992, was triggered by the passive smoking controversy (chapter 9).

Many more risk controversies—some contained in the agency, others public, but always reflective of the oppositions among audiences concerning the reality and acceptability of particular risks—influenced the definition of risk assessment: The internal dispute about PCE led to a revision of the scheme for carcinogen classification. The formaldehyde case led to a standardization of exposure calculation. The divergences among program offices on the risks of arsenic precipitated the invention of a notion of an "uncertainty factor" (chapter 7). Each time, these affairs accelerated the formalization of a mode of administration, which led to its installation in the EPA. It also inspired discourse by its leaders to represent how the agency generically addresses uncertain issues.

The Political Nature of Frameworks

Despite their apparent logic and mimicking of the scientific method, the risk assessment–risk management framework, and others that succeeded it, are fully political objects. They are a response to existing controversies, as well as a resource to restructure, if not resolve them. Science is instrumental to the EPA because it helps in analyzing, structuring, and responding to controversy. The networks of scientists—of this special kind of scientist that pertains to the broad area of bureaucratic design sciences, those who care to manage risks and learn from contexts of administration—have consistently been active, observing the limits of the administration of uncertain issues and adapting bureaucratic knowledge and technologies to ensure the assemblage of people and of their views. It is, in the deepest sense, a political science, which helps make decisions in the face of apparently intractable situations and respond to selected, idealized audiences.

As this book has illustrated at length, bureaucratic knowledge evolves through contests, counterpropositions, and political interventions. The risk assessment–risk management framework is a reflection of a time when the EPA tried to articulate demands for protection against chemical risks; a rising agenda of deregulation championed by the chemical industry and Republican presidents; and continued demands, inscribed in its statutes, to be science-based, and quantify risks. The framework articulated and ordered all of this at once. The scientization of risk assessment is inseparable from the aggression of various Republican administrations toward environmental regulation in the early 1990s and between 2002 and 2010. Once Republicans in Congress went in the same direction as the industry in the 1990s, this sealed the political fate of risk assessment, which came to be perceived negatively by those who wanted to advance the goals of health and environmental protection.

The environmentalist critique of risk assessment crystallized between 1994 and 2000 because Republicans in Congress embraced it to constrain and roll back environmental policy. The political disinvestment of risk assessment and risk-ranking under Carol Browner is also a testimony to this politicization of the design applied within the agency. Ecological risk assessment, exposure assessment, and cumulative risk assessment developed thereafter, we can assume, in order to restore capacities to make protective decisions as other technologies conservative decision-making, namely the linear extrapolation model in chemical risk assessment, started to be put in question. The invocation of sustainability as a broader imperative, taking over that of risk, probably reflects the failure of a risk-based paradigm to continue to protect the legitimacy of the EPA.

Despite constant talk of a paradigm, there is no clear shift from one paradigm to another at a particular time. There is an incremental invention of new forms, in networks that are continuously active across Washington, that observe and reconceptualize what is happening in a variety of environmental decisions. New knowledge representations and bureaucratic technologies emerge in these networks, connected to one or another office of the agency—not least its ORD—that administrators pick up on depending on the agenda that they wish to promote, to assemble and defend the identity of the organization in new ways. Problem formulation and public engagement are two bureaucratic technologies that are central to the sustainability paradigm that emerged more forcefully in the 2000s. These concepts already existed somewhere in the EPA, but they became instrumental when its leaders felt the need to advertise the complexity of environmental health issues, the interconnections among them, and the limits of a chemical-by-chemical approach. Again, there is a strong link between bureaucratic technologies for making decisions and the broader political context. The former are conceived and incorporated into grander administrative plans, depending on the evolution of the latter.

The fact that the risk paradigm ended does not mean that the whole apparatus of knowledge and technologies that crystallized to handle risk disappeared. The knowledge and technologies are still there, important and employed day after day. They still evolve, one by one, and incrementally. When the whole set of technologies seems to have changed, another name is given to it. But they are less central to the representation of what the agency does as a whole, to the images of its action that circulate in the polity.

"Not Invented Here": The Autonomy of the EPA

Throughout the various episodes recounted in this book, the EPA has consistently been, if not the sole agent of invention of rationalistic forms of decision-making, at least an important site for it. It is a rationalizing agency of some sort—an organizational space that stimulates and hosts design. It is an agency that rationalizes from within.

This can be observed in the fact that nearly all the concepts that the EPA administrators put forward and applied across the agency already existed in one corner of it. No framework emerged wholesale from the NAS, however laudable RAFG and its authors were. The authors of RAFG relabeled and reordered concepts and processes that had been proposed inside the EPA. Much of the history of the decision-making frameworks at the EPA is also the history of the codification of implicit practices and concepts that constitute the typical knowledge of the bureaucratic organization.

Guidelines for quantitative risk assessment and risk-ranking are other blatant examples. Only under Administrator Bill Reilly did the latter acquire greater public visibility, particularly with the release of *Unfinished Business*. Only then was it promoted to become a defining vocabulary for the agency as a whole. But people in the Toxics Integration program, starting in 1977, already knew the value of this methodological concept for the EPA. Risk characterization emerged in the early 1990s, ten years after both the authors of RAFG and Milton Russell established how instrumental an explication of uncertainties and uncertainty-reducing techniques was for legitimating the decisions of an agency. Risk communication crystallized through the "seven rules of risk communication" memo, under the pressure of the radon controversy, years after being conceptualized by Milton Russell and William Ruckelshaus. Problem formulation, finally, was already inscribed in the guidelines for regulatory impact assessment of the early 1980s, but it took on a far greater importance later on, in the context of the development of the practice of ecological risk assessment in the Superfund office and regional branches. They were elevated to become a more generic concept for application across programs by Administrator Carol Browner.

The idea that these designs were emerging inside the EPA is actually not lost on the officials and managers of the agency. They can always argue, when external designers try to impose a new approach or ask for its generalization, that the agency is already doing what is being asked of it (even it is doing so only in part, or only in a particular corner of its organization). This discourse arose in the 1990s when risk characterization emerged as a new imperative, or around 2010, when problem formulation was publicly floated.

Another implication of this rationalization from within is that people in the agency can always argue that it is learning and improving. The trajectory of these techniques—first confidential and not very sophisticated, then precise and generalized—forms the basis of a narrative of rationalization. Pointing to the difference between the times when a method was being experimented in only one part of the agency and the days of its generalization, the EPA will appear to have grown more competent and more dedicated to the implementation of a given approach. Most of the notions that define its way of operating and expertise are generic and can appear in multiple forms and versions. They thus tolerate, or even enable, this discourse about rationalization, of progression from a basic and local practice to a fully conceptualized, global one. Convergence across program offices and between human health and ecological risk assessors on problem formulation in the 2000s, convergence of very different NRC panels on building on the codification proposed in RAFG, and convergence, after some reflection, of Administrator Browner on the memo of the preceding deputy administrator, Hank Habicht: these are examples of references that, over time and depending on circumstances, gain support to become standard bureaucratic knowledge.

EPA and Its Design Networks

And so, from the politics of design standpoint used in this discussion, an agency like the EPA appears to have a fairly high degree of autonomy. This does not mean that it rationalizes alone and on its own terms, though. Rather, it means that the agency is a site in which the formation and activity of *design networks* are stimulated.

RAFG helped the agency combine two separate ideal designs. One was health risk assessment, which combines scientific experimentation of the risk with defined policy criteria to avert uncertainty and legitimize regulatory intervention despite any doubts—a form of precaution predating the principle. But the report also made space for the economists' passion for comparability, consistency, and control over decisions. This rationale found its expression in cost-benefit analysis and regulatory analysis. Through this framework, the vision and expertise of toxicologists, economists, policy analysts, and political managers were articulated. This has happened in a particular configuration, provoked by the extreme uncertainty of the early

1980s concerning the existence of the EPA. This was, in many ways, an exceptional moment. At other times, it is more common to hear that toxicologists and policy analysts in the agency do not manage to speak to each other, or they cannot quite establish routine cooperation (Powell 1999). Scientists, particularly when placed in the ORD, and decision-makers in regulatory offices often do not manage to align. This is a nearly constant condition at the EPA, as the repeated attempts to organize its research and make it more useful to regulatory agencies—in 1977, in 1983, in 1994, and other times—demonstrate.

But design networks are continually active in and around the agency. The enduring cancer risk assessment method cannot be disentangled from the existence of a network of health scientists that consistently defended this calculative-protective method of making decisions. A pivotal person in this group is Bernie Goldstein. Goldstein, academically, was the founding director of the Environmental and Occupational Health Sciences Institute, a joint program of Rutgers University and Robert Wood Johnson Medical School in Piscataway Township, New Jersey. He was trained by Roy Albert, the scientific leader of the early CAG, and a staunch defender of the cancer risk assessment and decision-making technology described in chapter 2. He advised the department of the environment of the state of New Jersey, then facing some of the highest levels of cancer in the United States and pollution of water sources by chemicals, many of which were carcinogenic. Goldstein became head of ORD in 1983, in a critical moment of adaptation of the risk framework in the agency. He recruited Peter Preuss, who had started his career working for the state of New Jersey. Preuss went on to have a long career at the ORD, for many years leading the center that performed risk assessments and working to improve the coordination between the ORD and regulatory offices, or between risk assessment and risk management, holding the ground of the more precautionary approaches embedded in the guidelines used in the agency for both cancer and noncancer risk assessment.

Preuss had worked with a young epidemiologist, Tom Burke, at the department of environmental protection of New Jersey. Burke's mentor was Reuel Stallones, the chair of the committee that produced RAFG. He learned risk decision-making thanks to Joseph Rodricks (a toxicologist involved in the IRLG effort [see chapter 2] and the production of RAFG as well as *Science and Decisions* [chapter 4 and 10]), during a national conference in San

Antonio that was set up to spread the framework to state officials. Burke later became the chair of the panel that produced *Science and Decisions* and then was deputy assistant administrator of the ORD, as well as the EPA's science advisor during the administration of Barack Obama. This group of people forms a loose network that spanned the boundaries of the agency, working at times inside the agency, at other times outside, advancing the same rationale of using science to engineer health-protective decisions. All were marked by the situation of New Jersey in the 1970s, and all started to find their place in the relevant administrations through the notions of risk assessment and risk management emerging from the RAFG.

Other networks have been active across the few decades covered by this book, focusing on other dimensions of uncertainty and advancing alternative designs. A network of policy analysts and economists, successively active in the agency or close to it in universities or think tanks such as Resources for the Future, has supported the various developments in terms of risk comparison and risk ranking in the agency. Alvin Alm, Richard Morgenstern, Terry Davies, and Adam Finkel, among others, advocated the rationale of strengthening the agency's cohesion through transversal, synoptic consideration of lists of risk. The 1990s saw the emergence of a coalition of actors with a scientific background, engaging with the methods of risk assessment applied by the agency to overturn them. The way that John Graham trained and mobilized scientists through the Harvard Center for Risk Analysis to engage on this front over two decades is highly representative of this movement. It helped a network supporting an alternative design—one that regulates based on calculation and prediction—to come together and contest the decisionistic rationale embedded in EPA guidelines, notably when one of Graham's close colleagues, George Gray, took on the position of head of the ORD in 2006.

All three networks documented in this discussion are, in their own ways, deeply concerned by and involved in the fate of the EPA and of the administration of the environment. Despite changing functions and organizational affiliations at the individual level, they are consistently engaged in environmental regulation collectively. They are cross-organizational networks, cutting across the boundaries of EPA offices and other organizations in the area—public academic organizations, think tanks, committees, and services of the executive or Congress, and more occasionally private organizations such as research institutions, consultancies, and industry lobby groups. This means that the networks are always present and active, either inside the EPA or through one or more of their members.

The relationships between people in the same network are maintained over time: All three networks are multigenerational. In an important sense, these networks harbor and perform a particular regulatory morality, more or less liberal or conservative, helping the EPA administration to sustain regulatory intervention, with others trying to optimize this regulatory intervention, and yet others actively working to roll regulations back. Design networks are sometimes explicitly connected to either the Democratic or Republican Party. This political affiliation is generally not made explicit, but it is sufficiently strong to explain that a given network will be more present in the agency, depending on the president and the chosen EPA administrator.

These networks comprise, at their heart, competences coming from this particular sort of science, which I have called *design science*. Their members are trained in and draw from a wide spectrum of disciplines, which go from quantitative health risk assessment to economic regulatory analysis (see chapter 1). They accumulate experience in the performance of analytical tasks for regulatory purposes, observing where and when it seems effective, or not, and delivering the kind of environmental standards that they prefer. Their action on the agency and environmental regulation is mediated by existing bureaucratic knowledge and technologies. Their influence is all the greater in that they work on established designs, from within, to subtly reformulate them. In this kind of administrative politics, it appears more effective to argue that established forms of bureaucratic knowledge have failed and should be adjusted than to put forward an altogether different, but unrecognizable model of administration.

Networks intersect and tie in certain places and moments, making the process of bureaucratic design suddenly more contentious. This book has shown the importance of a number of institutional venues where this happens. From the design perspective adopted here, the NRC has influence on the EPA not only because of its accumulated scientific prestige. It has power, insofar as it is the center of a design configuration—a place where various networks intersect, suitable for elaborating a common design and borrowing proposals from various networks and rationales. Because RAFG contributed to the institutionalization of a mode of administration at the EPA, it seems to be inevitable, for anyone wishing to alter this mode of administration, to work through the NRC. In actor network theory parlance, it has become an obligatory point of passage, the institutional-geographical point at which design networks can be brought together.

Judging by the recurrence of influential reports produced by the NRC and the number of reports concerning chemical risk assessment in the past decade alone, it seems clear that the Academies, the National Academy of Science and the National Research Council in the first place, have become an even more important site to forge institutional responses to controversial situations since the 1983 publication of RAFG. This observation is coherent with the number of participations in NRC panels of protagonists of the design process encountered in the various chapters. Joseph Rodricks, a toxicologist by training and former FDA official, participated in the panel that published RAFG, but also took part in the development of *Science and Judgment*, and of *Science and Decisions*. Gil Omenn, a geneticist and toxicologist, served in the OSTP under President Jimmy Carter, was a pivotal member of the RAC that produced RAFG, and later served as chair of the PCCRARM in 1997. Warner North, a decision scientist, was a member of the panels of RAFG, *Improving Risk Communication, Science and Judgment,* and *Understanding Risk*. The latter report was special, in that it was produced by a panel that included the science and technology studies scholar Sheila Jasanoff, the philosopher Kristin Shrader-Frechette, and the psychologist and risk perception specialist Paul Slovic. Slovic had been a member of the RAFG panel, while Sheila Jasanoff had been on the panel for a precedent NRC report, *Science and Judgment in Risk Assessment*. The importance of maintaining a good relationship with the Academies and the status of its recommendations, asserted by EPA officials, testifies to the centrality acquired by the NRC in the production and management of change at the agency.

The agency and these design networks are not evolving independent of each other. It is probably more accurate to say that the agency is a site of formation of such networks, and its goals a motive for their actions. All of these people were linked to the EPA through one or several of the positions they held during their careers as scientists, former officials of this agency or another, as consultants for the EPA, as members of one of its advisory panel, as members of a service of the White House overseeing the agency, and others. The source of the capacity of the EPA to rationalize itself is this use of networks and of the venues in which they take form and in which they try to design modes of administration. Tellingly, the interactions between the agency and the NRC have become somewhat more critical over the years. Discussing the charge and the selection of experts for panels, feeding panels with detailed information about what the EPA knows and does, advancing concepts to shape its thinking ... all of these pertain to the critical strategic work that EPA people and leaders consistently perform in order to take part in the shaping of the standards of rational decision-making.

The EPA's Rationalities

Risk assessment and risk management, as we are reminded here, is not a homogeneous rationality. This poses the question of what, fundamentally, is the underlying material rationality of the EPA—if it can be at all qualified in such a global, integrated fashion.

Some see the agency as home to a large population of committed environmentalists who are ready to ignore the science where and when it indicates a moderate level of risk. Others consider that the EPA, like other agencies, is captured by private interests—a situation that holds thanks to the injunction to be science based, to use scientific evidence, and to reduce uncertainty (Wagner 1995). This pressure mostly results in delays and the inability to decide. The sophisticated scientific methods and data that justify the pursuit of ever-more-refined risk assessments—such as for dioxin—are jointly defended by private interests and many scientists and decision-makers in the agency. Recently, an investigation traced the more frequent use of PBPK models in risk assessment to the continual influence of an industry-funded group of scientists (Brown and Grossman 2015). Quoting the proponents of this kind of modeling method, the paper claims that it has contributed to blocking processes of a special review of pesticides. The authors of the investigation find coherence between this outcome and the fact that the most well known proponents of the method were affiliated with a contractual research institute working for the chemical industry, in which a string of EPA scientists were trained. The case of PBPK modeling would thus illustrate the pressure exerted by networks of industry-linked scientists on the EPA to always make "use of the best available science" (McClellan and North 1994, 629) and reduce the use of defaults and protective assumptions.

But especially since the 1990s, predicting risk and reducing uncertainty through the search for more knowledge, and more accurate knowledge, appear to have become a strong rationale. It remains difficult to argue that the EPA in general became more scientistic, paralyzed by the uncertainties that an evolving science was supposed to lift, by following strictly deterministic, science-dictates-decisions kind of frameworks. What is certainly true, however, is that businesses and associated scientists are much more central, active, and influential in the configurations in which this bureaucratic knowledge is forged.

Attention to and pressure on the EPA has only increased, and the 1990s were a striking moment in that respect. The changes of the 1990s—the beginning of this discourse according to which the EPA had followed a model, and the fact that this model of separating risk assessment from risk management stemmed from a report dubbed the "Red Book"—took place in a much denser configuration, with cross-linkages and multiple workshops that intermingled EPA people with their opponents and supporters. These were days when workshops of the Silver Book succeeded at a fast pace to conferences and panel meetings at the NRC, and more and more frequent meetings of the Society for Risk Analysis. The Office of Technology Assessment and the GAO produced more reports about the EPA in those years. Think tanks, from Resources for the Future to the Carnegie Foundation, took their share as well. In the early 1990s too, the OMB was extremely active on the topic of EPA concepts and methods, as illustrated in chapter 9, as were the AIHC and the American Petroleum Institute, and the American Chemistry Council in the more recent past. Businesses may not be influential down to every minute decision of the EPA. But they may be influential in setting the course of the slow development of the agency's bureaucratic knowledge.

This pressure combines with the dominance, over the years, of what I have termed a *decisionistic design,* strongly ingrained in the agency. The strength of decisionism shows in the emergence of the risk assessment–risk management framework itself, which made the methods of debating values—risk evaluation, as discussed in chapter 4—redundant. Risk communication, once a central exercise during the second term of William Ruckelshaus as EPA administrator, lost political importance in subsequent years. Public engagement, in the latest frameworks for risk management (e.g., Presidential and Congressional Commission 1997), has become an important prescription. But as the many graphs depicting the new, holistic risk management methods show, it remains peripheral to the processes of risk calculation and regulatory analysis.

Rationalizing (in) Adversity: Institutional Innovation at the EPA

One of the puzzles that motivated this work, finally, lies in the fact that the risk assessment–risk management framework became a standard. As mentioned in the Introduction, risk assessment, risk management, risk characterization, and communication form part of a global gospel of risk governance—a repertoire of rules that frame the practices of bureaucracies worldwide (Jasanoff 1999; Jardine et al. 2003; Renn 2005; Renn and Walker 2008; Winickoff and Bushey 2010; Demortain 2011, 2012). A conventional narrative ties these rules to RAFG, as the source of the framework, and to the EPA experience of the early 1980s (see note 3, chapter 1).

This process of institutional innovation is puzzling if we accept the premises of new institutionalism, according to which organizations tend to incorporate those scripts that are standard among organizations in the same field (Di Maggio and Powell 1983). The authorization of the generic concepts of risk assessment and risk management by a panel of scientists of the NRC, of course, was a very important factor to explain that the EPA officially embraced these notions and extended them across the organization to restructure it. This ceremonial adoption resembles the sort of process that new institutionalists describe as part of the notion of isomorphism. And it helped the agency become a manifestly rational organization, with an apparent plan and purpose.

But this explanation does not fit the history recounted here. The EPA did not buy risk decision-making wholesale from another organization or institutional carrier. The framework gradually emerged through a process of definition of the agency's cross-office procedures for quantifying risk. It was not adopted because of its a priori legitimacy, but it gradually articulated it, as and when it appeared that the categories emerging in the configuration of the early 1980s effectively reduced the disagreements, misunderstandings, and conflicts, both inside and outside the agency, about what the various processes of risk assessment, risk analysis, risk evaluation, hazard or health assessment and the like meant and implied. Once embedded in the organization, the framework represented more than an empty ritual. Although there is some measure of decoupling between frameworks and the reality of practices inside the agency—in part because the agency needs what I have termed a *bureaucratic screen*, a function that the risk assessment–risk management framework fulfilled alongside other tools, such as the risk assessment guidelines—risk frameworks do correspond to the way people behave, the knowledge they acquire and use, and the roles and identities that they take on.

It would be an exaggeration, however, to argue that the EPA possesses a large, autonomous capacity of institutional innovation. First, the design of decision-making frameworks is motivated and constrained by the need to articulate the manifold audiences and criteria of legitimate environmental action—starting with the heterogeneous criteria of risk and of proof inscribed by lawmakers in environmental statutes, as well as in the work of the agency's offices. This rationalization took place in a complex environment at the EPA comprised of the multiple audiences with which it is constantly grappling. Being strategic, in such a configuration, arguably implies reducing the distance between your audiences and aligning the criteria of decisions that they favor and impose on the agency.

In a deep sense, a framework is such a multicriteria organizational plan—one through which various audiences, and approaches to risks, can be articulated together. The science of decision-making, one could assume, is animated by, and embodies, such an ambition to balance and order multiple criteria and public audiences. Further, the assembling of the framework was not done by EPA scientists and bureaucrats alone. It was forged within networks of professionals circulating in and around the EPA, working together in particular moments of questioning and refashioning of administrative institutions.

These elements point to the fact of drawing from an institution—that of science and its ambition to frame and avert dispute—so as to appear to be an organization that can respond to, and even reconcile, multiple audiences. This mode of rationalization concerns organizations whose legitimacy and indeed existence are regularly threatened because of the multiplicity and ambiguity of the audiences and demands that they have to satisfy. Simultaneously, it is accessible only to these organizations that are densely connected to these other parties, forming networks within which this adversity can be re-created, the source of the controversies explicated and analyzed, and solutions drawing on culturally accepted repertoires, like science, tested.

Both of these dimensions stand out in the case of the EPA. This agency, first, is caught in a complex web of evaluations and prescriptions, projected by the multiple audiences and principals on which its existence depends.

The White House and the OMB evaluate the agency, posing questions about the appropriateness of each and every one of its acts. A diverse network of audiences, NGOs, and industry groups attack all the agency's decisions or at least publicly disparage them, with hardly anyone left to defend them. Courts, therefore, question the benefits and meanings of the EPA's actions. EPA officials are very common invitees to Congress, too. The number of reports of the Academies concerning the EPA, its methods, and its assessments is impressive. This agency is subject to greater scrutiny than any other. Each of the landmark NRC reports studied in this book are now referenced outside the area of chemical risk assessment, but all originated in a review of the work of the EPA. It is also an agency of which people say that it is "asking the wrong questions" (Landy et al. 1994) or has "gone mad" (Fleming 1994), and one that Republicans do not unfrequently want to reengineer, if not shut down entirely. The EPA is caught in a dense set of relationships with people, principals, and organizations that all evaluate and contest parts of its action, attributing negative dispositions to it (Dowling and Pfeffer 1975).

Second, the EPA is also an agency that benefits from the engagement of its professionals and leaders in the networks in which the norms of scientific reasoning and the scientific method are defined, tested, and stabilized. What I have emphasized, through the description of design networks, is the web of relationships that tie the agency to the many places in which audiences forge their criteria of what an environmental agency should be doing, and infuse it (Selznick 1949). In the case of the EPA, this institutionalization is forced by even greater legitimacy problems, or existential threats. And this relation with the environment is particularly complex: Each mission and statute executed by the agency implies one or several audiences. The attribution of a unified character to the organization is constantly hindered by its internal fragmentation and the diversity of environmental issues it administers. But if and where science puts its ability to identify and articulate together the criteria of decisions, and if logic and objectivity translate into testing systems for ordering the diversity of knowledge and viewpoints to arrive at common positions, it seems in this case that its broad cultural authority can be put at the service of administration and of legitimacy.

The history of the EPA, at the end of the day, reveals a tension that characterizes most science-based bureaucracies. Science hardly protects, as such, these administrations from conflicts among their audiences and principals, or from the kind of political polarization that the adversarial American political system typically engenders (Jasanoff 1990). At times, however, if it enables analyzing conflict and testing solutions to avert it, then it does legitimize government. It is, in sum, because agency officials and the professional network of design sciences supporting the agency have consistently searched for ways to create the articulation between conflicting parties and opposed views of the environment and risks that it succeeds in instituting forms of decision-making, in the United States and beyond.

Archival Sources and Documents

Electronic EPA Archives, nepis.epa.gov.

Collection and analysis of reports, standards, policy statements related to risk assessment, management, characterization, risk assessment guidelines, etc.

National Archives and Records Administration, Records of the Environmental Protection Agency, Record Group 412.

Director's Correspondence Files, compiled January 1, 1985–December 31, 1986, ARC Identifier 6863743/MLR Number UD-11W 252, Textual Records from the Environmental Protection Agency (EPA) Office of Pesticides and Toxic Substances (1980–1990).

Interagency Agreement Management Files, compiled January 1, 1983–December 31, 1997, ARC Identifier 6862252/MLR Number UD-11W 338, Textual Records from the Environmental Protection Agency (EPA) (December 2, 1970–), National Archives at College Park—Textual Reference (Civilian), College Park, MD.

National Archives at College Park—Textual Reference (Civilian), College Park, MD, Director's Correspondence Files, compiled January 1, 1988–February 28, 1989, ARC Identifier 6863741/MLR Number UD-11W 250.

National Archives at College Park—Textual Reference (Civilian), College Park, MD. Director's Correspondence Files, compiled March 1, 1989–February 28, 1990, ARC Identifier 6863740/MLR Number UD-11W 249, Textual Records from the Environmental Protection Agency (EPA) Office of Pesticides and Toxic Substances. (1980–1990).

Records of the Office of the Administrator, Textual Records, 412.5.1 : General correspondence, 1971–1983. Letters sent, 1977–1982. Intra-agency memorandums, 1977–1984.

Textual Records from the Environmental Protection Agency (EPA) Office of Pesticides and Toxic Substances (1980–1990), National Archives at College Park—Textual Reference (Civilian), College Park, MD.

National Research Council archives

Files related to the Committee on Institutional Means for Risk Assessment, 1981–1983.

Files related to the Committee on Risk Assessment of Hazardous Air Pollutants, 1992–1994.

Russell, Milton. Special Collection, Howard Baker, Jr. Center for Public Policy, University of Tennessee, Special Collections Library, Knoxville.

Speeches, memos, internal and external correspondence concerning risk assessment and risk management, regulatory and cost-benefit analysis, risk communication, management of research, 1983–1987.

Acronyms

ADI	acceptable daily intake
AIHC	American Industrial Health Council
BBDR	biologically based dose response
CAA	Clean Air Act
CAG	Carcinogen Assessment Group
CASAC	Clean Air Scientific Advisory Committee
CORADM	Committee on Risk and Decision Making
CRAM	Committee on Risk Assessment Methodology
CSPC	Consumer Safety Product Commission
DEHP	diethylhexyl phthalate
EDB	ethylene dibromide
EPA	Environmental Protection Agency
ETS	environmental tobacco smoke
FDA	Food and Drug Administration
FIFRA	Federal Insecticide, Fungicide, and Rodenticide Act
IEMD	Integrated Environmental Management Division
IRIS	Integrated Risk Information System
IRLG	Interagency Regulatory Liaison Group
LMS	Linearized Multistage (model)
NAAQS	National Ambient Air Quality Standards
NAS	National Academy of Sciences
NCEA	National Center for Environmental Assessment
NGO	nongovernmental organization
NOAEL	No Observed Adverse Effect Level
NRC	National Research Council
OHEA	Office of Health and Environmental Assessment

OIRA	Office of Information and Regulatory Affairs
OMB	Office of Management and Budget
OPPE	Office of Policy, Planning, and Evaluation
OPRM	Office of Policy, Resources, and Management
OPTS	Office of Pesticides and Toxic Substances
ORD	Office of Research and Development
OSHA	Occupational Safety and Health Administration
OSTP	Office of Science and Technology Policy
OTA	Office of Technology Assessment
OTS	Office of Toxic Substances
PBPK	physiologically based pharmacokinetic modeling
PCCRARM	Presidential and Congressional Commission on Risk Assessment and Risk Management
PCE	perchloroethylene
PPBS	Planning Programming Budgeting System
RAC	Risk Assessment Committee
RAF	Risk Assessment Forum
RAFG	*Risk Assessment in the Federal Government: Managing the Process* (NRC 1983)
SAB	Science Advisory Board
SRA	Society for Risk Analysis
TCE	trichloroethylene
TSCA	Toxic Substances Control Act
UF	uncertainty factor
WOE	weight of evidence

Notes

Introduction

1. The NRC is part of the complex of US national scientific academies, which include the NAS (the oldest academy, founded in 1863), the Institute of Medicine (IOM; recently renamed the "National Academy of Medicine"), the National Academy of Engineering, and the NRC. The NRC is the operational arm of the Academy complex and most frequently delivers the reports that customer organizations (often governmental) request and fund.

2. I am using the acronym *RAFG*, and not "the Red Book," in order to remind the reader of the content of this report—originally a report about the organization of risk assessment in the federal government, with nuances and ambiguity even as concerns this very issue (Jasanoff 1992). Using the nickname "Red Book" tends to simplify the representation of what is inside the report, and let people believe that it offers a ready-made administrative model for the use of science in decision-making. As this book shows, it is an oversimplification of the historical process by which risk analysis *became* a model.

3. It is also sometimes called *the NRC paradigm, the NAS paradigm,* or *the risk assessment–risk management framework.* It is difficult to overestimate the number of reports, articles, or books that call this a paradigm, with *Risk Assessment in the Federal Government* as its sole origin (Newman 1994, 2014; Cothern 1995; Brownson and Petitti 1998; Holgate et al. 1999; Gardner et al. 2000; Jayjock et al. 2000; Aron and Patz 2001; IOM 2001; Thornton 2001; Vallero 2003; Kassim and Williamson 2005; WHO 2005; Calow 2009; McQueen 2010; Sigel 2011; Driver et al. 2012; Friis 2012; Linders 2012; Gulis et al. 2013; Simon 2014; Doern et al. 2016; Fabiansson and Fabiansson 2016; Frantzen 2016). The Scopus/Elsevier database indicates that the report had been cited more than 1,700 times (as of November 30, 2018).

4. The list of administrators of the EPA since its foundation is as follows:

William Ruckelshaus: 1970–1973 (under Richard Nixon)
Russell Train: 1973–1975 (under Gerald Ford)

Douglas Costle: 1975–1981 (under Jimmy Carter)
Ann Gorsuch: 1981–1983 (under Ronald Reagan)
William Ruckelshaus: 1983–1985 (under Ronald Reagan)
Lee Thomas: 1985–1989 (under Ronald Reagan)
Bill Reilly: 1989–1993 (under George H. W. Bush)
Carol Browner: 1993–2001 (under Bill Clinton)
Christine Todd Whitman: 2001–2003 (under George W. Bush)
Michael O. Leavitt: 2003–2005 (under George W. Bush)
Stephen L. Johnson: 2005–2009 (under George W. Bush)
Lisa P. Jackson: 2009–2013 (under Barack Obama)
Gina McCarthy: 2013–2017 (under Barack Obama)
Scott Pruitt: 2017–2018 (under Donald Trump)
Douglas Wheeler: 2018–present (under Donald Trump)

5. Ellen Silbergeld, interview with the author.

6. The Society for Risk Analysis (SRA) never succeeded in agreeing on a common glossary of terms, including a definition of *risk* (Gratt 1987).

7. Labeling literatures and perspectives always involves a simplification of the often-vast set of studies published, as well as their nuances. The opposition between a strategic and cultural view of rationalization, however, helps set the terms of the debate and stimulate the search for alternative explanations of the process of organizational rationalization through the incorporation of analytical knowledge. For a similar opposition between strategic and cultural (or new institutionalist) views, see Brunsson (1985), Oliver (1991), Dobbin (1994), Suchman (1995).

8. Historians of science, for their part, have shown that even the best-known incarnations of rationality and rational tools, such as game theory, have diversity (Erickson et al. 2013). Berg's work on the tools of rational decision-making in medicine is also illustrative of the internal diversity and dynamics of rationalization (Berg 1997).

9. On the notion that design is a dialectical process between the shaping of an object outside use settings and inside situations of use, see Storni (2012) and Bjogvinsson et al. (2012).

Chapter 1

1. Hoos (1983, 294) later wrote that technology assessment was the umbrella term for the administrative use of many of these techniques.

2. For Cooke (1991, 41), "Policy analysis is really a catchall for everything that does not fit into the areas discussed [scenario analysis, systems analysis, probabilistic risk analysis]. It will be understood to include macroeconomic modeling, economic

forecasting, energy planning, project management, environmental impact studies, etc." William Thomas, a historian of operations research, more recently identified "sciences of policy" (Thomas 2015).

3. "Operations research is a scientific method for providing executive departments with a quantitative basis for decisions" (Kittel 1947, 150). Its generic process "begins by carefully observing and formulating the problem and then constructing a scientific (typically mathematical) model that attempts to abstract the essence of the real problem" (Hillier and Lieberman 1980).

4. That redefinition of systems analysis provided a platform for statisticians and mathematicians to start dealing with public programs in various areas, connecting to the ongoing but separate development of a policy science by Lasswell (Lasswell 1951; Farr et al. 2006). The two editors of the book *Pitfalls of Analysis,* both of whom had formerly worked for RAND, approached systems and policy analysis as equivalents: "In this volume, systems analysis and policy analysis are used as essentially synonymous terms for the same activity: research that attempts to bring modern science and technology to bear on society's problems, seeking ameliorative solutions" (Majone and Quade 1980, 6). Like systems analysis, Laswell's policy science was a science that optimized the efficiency of military and governmental organizations' decisions, strategies, and programs. But his platform for the development of policy sciences had other facets as well. One of these was training the practitioners of "the science of democracy," (Lasswell 1942) for the purpose of explicating and clarifying the goals of policies, their relation to democratic values, and their capacity to solve collective problems. This noninstrumental public policy analysis later developed through A. Wildawsky, H. Heclo, and C. Lindblom (see Hoppe 2005; Spenlehauer 2013).

5. The preference for viewing the war in numbers, including during field visits by Washington-based strategists and by McNamara himself, prevented many from contemplating the possibility of a defeat in the Vietnam War (Halberstam 1988).

6. See the title of her obituary in the *New York Times*: "Ida R. Hoos Is Dead at 94; a Critic of Systems Analysis."

7. Technology assessment took shape in 1972 with the US Congress Office of Technology Assessment, headed by Emilio Daddario. In a bill that he tabled in the House of Representatives in 1967, technology assessment announced risk assessment: It was a way of "identifying the potentials of applied research and technology and promoting ways and means to accomplish their transfer into practical use, and identifying the undesirable by-products and side-effects of such applied research and technology in advance of their crystallization, and informing the public of their potential danger in order that appropriate steps may be taken to eliminate or minimize them" (cited in Wynne 1975, 117).

8. The Windscale Inquiry was a public hearing series under the Town and Country Planning Act of 1971 in the United Kingdom, over the application by British Nuclear Fuels Limited to establish a plant for processing irradiated fuels.

9. This paper was republished in his book *Normal Accidents* (Perrow 1984).

10. Lester Lave is called a "self-styled right-wing economist" and a "liberal academic" found to appear "to be to the right of Sierra Club interveners" (Perrow 1982, 299)

11. Wildavsky followed a different path. In *Searching for Safety*, he shows that the calculation of risks pertains to a generic strategy of anticipation and avoidance of potential harm. Anticipation is a much less effective strategy than its generic opposite—accepting that risk will materialize anyhow, and preparing to counter it—which he calls "resilience." Learning from experience is objectively better than calculating subtle variations in probabilities through experiments and models that cannot anticipate or even imagine "major surprise, a change in kind, a change like acquired immune deficiency syndrome that central decision makers could not imagine might occur" (Wildavsky 1988, 8).

12. This does not apply as strictly to the work of a philosopher such as Kristin Shrader-Frechette, who forensically deconstructs the scientific methodologies and values of risk assessors as observed in particular cases (see Shrader-Frechette 1985, 1991).

13. One of the explanations for why Beck omits the term and makes no reference to the SRA is that at the time that he wrote this, risk analysis was only slowly emerging as a discipline, and mainly in the United States. By 1985, for instance, the SRA membership had just reached 1,000 people. Although there were many Europeans in the SRA, it was predominantly US-based. Another reason for the absence of the term is that it was not a relevant phenomenon in the framework of Beck's sociological program. His work was on social change, individuation, and new modes of social stratification (as well as new forms of politics), so the emergence of probabilistic sciences and instruments was of far more interest to the other facet of social science research on risk, which applied a Foucaldian governmentality perspective, in which "risk is analysed as a component of assemblages of practices, techniques, and rationalities concerned with how we govern" (Dean 1999, 132).

14. There were repeated efforts in the 1970s to structure the field and to identify the common concepts and methodologies that specialists of variegated sources of hazards (plants, chemicals, natural disasters, etc.) could claim to have in common (Boudia 2014). In the late 1970s, the National Science Foundation (NSF) founded the Technology Assessment and Risk Analysis (TARA) group to manage the impetus and funding for research on these new perspectives on probabilistic and quantitative risk assessment, the calculation of the benefit-risk ratio and the perception of risks. The NSF's arrival on the scene resulted in an expansion of research devoted to the issue and contributed to structuring a professional community dedicated to risk analysis and management (Golding 1992). This was the context in which the SRA took shape.

It was created in 1981 in Washington, D.C., and the journal *Risk Analysis* published its first issue that same year. The aim was to create a common home for geographers, toxicologists, and engineers who were motivated to turn safety and risk into a sui generis specialty rather than a peripheral object for their initial discipline (Rip 1986).

15. Starr was an engineer and physicist. He was a member of the Manhattan Project, working on the delivery of uranium-235 material to Los Alamos for the production of the atomic bomb. After the war, he worked at Oak Ridge National Laboratory in Tennessee on the development of nuclear propulsion for rockets and ramjets, miniaturized nuclear reactors for space, and the design of atomic-power electricity plants. In the report and in a paper published in *Science*, risks were quantified in terms of the number of deaths attributable to the use of this technology, while its benefits were measured using the proxy of dollar amounts of consumption of the technology. The paper introduced multiple risk-benefit comparisons and inspired the subsequent development of so-called comparative risk assessment. Starr also introduced the quantification and monetary valuation of risks and benefits, paving the way for the application of cost-benefit and risk-benefit analysis to environmental issues. The third opening concerned the measure of social opinions on hazards. Starr borrowed from economists the method of revealed preferences to measure this perception of the risk. He saw this dual quantification (of real and perceived risks) and the reconstruction of relative risk levels as means to show that citizens overestimate numerous risks, particularly those that are involuntary. The paper is often cited by people both within and outside the field as the origin of the field of risk analysis (Short 1984; Hacking 2003; Burgess 2006).

16. Starr was an outspoken supporter of the development of nuclear energy. He founded the Electric Power Research Institute in 1973 as an industrywide cooperative program for electricity and environmental research. In the 1990s, Starr joined the ranks of climate-change deniers, signing onto the 1998 Global Warming Petition Project opposing the Kyoto Protocol.

17. Two American political scientists launched a survey of "risk professionals" (Dietz and Rycroft 1988); that is, those "who spend most of their time dealing with environmental risk concerns" (ibid., 8) and who self-identify with this label. The survey itself showed that the population of risk specialists was increasing at the time and becoming increasingly visible and active. Its results showed that the field, though diverse, was becoming increasingly well structured. Risk professionals, the two authors found, were generally highly educated in specific disciplines, ranging from physical sciences, engineering, mathematics, and statistics (about a third of them), to biological sciences (13%), including medicine and public health (6%). Risk professionals worked within a wide variety of organizations, but federal organizations hosted nearly 30% of them (including 15% at the EPA alone). Risk professionals shared an ideology that combined scientism and environmentalism. These professionals typically adhered to the views that the environment needed to

be protected and that the diffusion of scientific information was necessary and sufficient to clarify concerns and find solutions. This was combined with a strong belief in the virtue of policy, decision or risk analysis for improving or optimizing governmental intervention, and with a preference for working with and helping decision-making by large administrative organizations in government, or corporations.

18. This claim of relevance to public decisions is repeated in the first issues of *Risk Analysis*, by the founders of the SRA: "Decision-making is an essential feature of human society and anything which impedes it must be regarded as non-adaptive. Thus, mechanisms must be developed to allow the decision-making process to go on in the face of risk and uncertainty. Such mechanisms can be collectively called 'risk management' … The risk analyst should look at the particular risk problem in its context and finally consider the decisions that must be made, the options involved, and not only the risks" (Cumming 1981, 97–99). Weinberg (1981, 7) called for risk analysts to deal with decision and administrative action: "I hope those who work on the scientific side of risk analysis will interest themselves in the formulation of policy that flows from their scientific findings."

Chapter 2

1. Assistant administrators manage the various offices of the EPA, from the ORD to the Office of Policy, among others, and the various "program offices" (Office of Air and Radiation, Office of Water, etc.). Assistant administrators are political appointees, not career staff. They are chosen by the incoming EPA administrator, after approval by the White House, and must be approved by the Senate.

2. The committee and report were the latest moves in a battle of opposing factions of medical scientists on the issue of carcinogens. The Saffioti report was a reaction to a report (NRC 1969) that suggested that one could compute "safe" levels of carcinogens in food, in an attempt to scientifically justify moving beyond the precautionary, zero-tolerance Delaney amendment (McGarity 1983). Public Citizen Health's Research Group accused the NRC panel of being biased toward the views of the chemical and food industries (Vogel 2013), which then were actively at war against the Delaney amendment.

3. This office has changed names several times during the period covered by this book. It was first called the Office of Planning and Management (1970–1975), then the Office of Policy and Resources Management (OPRM) (1976–1981). It then became the Office of Policy, Planning, and Evaluation (OPPE). What this text really discusses is the activity of the latter, in the 1980s and 1990s. Currently, the office is called the Office of Policy. Its activity is described in the following way on the EPA website (accessed in February 2017): "the Office of Policy (OP) is the primary policy arm of EPA. We work with our EPA colleagues to support Agency priorities and enhance decision-making. We provide multi-disciplinary analytic

skills, management support, and special expertise in five areas: regulatory policy and management, environmental economics, strategic environmental management, sustainable communities, and climate adaptation" (https://www.epa.gov/aboutepa/about-office-policy-op).

4. Source: Economist, Office of Policy Analysis, interview with the author.

5. See, for instance, Anderson's testimony before Congress in 1983: "In assessing the possible risk of carcinogens posed by individual chemical substances, EPA follows a two-step approach. Risk assessment attempts to answer two questions: 1) how likely is the event to occur? 2) on the assumption that the event does occur, what is the magnitude of the public health impact? Since only rarely do we know for sure that an agent is indeed a human carcinogen, the first step involves an evaluation of all the biomedical data to determine the weight-of-evidence that an agent might be a human carcinogen. The second step involves the quantification of risk, that is public health impacts, in terms of rough estimates for current exposures as well as estimated exposures for various regulatory options" (US Congress 1983a, 372). See also Anderson (1983).

6. *Ethyl Corp. v. EPA,* United States Court of Appeals for the District of Columbia Circuit 541 F.2d 1 (1976).

7. The first quantitative risk assessment performed in the EPA was the "Quantitative Risk Assessment for Community Exposure to Vinyl Chloride," completed in 1975 in the Air Office (Kuzmack and McGaughy, 1975).

8. The linear model, as its name indicates, is a mathematical model that forces the computation of dose effects that, plotted on a line, will produce a no-threshold, nearly straight line. Probit is a model that produces a supralinear or sublinear curve (half of a bell shape). Other models available at the time include K-hit, Armitage-Doll, and Weibull.

9. In February 1983, in a testimony before the House Committee on Agriculture, Albert argued that the only thing that could realistically be done to counter cancer was to regulate the excess cases of cancer in the population across all modes of exposure, instead of taking on substances individually, at the risk of losing each of these judicial battles. An overall target of reducing cancer, following the principle of protection against ionizing radiation, ALARA, was the only way forward. This became the policy of the agency: "Risk assessment of human exposure to radioactivity offered the conceptual framework for the assessment of risk from exposure to chemicals" (EPA 1985c, 2).

10. Landy et al. (1994) cite Anton Keller, an OSHA official, describing the 1976 EPA guideline as a "Sistine Chapel" approach: that is, experts go into the chapel and emerge with divinely expert judgment.

11. Scientific coordination among offices was feasible where there were programmable analytical tasks. The Air and Water offices, for instance, duly used the forces of OHEA to produce the series of "criteria documents" that had to be established for legally defined, definite lists of chemicals. As mentioned earlier in this chapter, the Air Office was the first to endorse ORD's risk assessment guideline of 1976 for application to the assessment of the six criteria chemicals designated under the Clean Air Act. The Air Office was the first "client" of the OHEA (around 1980, 45 percent of OHEA was dedicated to providing scientific advice to this program office). OHEA routinely did cancer and noncancer assessments for this office, limiting the risks of inconsistent opinions. In 1980, the Water Office announced the availability of water quality criteria documents for sixty-four contaminants. It published, along with the criteria, the guidelines that were used to calculate these numerical concentration limits, below which human health and aquatic life are believed to be safe, in terms of cancer risk, noncancer risks and organoleptic effects (Stephan et al. 1983). These guidelines are dubbed "the first EPA document describing quantitative procedures used in risk assessment" by the EPA staff paper on risk assessment principles and practices of 2004 (EPA 2004, 4).

12. When the EPA was created in 1970, the ORD was totaling 2,000 staff, out of the 6,000 employees of the agency. Ten years later, the ORD had 2,400 staff, out of 13,000 agency employees (the administrator's office counted around 1,100; Office of Water, 800; Solid Waste, 300; Air: 1,100; Pesticides: 1,360; ORD: 2,400; and the ten regional offices has more than 5,000 staff).

13. Partly as a consequence of the 1977 NRC report, Congress reestablished the Scientific Advisory Board in its 1978 Science Advisory Act (it had been created in 1974 to reunite various advisory committees inherited from the departments that preexisted EPA). After 1978, the SAB counted 100 external scientists and 300 consultants. Its mission was to provide extramural information and advice to the administrator and other officials of the EPA (Jasanoff 1990).

14. The Environmental Biology Research Laboratory; the Environmental Chemistry and Transport Research Laboratory; the Environmental Engineering Research Laboratory; the Environmental Measurements Research Laboratory; the Health Effects Research Laboratory; the Health and Environmental Assessment Center (tentatively renamed, but it would keep the name of OHEA eventually, until the reform of ORD in 1995).

15. Eight areas of potential cooperation had been identified in this agreement, including risk assessment, toxicity testing, joint research planning, information exchange, coordination of inspection efforts, uniform responses to problems of concern for more than two agencies, communication and education, and policy for the review of epidemiological data.

16. Omenn graduated from Princeton. He received the Freshman First Honor prize at Princeton University in the winter of 1961. He then obtained his medical

degree and PhD from Harvard. In September 1973, then an assistant professor at the University of Washington, Seattle, he took a year's sabbatical to serve as a White House fellow. He worked as special assistant to the Chairman of the Atomic Energy Commission. Omenn recounts this episode, as well as many others, to discuss his involvement in high-level policy issues throughout his career, as well as his position within both policy and medical elites (Omenn 2011).

17. A first step had been made in 1978: The Department of Health, Education, and Welfare moved in this direction when it centralized its research and evaluation work on carcinogens in a National Toxicology Program, partnering in this with the FDA, the National Cancer Institute, the National Institute for Environmental Health Sciences, and the National Institute for Occupational Safety and Health.

18. In 1982, the OHEA proposed additional guidance for cancer assessment. In this document, it opened up to considering mechanistic knowledge—understanding of the modes of action of chemicals in the body, to get to more nuanced judgment about their carcinogenicity. The guidance distinguished between mutagenic and non-mutagenic carcinogens and recommended the application of a conventional toxicological method to these (using the NOAEL and applying a safety factor, to determine a safe level of exposure). Roy Albert received dozens of letters of comments, with those from industry pleading for even more flexibility in the guidelines. Gehring, then-director for health and environmental sciences at Dow, explained that "sound scientific principles" demanded an "even more scientifically flexible guideline than currently proposed," away from the "inflexible predesignated 1000x safety factor," and the "assumed linearity at low doses" (Gehring, P. J., Letter to Dr Roy Albert, dated June 28, 1982, with enclosure). The scientists whom Dow recruited to comment on the guidance made it clear that the statistical model borrowed from Kenny Crump by the EPA to found the multistage model originally did not include any assumption of linearity. Dow's scientific consultants pointed out that the EPA had made a policy—not scientific—decision to incorporate this assumption in the model, and that it should assume the responsibility of this policy decision and agree to put it up for debate rather than presenting it as fact. David Hoel, from the US Department of Health and Human Services, argued against what he saw as a cookbook approach, while Ronald Hart, director of the National Centre for Toxicological Research, found fault with the logic of applying safety factors even in cases where human data were available, thereby suggesting that the guidance should make space for a case-by-case application of the safety factor and linear extrapolation methods (EPA 1980c, 55).

19. The IRLG document was published in the Federal Register, without endorsement by OSHA, which objected to this quasi-official status. Costle insisted on having the document endorsed by the White House Regulatory Council, which he chaired. This was done (at least somewhat) when the council published a statement on carcinogens, citing the IRLG guidelines in an appendix. But the document, as acceptable as it may have been to all those who participated in drafting it, was never reviewed

internally by the agencies or approved by them. After President Ronald Reagan's election in 1980, the IRLG group was disbanded. (For details, see Landy et al. 1994, 194–200.)

20. Justice Stevens stipulated, "The burden was on the Agency to show, on the basis of substantial evidence, that it is at least more likely than not that long-term exposure to 10 ppm of benzene presents a significant risk of material health impairment ... In this case OSHA did not even attempt to carry its burden of proof." The closest that it came to making a finding that benzene presented a significant risk of harm in the workplace was its statement that the benefits to be derived from lowering the permissible exposure level from 10 to 1 ppm were likely to be appreciable. In truth, the agency did not draw a dose-response curve to extrapolate the effects of the substance at 10 ppm from the observed effects at higher concentrations. It stuck to a generic, qualitative consideration that no thresholds exist for carcinogenic substances, categorizing chromosomal effects as an adverse biological event of serious concern, which may pose or reflect a potential health risk. The Court recognized that imposing a proof of certainty before issuing a regulation to the agency would equate with paralyzing it, and that the Act requires no such proof. It also acknowledged that the rules for judging carcinogenicity, in the absence of the possibility of scientific certainty, were a matter of policy. In the words of Justice Stevens, "the requirement that a 'significant' risk be identified is not a mathematical straitjacket. It is the Agency's responsibility to determine, in the first instance, what it considers to be a 'significant' risk."

21. A subsequent court ruling in another case upheld this principle: *American Textile Mfrs. Inst., Inc. v. Donovan,* 452 U.S. 490 (1981).

22. *Gulf South Insulation v. United States Consumer Product Safety Commission,* 701 F.2d 1137 (5th Circuit 1983). The CPSC had used a quantitative risk assessment, evaluating the excess cancer risk in the range of 0–51 per million, based on a single study.

23. *Ethyl Corp. v. EPA.*

24. One illustration of that is the fact that major judicial setbacks inflicted on certain agency programs, such as the "*Corrosion-Proof Fittings v. EPA*" ruling of 1991 concerning asbestos did not result in alterations of the preferred frameworks. Fisher et al. (2015) demonstrate this with regard to the NAAQS program: EPA analytical benchmarks are used by courts and generally upheld.

Chapter 3

1. A legal battle ensued to determine whether the Act superseded the decree, and whether the program launched by the EPA to establish criteria should be pursued—with the industry lobbying for its interruption and Gorsuch likewise leaning in this direction.

2. Clean Air Act, 42 U.S.C. § 7412(a)(1) (1982). The Act stipulates that there is a safety threshold, perhaps to compensate for the difficulty (or impossibility) of attaining zero risk. But as Powell (1999) explains, this assumption is incorrect in the case of air pollution because biological effects were found in conditions of exposure to ambient air pollution.

3. These were the major Acts for which the agency had full responsibility. Table 3.1 does not mention Acts of which the EPA implemented only a particular section. Some of the Acts indicated here contained more than one regulatory program. The Clean Air Act included a program that applied to criteria air pollutants, and another to hazardous air pollutants. The agency would generally create separate offices for each type of pollutant.

4. After Nixon's reelection, Costle spent some time at the congressional office of Budget. When Jimmy Carter appointed him, he strategically chose to espouse the presidential agenda in an attempt to foster the relationship with the White House and continue to amass support from the president (Landy et al. 1994).

5. On the one hand, Train won from the White House an ascendance over environmental decisions, notably through the written confirmation that the EPA, rather than the OMB, had the last word over the substance of its standards and regulations. On the other hand, Congress gained confidence to create new environmental laws and oversee environmental policy.

6. The EPA designated six criteria air pollutants: particulates, sulfur dioxide, nitrogen dioxide, carbon monoxide, ozone, and hydrocarbons (with the latter soon delisted and replaced by lead).

7. For instance, four people (one in OHEA, one in the Air Office, and two in separate parts of the Water Office) were working on formaldehyde. No fewer than six groups were dealing with benzene for technical assistance, regulation of environmental effects, the revision of the regulation of environmental effects, and the establishment of water quality criteria under the CWA; for regulation of its environmental effects under FIFRA; and for its preregulatory assessment of its health effects under the SDWA.

8. The EPA gained its independence from constant White House oversight only when its administrator, William Ruckelshaus, threatened to quit unless his authority was acknowledged (Eads and Fix 1984). Train, who had chaired the Council on Environmental Quality, had to do the same after he took the reins of the agency in 1973.

9. In his capacity as head of the Regulatory Council, Costle wrote to other heads of regulatory agencies and lawyers in charge of the development of regulations in these administrations in July 1980. He asked his counterparts and regulation writers in each agency to duly consider all regulatory options and alternatives as part of rule-making. His letter was founded on the work of Michael Levin, the chief of the

regulatory reform staff, a group created inside the Office of Planning and Evaluation to work toward the implementation of Executive Order 12044 for efficient and effective regulatory strategies. Levin had developed a checklist of all possible ways of regulating technologies and products and industries, ranging from direct regulatory mechanisms (technology-based standards, performance-based standards, prohibitions, and limitations) to innovation incentives, through information mechanisms. Toxics integration was fully in line with the process, given its recommendations to include cost considerations to help decide whether to regulate a substance.

10. Daniel J. Fiorino, Chief Regulation Management Staff, Memorandum to C. Ronald Smith, Director of the Office of Standards and Regulations, "Survey of agency practices regarding OMB review," Office of Policy and Resource Management, November 30, 1982, Milton Russell Special Collection.

11. Beyond cost-benefit analysis, the order also aimed to increase agency accountability for regulatory actions, minimize duplication and conflict of regulations, and ensure presidential oversight of the regulatory process. Therefore, while the OMB had neither the authority nor the competence to review the EPA's risk assessments, potentially it could still severely restrict the agency's authority and legitimacy for major decisions affecting highly toxic and widely used chemicals, as it did by regulating dioxin, benzene, lead, arsenic, or formaldehyde. Under that order, the OMB gained the right to review every single individual cancer regulation drafted by those agencies to examine the way that EPA staff considered the costs and benefits of the rules. It set up a dedicated task force to do so.

12. The passback was the OMB's communication (written or oral, in any case generally confidential) to the agency of its budget for the coming year, following decisions by the head of the OMB and the president.

13. Morgenstern had become director of the Office of Planning in mid-1982, on special assignment from the Urban Institute under the Intergovernmental Personnel Act. He was director of the Urban Institute's Energy Program from 1980 to 1982, senior legislative assistant to Senator J. Bennett Johnston from 1979 to 1980, and deputy assistant director for energy, natural resources, and the environment at the Congressional Budget Office from 1976 to 1979. From 1971 to 1976, Morgenstern was a tenured associate professor of economics at Queens College of the City University of New York. Prior to that, he taught for a year at the American University in Washington, D.C., and obtained a PhD from the University of Michigan in 1970.

14. Member of the Integrated Environmental Management Division staff, interview with the author.

15. Just as in 1977, these options differed widely, notably in terms of authority granted to functional offices, whose role has been difficult to define ever since the beginnings of the agency. The first option was to give a priority-setting task to the Office of Toxics Integration in the OPRM. The Office of Toxic Substances would

list substances for multimedia risk assessment and then coordinate the work by program offices. The second was to assign to a new office in the OPRM the task of coordinating and reviewing assessments done in separate offices, and of preparing guidelines for application by program offices. The third option was to combine all hazard, exposure, and risk assessments with economic analysis in a single organizational unit in OPRM. From option 1 to option 3, the degree of centralization obviously increased to the point of becoming politically unrealistic. Program offices were unlikely to give up their resources and mandates in risk assessment. Furthermore, options 2 and 3 outlined a system in which the OPRM effectively mediated among program offices and the top level of the agency, the deputy administrator and administrators, who signed regulatory decisions. In the eyes of its staff, the OPRM was to prepare the file for final decision-making and arbitration at higher levels. Nowhere in the regulatory regimes that the agency operated was it indicated that the OPRM would fulfill this role. Beardsley thus outlined several options and let the administrator decide.

16. The "conservative" (i.e., protective) policies were rooted in statements of individual risk, of the following kind: One (theoretical) person has 1 in a million chance of developing cancer from being exposed to substance *x*. The chemical industry was accustomed to criticizing this criterion, saying that, in the aggregate, the kind of individual risks that the EPA calculated amounted to no more than a dozen deaths per year for the whole US population—a very small number compared to casualties of automobile accidents, for instance. The OHEA always had difficulty countering these arguments. In 1981, in the era of systematic cost-benefit analysis, OMB supervision, and environmental skeptics at the top of the agency, cancer risk assessors conceded that indications of total risk should be provided alongside individual risk levels.

17. Arnold M. Kuzmack, Director, Office of Program Development & Evaluation, and Elizabeth L. Anderson, Director, Office of Health and Environmental Assessment, Memorandum to TSPC, "Request for Review of Draft Cancer Policy Statement," Office of Water, US EPA, July 22, 1981.

Chapter 4

1. Parts of this chapter have appeared in Demortain (2016). I thank the publisher for authorizing me to use parts of this chapter here.

2. For example, step 5, to be performed by the expert panel, concerned the preliminary estimation of carcinogenic hazards to determine the relative potency or severity of effects and to permit a reasoned and consistent setting of priorities. Step 6, a task of the regulatory agency, consisted of establishing preliminary priorities for regulation, if warranted by the outcome of scientific determination in step 5, and assembling data on actual or potential exposure. Then the regulatory agency would

continue to step 7, determining the need for and priority of regulation. The whole process ended with step 11, risk evaluation, including the identification of social valuation of the risk as derived through a risk-benefit evaluation, à la Chauncey Starr; and step 12, providing the most cost-effective method of achieving or maintaining a level of control.

3. See the comparable propositions put forward by the political scientist Allan Mazur (1973, 1977), who ultimately collaborated with Kantrowitz.

4. Raiffa is the author of a best-seller on negotiation games (Raiffa 1982).

5. "The policy maker has to consider, formally or informally, the alternative actions he or she might pursue, the institutional and political constraints, value and ideological judgments, and so on. This is risk evaluation. It can be viewed as a subset of what some people call policy evaluation or policy analysis" (NRC 1982, 33).

6. In a book that was published during CORADM's mandate, based on this committee's experience (Cohen and Lindblom 1980), Lindblom explained that the power of professional analysis to decide and set rational courses of action was a myth. The knowledge that was used in decisions was not professional, formal, or analytical knowledge.

7. A longer, 700-page report was written by Raiffa, but never actually published. The official report from the committee was a shorter, 82-page document entitled *Risk and Decision Making: Perspectives and Research* (NRC 1982).

8. Risk Analysis Research and Demonstration Act of 1981, H.R.3441—97th Congress (1981–1982), 11.

9. Projects Proposed: Risk Assessment and Federal Regulation Policies, Report No. 96–1030, Agriculture, Rural Development and Related Agencies Appropriation Bill 1981, Archives NAS-NRC Executive Offices Organization.

10. On November 18, 1981, the *Journal of Commerce* published a paper entitled "Gov't Reviewing Cancer Policy." The article quoted the executive director of the synthetic organic chemical manufacturers association as saying that a "pending study from the NAS [is expected] to fuel congressional investigations," especially on whether and how science could be separated from politics. See also *New York Times*, October 11, 1982, "The Calendar": "Risk assessment: How effectively are regulatory agencies working with scientists in assessing health and technology risks? That will be the focus of a three-day conference sponsored by the American Chemical Society and the National Bureau of Standards beginning today at the bureau's headquarters in Gaithersburg, Md. Wednesday."

11. The three academies (of science, engineering, and medicine) have standing committees of scientists, who are in charge of establishing the program of studies and events of the academies in a particular domain of competence. These boards

oversee the completion of studies, the establishment of the panels that will conduct the study. Board members may themselves sit on these panels.

12. NAS does not strictly require experts to be independent, but it selects people whose biases and attachments cancel each other out. Authority and neutrality are less properties of individuals than outcomes to be achieved through collective deliberation. The committee as a whole must appear as credible, authoritative, and speak with one voice. This subtle functioning gives committee chairs a key role. And while there are recipes for a successful committee, it is hard to predict whether a committee will succeed in producing a report consensually.

13. At the same time, Tardiff was playing an instrumental role in prefiguring the Society for Risk Analysis.

14. Ted Greenwood was then an assistant professor at MIT, working precisely on the question of regulatory agencies' use of science. His developing work proved useful to the committee to appreciate what he termed the interplay of "knowledge and discretion." He published a book called *Knowledge and Discretion in Government Regulation,* soon after the committee disbanded (Greenwood 1984).

15. Frank Press, before becoming president of the NAS in 1981, was President Jimmy Carter's science advisor and the administrator of the OSTP.

16. Fred Robbins, Memorandum to Frank Press, "Committee on Institutional Means for Assessment of Risk to Public Health," July 23, 1981, NAS-NRC Archives, RAC 1981–1983 files.

17. With both Omenn, the initial choice for the chair, and Stallones being in some way related to the AIHC or to the positions it advocated, the inference can be made quite safely that the AIHC was involved in suggesting names of experts to the NRC.

18. Interview with the author.

19. The Stanford Research Institute was Stanford University's contract research branch.

20. "Public" and "Industry" are the categories used in NRC documents.

21. Interview with the author.

22. It was clear, moreover, that the leaders of NAS did not want to create this panel within its walls. The panel would have given a quasi-regulatory role to NAS, irreconcilable with the institutional and intellectual independence that was the source of its great reputation and continued success in attracting study requests from the government or elsewhere. In September 1982, six months before the report was released, Stallones and Omenn briefed Lazen about some of the recommendations that were being sketched in the report, notably that of creating a "board for risk assessment methodology" within NAS/NRC, to get Frank Press's feelings about it. They felt that

this recommendation could embarrass the academies. First, it was not within the committee's ambit to develop such a line of recommendations. Second, a board for risk assessment methodologies would certainly represent a shift in the traditional activities of the NAS and the NRC, toward a more active regulatory role. It would necessarily involve some institutional and legal clarifications before being set up.

23. Moffett stated, "I must say in all candor that my reading last evening of the testimony that will be presented here this morning left me more than a little disturbed and disappointed, if not angry, about the performance of EPA during the past several months. There is apparently an awesome litany of retreat from serious environmental problems. […] Oh, yes, we are all in favor of regulatory reform. We all subscribe to the need for regulatory reform. There is a broad coalition indeed for getting rid of unnecessary and unproductive regulations, but what appears to be taking place is something more than that; something more than reform; something more than revision. This seems to be a radical departure. Yes, indeed, a radical or extreme departure from what the American people want and what their elected representatives have expressed legislatively. Perhaps the most important question that we need to explore today is what kind of regulatory mentality now exists at the lead environmental agency?" (US Congress 1982, 1–2).

24. John Todhunter, Memorandum to Administrator Gorsuch through Dr. Hernandez, "Review of Data Available to the Administrator Concerning Formaldehyde and Di(2-ethyl-hexyl)phthalate (DEHP)," US EPA, February 10, 1982 (US Congress 1983b, 248–249).

25. This separation played an important role in decision-aiding science from the outset. The prescription that problems should be clearly formulated before any procedure can be found in most handbooks of policy or systems analysis. According to Fortun and Schweber (1993), the physicists who founded operational research (the first to become involved in military affairs in the United Kingdom, and then in the United States) drew on a discourse of objectivity and neutrality, articulated with a practical reluctance to take the place of decision-makers.

26. Interview with the author.

27. McCray uses the language of "discretion" much more than the report will eventually do, in resonance with the work of Ted Greenwood (1984).

28. The group outlined a set of performance considerations and selected for case analysis: the relationship between the National Institute of Occupational Health and the OSHA, EPA's Scientific Advisory Panel (pesticides), the NRC, the CAG, EPA's Clean Air Scientific Advisory Committee (CASAC), OSHA's division of health standards, and the FDA's public board of inquiry.

29. The review showed that there were considerable variations across regulatory agencies, both in the kind of science they used and in the way they used it. Some

used in-house scientific departments' expertise, while others drew on that of external national research institutes. No one organization seemed better than the others in terms of producing more credible regulatory decisions, and there seemed to be no single best organizational solution for the use of science in regulatory decision-making processes.

30. The contract between the FDA and the NRC, dated September 18, 1981, No. 282-81-8251, stipulates that the committee should delineate the process of risk assessment "in terms of its individual components, identifying and distinguishing those that are scientific in nature from those that are value judgments or policy. In addition, an effort will be made to identify and describe those components that are neither strictly science or policy but a hybrid consisting of elements of both." It should be noted that the NRC was already thinking in terms of a hybrid category of "risk assessment policy" because this was used as a category of expertise to be populated by a nominee, back in June, as the council started to contact potential members. In short, the committee did not invent the hybrid notion of risk assessment policy. It was already available at the start of the process.

31. The document, for instance, stated that the purpose of the recommendation was to ensure consistency in risk assessments within the Office of Water. It also spoke of "risk assessment documents" (not "health effects documents" or "health assessment documents"), as comprising a health assessment, an exposure assessment, and ... a risk assessment.

32. The word *value* appeared only three times in this 191-page report. RAFG went for a different, arguably less prescriptive term: *assumption(s)* (22 occurrences). For Shrader-Frechette, the report could very well have used a more normatively explicit term, like *methodological value judgment* (Shrader-Frechette 1995).

33. Subsequent discussions in the field of risk assessment and risk management revolved around the need to attend to the political aspect of risk decisions through more explicit communication with audiences and the public, or incorporation of the public in the process of making decisions, through deliberative arrangements (NRC 1989, NRC 1996, Stern 2009).

34. Warner North, interview with the author.

35. The article covered the positions of a number of attendees, such as Warren Muir of John Hopkins University, former director of the Office of Toxic Substances at the EPA (guidelines are beneficial; toxicologists want to exert professional judgments, but use unspoken assumptions); Nathan J. Karch, former staff member of the council on environmental quality (explicit criteria of risk will ensure consistency and lead to a more rational process of decision-making); Sherwin Garder of the Grocery Manufacturers of America (generic guidelines are possible for risk assessment; key is peer review); and Thomas Grumbly, former staff director of the House Science

Investigations and Oversight Committee (too many differences between agencies; focus away from a central board for risk assessment and toward a "let 100 flowers bloom" attitude in regulatory science; avoid separation). Karch and Swanson, of the American Petroleum Institute, concurred, demonstrating that the AIHC proposal was not so widely shared and that there were many people around Washington with a positive view of how agencies were doing their job.

36. Anonymous, "Schedule: Risk Assessment Study," 1982, NAS-NRC Archives, RAC 1981–1983 files.

37. Jasanoff (1992) also shows that the report was ambiguous as to its practical recommendations for handling the relation between risk assessment and risk management.

38. Terry Davies, interview with the author.

39. Academic panel members have often described this review process as the most rigorous that they have ever experienced.

40. Reuel Stallones, "Letter to Fellow Committeepersons," February 1, 1983, NAS-NRC Archives, RAC 1981–1983 files.

41. Al Lazen, "Letter to Frank Press," September 9, 1982, NAS-NRC Archives, RAC 1981–1983 files.

42. Various committee members, interviews with the author.

43. David Lazen, "Letter to Reuel Stallones," April 18, 1983, NAS-NRC Archives, RAC 1981–1983 files.

Chapter 5

1. See footnote 2 of the Introduction.

2. Larry McCray, Memorandum to Members of the Committee on the Institutional Means for Assessment of Risk to Public Health, "Recent events," April 18, 1983, NAS-NRC Archives, Risk Assessment Committee 1981–1983 files.

3. Joe E. Penick, letter to Philip Smith, executive officer of the National Research Council, March 31, 1983, NAS-NRC Archives, Risk Assessment Committee 1981–1983 files.

4. See also Ted Greenwood, letter to Larry McCray, April 1983, NAS-NRC Archives, Risk Assessment Committee 1981–1983 files.

5. After leaving the EPA in the spring of 1973, as the Watergate scandal was mounting, Ruckelshaus was called by Nixon to become acting director of the Federal Bureau of Investigation (FBI). He then moved to the second-highest position in the

US Department of Justice, but he was soon forced to resign after he refused to fire special prosecutor Archibald Cox. While Gorsuch was head of the EPA, between 1981 and 1983, Ruckelshaus worked at the timber company Weyerhaeuser as Senior Vice President of Legal Affairs.

6. William Ruckelshaus, interview with the author. Linda Nash mentions that Ruckelshaus's law firm had the AIHC as a client at the end of the 1970s (Nash 2017).

7. Bernard Goldstein, interview with the author. As mentioned by a member of EPA staff who worked closely with him in the early 1980s, Ruckelshaus "was a very very skillful, a very very influential thinker, and he would have been familiar with some of these conversations and debates already, both because of his work in public service but also in the private sector" (Terry Yosie, interview with the author). Ruckelshaus confirmed that he "had done some thinking about that problem for some time, because early on at the EPA, we did that kind of rigorous analysis" (William Ruckelshaus, interview with the author).

8. William Ruckelshaus, interview with the author.

9. In December 1983, the *National Journal* mentioned the June speech and the "Ruckelshaus approach" (Mosher 1983): "Specifically, Ruckelshaus wants EPA and other regulatory agencies to beef up the science of their risk assessment procedures. But he has also pressed for legislation that would give the regulatory agencies 'a common statutory formula' for clarifying how risks are to be managed.... The EPA chief signaled his intention to press for such changes in a speech to some 150 members of the National Academy of Sciences on June 22."

10. The speech was republished in several journals over time, including in *Science* (Ruckelshaus 1983). This version has been cited 200 times to date.

11. See also this excerpt of another speech delivered to the chemical industry: "We have shifted our attention and concern from problems that are relatively easy to see and solve, to those that are subtle and vexing, from smoke and sewage to the attempt to eliminate toxic substances from the human environment. I can tell you that this shift has caused substantial problems for EPA ... controlling toxics is an entirely different sort of business. Where we once dealt with a dozen or so pollutants, we now must consider hundreds. Before, we concentrated on removing familiar substances by the ton; now, we often must cope with the exotic and worry about micrograms because we are confronted with materials that may be able to cause serious human health damage in vanishingly small concentrations," William Ruckelshaus, "Our Challenge," Remarks at the Semi-annual Meeting of the Chemical Manufacturers Association in New York City, November 8, 1983. See also Anonymous (1984a).

12. Anonymous ORD risk assessment scientist, interview with the author.

13. There is evidence that Ruckelshaus was modifying his messages to suit his specific audiences, drawing on the variations that the risk assessment–risk management framework allowed. Elsewhere, he had stressed that ignorance, rather than science, characterized the conditions in which the EPA had to make decisions: "We may ask, why is rational argument less than convincing in discussions about toxic chemicals? Why aren't the judgments more in line with our calculations? I believe it is because public concern is centered on those dreaded diseases that are plausibly connected with low concentrations of toxic substances; that is, cancer and the genetic and reproductive disorders, and because, at the heart of our risk assessment, there is, undeniably, a hollow place. I think people sense that we really don't know how and under what circumstances chemicals cause cancer; they're right, we don't." He went on to liken his action on cancer to British water hygiene policies in the nineteenth century, citing them as two cases of "action in the face of ignorance." This was certainly a daring comparison to make before proponents of risk assessment, which generally decry any sort of action that goes beyond the science (or what is now known under the rubric of precaution). "I think people sense that we really don't know how and under what circumstances chemicals cause cancer; they're right, we don't" (Ruckelshaus 1984; see also Anonymous [1984a]).

14. In the same speech, he declared the following: "At EPA we have tried to disentangle risk assessment, as a process, from the policy considerations that go into making a final decision about regulating a substance, which we call risk management. I realize there is not an obvious bright line between the two; still, I believe that good public policy obliges us to make it as bright as we can." So, from the point of view of the due administrative process of dealing with risks, the subtle definitions crafted in RAFG were "somewhat of a nuance" (William Ruckelshaus, interview with the author). From a public standpoint, in the aftermath of the Gorsuch crisis, it was simply "important to let people know that there was a scientific part of trying to determine what the nature of the risk was, that it should in no way be subject to political interference, or there should be no pressure put on scientists" (ibid.).

15. Lee Thomas, memo to Milton Russell, draft final report on risk assessment/risk management. Office of Solid Waste and Emergency Responses, July 31, 1984.

16. The EPA webpage summarizing the history of the reduction of air pollution from transportation, and the EPA's actions in the area, state that "EPA vehicle emissions standards directly sparked the development and implementation of a range of technologies. The automotive catalytic converter, in particular is considered to be one of the great environmental inventions of all time." See epa.gov/air-pollution-transportation/accomplishments-and-success-air-pollution-transportation (accessed on June 10, 2016).

17. In August 1983, Ruckelshaus was interviewed by the *National Journal*. The published interview was preceded by an introduction stating that "his [Ruckelshaus's] most cherished goal is to streamline the assessment and management of

environmental risks throughout government, and that will probably include introducing considerations of cost in setting air quality standards." Ruckelshaus reiterated that "the difference between assessing the risk and managing must be carefully made," applying that principle right away to issues such as dioxin and risks from fine particles in the air, and then on the agenda of the EPA, the press, and others. On dioxin, he stated: "There are two issues. The first is to define the risk; what are the health effects of dioxins? This is a separate scientific process and should have nothing to do with determining what to do about it. If we find dioxins in the soil in some remote site, that may suggest one solution. If you find it in the middle of a city where many people are exposed to it, and there are ways of disposing of it that are safe and relatively cheap, that suggests another strategy. We simply have to have a more flexible management strategy than just saying a certain level is unacceptable under every condition." Clearly, here, risk management implied that policy decisions were informed by a judgment of what an acceptable risk actually was, not by an absolute level of risk protection. Moreover, solutions and strategies could vary depending on the conditions. The problems were to be managed according to the way they emerged in a given context, much more than being sorted out in absolute terms.

18. Milton Russell, Memorandum to the Administrator, "Interagency Liaison," EPA Office of Policy, Planning, and Evaluation, September 2, 1983, Milton Russell Special Collection.

19. William Ruckelshaus, letter to other heads of agencies, "Proposal for an Interagency Coordination on Risk Management," September 21, 1983, Milton Russell Special Collection.

20. It took less than a year for the White House to kill the initiative once more, by launching a competing Cabinet Council Working Group on Risk Assessment under the aegis of the Council on Environmental Quality. Ruckelshaus was offered the chair of this council as well. See the hearing of 1984 (US Congress 1984) for positive comments from various agencies on the Interagency Risk Management Council project.

21. Monte C. Throdahl, chairman of AIHC's science policy task force, wrote to Ruckelshaus in November 1983 to ask the EPA to lend support to the bill.

22. H.R.4192—"A bill to establish coordinated interagency research and demonstration projects for improving knowledge and use of risk assessment by those Federal agencies concerned with regulatory decisions related to the protection of human life, health, and the environment, and to provide for the establishment of a Central Board of Scientific Risk Analysis as a means of improving the scientific review and evaluation of risk analyses made by Federal agencies, with particular emphasis upon risk analyses involving issues of chronic health hazards." 98th Congress (1983–1984), https://www.congress.gov/bill/98th-congress/house-bill/4192.

23. Ted Greenwood, letter to Larry McCray, April 1983, NAS-NRC Archives, Risk Assessment Committee 1981–1983 files.

24. Following her assessment of the initial Ritter Bill, dated June 3, 1982, and in her testimony on June 23, 1981, at the House's Subcommittee on Operations, Research and Foreign Agriculture, she defended views that were highly concordant with those of the RAC. She criticized the definitions laid out in this bill, which distinguished between risk assessment and risk evaluation, and was not in line with emerging definitions of RAFG (even though it was not yet published, which meant she followed developments of the committee's work). She was happy with the general support that the bill gave to the development of risk assessment. But her comments consisted of putting the bill back into the context of what the EPA did and current developments of risk assessment in the federal government at large, which the bill seemed to ignore. The bill suggests that the OSTP should coordinate a program for improving and facilitating the use of risk analysis—a system that Anderson deemed unclear and burdensome for agencies and the OSTP. Elizabeth Anderson, memorandum to Bernie Goldstein, assistant administrator for the ORD, "Comments on the Ritter Bill," March 6, 1982.

25. Arnold M. Kuzmack, director, Office of Program Development and Evaluation, Office of Drinking Water, memorandum to Peyton Davis, Policy and Strategic Analysis staff, Office of Water, "Draft Bill to Create an Independent Science Panel within the NAS," EPA, November 4, 1983, Milton Russell Special Collection.

26. Richard Hill, Science Advisor to the assistant administrator for OPTS, "Amendment to HR3840 offered by Mr Martin," October 26, 1983, Milton Russell Special Collection; Milton Russell, special assistant to the administrator, memorandum to Josephine Cooper, assistant administrator for external affairs, "Martin Amendment to Ritter Bill (HR3840) on Central Board for Scientific Risk Analysis," Office of Policy, Planning and Evaluation, US EPA, October 20, 1983, Milton Russel Special Collection.

27. Don Ritter and David Martin eventually withdrew Title II of their bill, the most controversial part of it, instituting a board for scientific risk analysis. The White House also opposed the bill and lent support to interagency coordination initiatives.

28. Milton Russell, memorandum to Alvin Alm, deputy administrator, "First Thought Piece on What EPA Could or Should Do on Risk Communication," January 17, 1984, Milton Russell Special Collection.

29. Milton Russell, "Communicating with the Public on Issues of Environmental Risk: Issues and Options," Draft paper prepared for Deputy Administrator Alvin Alm, January 17, 1984, Milton Russell Special Collection.

30. Roger Gale, special assistant to the administrator, memorandum to EPA Administrator William Ruckeslshaus, through Milton Russell, "Recommendations of the Risk Working Group," Office of the Administrator, EPA, March 27, 1984, Milton Russell Special Collection.

31. Arsenic was one of these high-visibility chemicals that posed recurrent problems to the EPA. The regulation of this substance was forced on the agency by the result of a suit opened by the Environmental Defense Fund, which wanted the EPA to list

arsenic as a hazardous air pollutant under the Clean Air Act. In June 1980, the Air Office followed suit, making arsenic the seventh substance on that list. By January 1983, however, the agency was ordered by the Federal District Court of Manhattan to set a standard for arsenic within six months (the agency had failed, purposely or not, to meet several past deadlines). When he joined the agency in May 1983, Ruckelshaus found that dossier on his desk and quickly decided to make a decision on this issue that was so symbolic of the governing difficulties generated by these disputed chemicals.

32. Handling the meeting proved extraordinarily costly and difficult for an EPA staff that had little training in direct deliberation with the public. Experience was limited to formal hearings, with prepared testimonies and limited on-the-spot responses. The hearings also proved very resource intensive for the regional branch in charge. Exchanging with the public appeared difficult and did not unroll as Ruckelshaus had imagined. Eager to apply a risk assessment/risk management distinction, where the public would mostly discuss risk management and possible future solutions to apply, he did not allow a full discussion of the EPA's scientific assessment of the risk levels or of the accuracy of the estimates derived from dispersion models used by engineers. At the same time, risk management was not much of an issue either. The public represented there was concerned about deindustrialization, the anticipated closure of industries, and the need to attract jobs. The discussion soon turned to local industrial policy and the need to gain more jobs from less-polluting industries in the area—issues for which the EPA was hardly competent (Reich 1985; Gutmann and Thompson 1998; Heifetz 2009). After Tacoma, Ruckelshaus's special assistant on risk, Roger Gale, tried to moderate the use of risk-related categories in public circumstances. Claims about risk inherently derived from considerations about the value of life, which he felt was slippery ground for the EPA. Moreover, the agency would not be able to handle such a generic issue alone, given its statutes and the role of other agencies. Finally, risk issues are fundamentally complicated and unappealing to the public and the business community. This notion of risk, he felt, overintellectualized problems, leading to near blackmail of the public, as in Tacoma. Roger W. Gale, special assistant to the EPA administrator, "Risk Assessment and Risk Management," December 6, 1983, Milton Russell Special Collection.

33. Roger Gale, special assistant to the administrator, memorandum to EPA Administrator William Ruckelshaus, through Milton Russell, "Recommendations of the Risk Working Group," Office of the Administrator, EPA, March 27, 1984, Milton Russell Special Collection.

34. Roger W. Gale, special assistant to the EPA administrator, memorandum to the members of the Risk Working Group, "Review of Revised Draft of Risk Working Group Memo," March 14, 1984, Milton Russell Special Collection.

35. Derry Allen and Richard Morgenstern, Office of Policy Analysis, to Milton Russell, assistant administrator for the Office of Policy Planning and Evaluation,

"Briefing Memo for This Afternoon's Meeting with the Administrator on Risk Communication Activities," April 1, 1984, Milton Russell Special Collection.

36. Paul Slovic, Memorandum to Milton Russell, "Putting Risks in Perspective," Decision Research—A Branch of Perceptronics, March 1984.

37. The report *Improving Risk Communication* did not itself present such a graph, but an amended framework was soon to appear with the "risk communication" module incorporated, a first occurrence of which can be found in subsequent publications of the World Health Organization (WHO).

Chapter 6

1. Richard Morgenstern, memorandum to Alvin Alm, deputy administrator, and to Milton Russell, special assistant to the EPA administrator, "Agenda of the First Meeting of the Toxics Integration Task Force," June 2, 1983, Milton Russell Special Collection.

2. Ibid.

3. Ruckelshaus also brought back with him a number of advisors and colleagues that he knew he could work with efficiently, such as Jim Barnes, Howard Messner, Jack Raven, and Phil Angel.

4. John Todhunter, memorandum to Richard Hill, "Adoption of NAS Recommendations with Regard to Risk Assessment," March 11, 1983, NAS-NRC Archives.

5. Interview with the author.

6. Richard Hill, Office of Toxic Substances, memorandum to John Moore, assistant administrator for the OPTS, "A Regulatory Decision-Making Nosology," US EPA, January 12, 1984, Milton Russell Special Collection.

7. By his own admission, Goldstein did not handle the interview at the White House ideally, failing to demonstrate the Republican credentials that he did not have anyway (interview with the author).

8. In May 1984, he spoke once again with Gil Omenn at a conference, discussing in depth the various recommendations of the report as it concerned guidelines and a methodological risk assessment board.

9. Quite emphatically, the same staff member considered that Ruckelshaus and Alm were "very charismatic, thoughtful, brilliant leaders," the kind of people who "surround themselves with very capable people of independent stature, and when you put that caliber of independence and stature together, what you get is the ability to think through decisions, not just in terms of the next decision, but also more strategically, over longer periods of time, and I think that's what Ruckelshaus and his team did" (interview with the author).

10. In April 1983, Larry McCray had briefed senior people of the ORD and the Office of Resources and Policy Management of the agency. Later, on June 10, 1984, Corn, North, and McCray briefed the Environmental Health Committee of the EPA's SAB. In July 1983, the representatives of the AIHC and Terry Davies met Ruckelshaus, the EPA administrator, to describe a legislative proposal for application of scientific peer review to agency risk assessment.

11. After leaving the EPA in 1985, Alm served in top positions in a number of management consultancies. The *New York Times* obituary for Alm mentioned that he "operated just under the bureaucratic rank where public service translates into renown"—another way of saying that his reputation did not quite match the successes that were attributed to him and his actions in the various organizations he served (Cushman 2000, A19).

12. Alvin Alm, memorandum to the general counsel, assistant administrators, and regional administrators, "My Preferred Approach to Management at EPA," EPA, August 18, 1983, Milton Russell Special Collection.

13. Ibid.

14. He became assistant administrator after clearance from the White House, and after testifying before Congress in October 1983, like the other people picked by Ruckelshaus and Alm, such as Bernard Goldstein and Jack Moore.

15. Milton Russell, interview with the author.

16. Ibid.

17. Ibid.

18. Ibid.

19. A similar process had been instituted at the White House during the Ford administration, with the participation of Alm. The then called "options paper process" was managed by the White House chief of staff, Dick Cheney, known at the time as "Mr. Straight" (Milton Russell, interview with the author).

20. Interview with the author.

21. Goldstein stayed as head of ORD until 1986, and Russell remained at the OPPE until 1988. Moore stayed at the EPA until 1989. He also acted as deputy administrator that year.

22. John Moore, assistant administrator for pesticides and toxic substances, memorandum to Alvin Alm, deputy administrator, "Candidates for Options Selection Process," November 10, 1984, Milton Russell Special Collection.

23. Milton Russell and Roger Gale, memorandum to William Ruckelshaus and Al Alm, "Report of the Risk Working Group," March 28, 1984, Milton Russell Special Collection.

24. Both phrases were articulated by Alm. Milton Russell, memorandum to Alvin Alm, "Options Review Process," March 28, 1984.

25. Richard Morgenstern, "Toxics Integration Task Force: Final Report Outline," Office of Policy Analysis, EPA, April 1984, Milton Russell Special Collection.

26. Alvin Alm, memorandum to Milton Russell, "Draft Memorandum on Criteria and Guidelines for Review of Agency Actions," September 16, 1983, Milton Russell Special Collection.

27. The guideline continued the substance of what had emerged during the Carter administration through the Regulatory Quality Council (for more information, see chapter 3, on Costle's regulation development memo).

28. The contents of the format were organized by the following rubrics: Name and type of regulation/action; statutory decision/criteria; target of control action; primary object protected (human health, aquatic organism, etc.); economic impact; major uncertainties; and other important considerations governing decisions. Three annexes with data are cost-effectiveness estimates, hazard assessment, and exposure assessment. The cost-effectiveness key includes control options, data on maximum individual risk, data on aggregate population risk, and benefits other than risk reduction.

29. Milton Russell, assistant administrator, OPPE, memorandum to Alvin Alm, deputy administrator, "Procedures Regarding OMB Contacts," January 9, 1984, Milton Russell Special Collection.

30. For instance, the OPPE sided with the OPTS to defend the "New Chemicals" program against recurrent OMB accusations of excessive conservatism and bias with regard to new chemicals.

31. Interview with the author.

32. In this case, much like the formaldehyde case, Todhunter had asked an EPA statistician to alter her estimates for EDB using a theory that she had never seen before—namely, the fact that risk levels decline exponentially with decreasing exposure time.

33. Milton Russell, memorandum to Alvin Alm, "Regulatory Development," October 1, 1984, Milton Russell Special Collection.

34. Ibid.

35. Don R. Clay, acting assistant administrator for pesticides and toxic substances, memorandum to Dan Beardsley, director, Integrated Environmental Management Program, "Interagency Chemical Strategies," November 14, 1983, Milton Russell Special Collection.

36. Dan Beardsley, director, Integrated Environmental Management Program, memorandum to Don Clay, director of the Office of Toxic Substances, "Your Memo

on Interagency Chemical Strategies," November 30, 1983, Milton Russell Special Collection.

37. Milton Russell, memorandum to William Ruckelshaus, through Alvin Alm, *Report on Risk Assessment and Risk Management,* August 13, 1984, Milton Russell Special Collection.

38. Milton Russell, memorandum to all assistant administrators, *Report on Risk Assessment and Risk Management,* July 6, 1984, Milton Russell Special Collection.

39. John Moore, assistant administrator for pesticides and toxic substances, memorandum to Milton Russell, assistant administrator for policy, planning and evaluation, "Comments on the Draft Final Report on Risk Assessment/Risk Management," July 26, 1984, Milton Russell Special Collection.

40. Milton Russell, interview with the author.

41. Ibid.

42. John A. Little, deputy regional administrator for region IV, and Peter Preuss, director, Office of Regulatory Support, ORD, "Risk Assessment Review," December 12, 1985, National Archives and Records Administration, Records of the Environmental Protection Agency (Record Group 412).

43. Ibid.

Chapter 7

1. Ernest F. Cloyna, memorandum to the administrator, William Ruckelshaus, "Report of the Laboratory Organization Review Group," Science Advisory Board, July 28, 1983, National Archives and Records Administration, Records of the Environmental Protection Agency.

2. Bernard Goldstein, memorandum to ORD laboratory and center directors, "Research and the Regulatory Agenda," April 6, 1984, National Archives and Records Administration, Records of the Environmental Protection Agency.

3. Ibid.

4. The SAB is comprised of external scientists, but the agency also has a staff director who is in charge of coordinating the work of the board with the agency's priorities.

5. "Historically, ORD program managers and the OMB have viewed the short-term and long-term components of EPA research as pursuing incompatible or unrelated objectives. In reality, both can serve to support EPA's fundamental mission: to identify, assess and abate the risk of pollution to public health and the environment. Viewed in this context, the strategic mission of ORD's research program, in both the current fiscal year and over a longer time frame, is to advance and develop

the scientific and technical basis for risk assessment and risk management." Ernest Cloyna, *Report of the Laboratory Organization Review Group to the Administrator,* July 28, 1983.

6. Terry F. Yosie, memorandum to the administrator, William D. Ruckelshaus, "Making Effective Use of the Science Advisory Board," November 2, 1983, National Archives and Records Administration, Records of the Environmental Protection Agency.

7. Ibid.

8. "The Agency's referral of studies and assessments to the SAB for peer review preceded, but is consistent with, the recommendations of the National Academy of Sciences in its report on risk assessment in the Federal government. A major recommendation of this report was for regulatory agencies to create independent peer review panels to review scientific studies that form the basis for major agency regulatory decisions" (Yosie 1987, 6).

9. Alvin Alm, memorandum to assistant administrators, general counsel, inspector general, regional administrators, "Risk Assessment," May 21, 1984, Milton Russell Special Collection.

10. Interview with the author.

11. Milton Russell, assistant administrator, Office of Policy, Planning, and Evaluation, memorandum to Alvin Alm, deputy administrator, "Problems in Risk Assessment, Risk Management, and Risk Communication," October 5, 1983, Milton Russell Special Collection.

12. Anonymous ORD risk assessment scientist, interview with the author.

13. Alvin Alm, deputy administrator, memorandum to assistant administrators, general counsel, inspector general, and regional administrators, "Risk Assessment," May 21, 1984, Milton Russell Special Collection.

14. Rodrick's business is now a large, multinational risk assessment consultancy called Environ.

15. Alvin Alm, deputy administrator, memorandum to assistant administrators, general counsel, inspector general, and regional administrators, "Risk Assessment," May 21, 1984, Milton Russell Special Collection.

16. In the internal newsletter called the *Risk Assessment Review,* Peter Preuss of the OHEA (one of the main risk assessment shops of the agency) wrote a short article about this new Risk Management Council, noting, "Recently the Administrator has established a number of 'Risky' structures at EPA whose profusion can bring confusion to the eyes of even the most battle-hardened bureaucrat." For Lee Thomas, the main objective of the Risk Management Council was to have a place in which to

discuss and institute a cross-media perspective in the normal regulatory operations of the agency. The council was to identify such cross-media, multioffice issues and promote common management strategies.

17. Alan Ehrlich, 1988. "Advances in Risk Assessment Guidelines." Presentation at the 34th Anniversary Technical Conference Mid-Atlantic States Section Air Pollution Control Association, Milton Russell Special Collection.

18. An ADI is a computation made for regulatory purposes, consisting of dividing by a safety factor of 10 or more, the lowest possible dose at which a chemical substance shows no adverse activity in the body of an experimental animal. The computed dose is deemed acceptable as it is presumed that a human person can be exposed to it on a daily basis during his or her lifetime without developing any serious adverse effects.

19. Donn J. Viviani, Chief Regulations Analysis Branch, memorandum to Al Jennings, director, Chemical and Statistical Policy Division, "Haven't We Met Before?," August 20, 1985, Milton Russell Special Collection.

20. Anonymous ORD risk assessment scientist, interview with the author.

21. IRIS is an information technology database containing agreed-upon risk values for a wide range of chemicals, accessible by all regulatory offices, initially by internal email, aiming at improving the consistency of risk assessments across the agency. It has become one of the international databases of risk estimations of reference worldwide.

22. Guidelines provide an explicit point of reference, against which discrepancies between regulatory offices can be measured and explained too, though not necessarily reduced: "[P]reparation of the guidelines was an instrument for addressing both an internal administrative problem and an external political problem. Within the agency, they could serve as an instrument to control how various offices used scientific data and how they recommend policy choices to the administrator. Externally, they represented one of Washington's most venerable principles: be sure that your potential adversaries (be they industrial firms, environmentalists, Congress or the Office of Management and Budget) debate your ideas, and your agenda" (Yosie 1989, 3).

23. Dorothy Patton, interview with the author,

24. Milton Russell, special assistant to the administrator, memorandum to Alvin Alm, deputy administrator, "Expected Reactions to Proposed Memorandum on Criteria and Guidelines for Agency Review," September 16, 1983, Milton Russell Special Collection.

25. Paul Ehrlich, memorandum to Bernard Goldstein, assistant administrator for the Office of Research and Development, and Elizabeth Anderson, director of OHEA, "Progress report on guidelines," June 21, 1984, Milton Russell Special Collection.

26. Richard Nill, memorandum to Jack Moore, assistant administrator, Office of Pesticides and Toxic Substances, "A Regulatory Decision-Making Nosology," January 12, 1984, Milton Russell Special Collection.

27. Ibid.

28. Ibid.

29. Milton Russell, memorandum to the administrator, "Equity and the Choice of Risk Measures," Office of Policy Planning and Evaluation, EPA, June 7, 1984, Milton Russell Special Collection.

30. Milton Russell, memorandum to all assistant administrators, "Requirements for Exposure-Related Analysis," May 8, 1984, Milton Russell Special Collection.

31. Alvin Alm, memorandum to all assistant administrators, regional administrators, and general counsel, "Accuracy in Risk Assessment," January 1, 1984, Milton Russell Special Collection.

32. Anonymous ORD risk assessment scientist, interview with the author.

33. This means that the experiments in animals did not show cancer in all species tested, or only at certain doses, and tumors were difficult to analyze as malignant or benign.

34. Milton Russell, memorandum to Alvin Alm, deputy administrator, "Carcinogenicity, Strength of Evidence and the Need for Agency-wide Consistency," Office of Policy, Planning, and Evaluation, March 16, 1984.

35. Ibid.

36. Paul Milvy, Office of Drinking Water, memorandum to Al Jennings, Office of Policy, Planning and Evaluation, "The Three Conclusions Reached by CAG on the Carcinogenicity of Perchlorethyelene (PCE) and the 3/8 Memorandum Entitled 'Strength of Evidence of Carcinogenicity—a Need for a Consistent, Agency-Wide Policy,'" March 9, 1984, Milton Russell Special Collection.

37. Russell to Alm, March 16, 1984, attachment I.

38. Ibid.

39. Elizabeth Anderson, director of OHEA, memorandum to Richard Morgenster, director of Office of Policy Analysis, "Weight of Evidence in Decision-making," US EPA, March 28, 1984, Milton Russell Special Collection.

40. Russell to Alm, March 16, 1984, attachment I.

41. Bernie Goldstein, memorandum to the administrator, William Ruckelshaus, and deputy administrator, Alvin Alm, "The Presence or Absence of Thresholds in

Chemical Carcinogenesis: Scientific Issues," October 11, 1983, Milton Russell Special Collection.

42. Terry Yosie, interview with the author.

43. Elizabeth Anderson, director of Office of Health and Environmental Assessment, memorandum to Alvin Alm, deputy administrator, "Characterizing Cancer Risk Quantitatively," June 8, 1984, Milton Russell Special Collection.

44. William Ruckelshaus, administrator, "Letter to EPA Program Managers," November 23, 1984, Milton Russell Special Collection.

45. The guidelines, as agreed between Anderson and the SAB, would be reviewed later by the board, but they were to be applied right away by each office.

46. During the press conference for the launch of the guidelines, Bernard Goldstein said, "The overall theme is to improve accuracy" (Anonymous 1984b).

Chapter 8

1. Ruckelshaus joined a company in an industry that the EPA regulated. In 1987, he joined the waste disposal company Browning-Ferris as director, soon after the EPA announced that it would sue the company for violations at a Louisiana landfill. This led to a settlement between Browning-Ferris and the EPA, which was considered controversial by some commentators. After the settlement, Ruckelshaus was named president and CEO. Phil Angell, who had been his close adviser during both of his mandates at the EPA, was Browning-Ferris's spokesman.

2. Milton Russell, memorandum to the administrator, Lee Thomas, and to the deputy administrator, James Barnes, "OPPE Role in Decision Process," October 21, 1985, Milton Russell Special Collection.

3. Lee Thomas, "Assessing and Managing Risks in the Real World," address before the National Petroleum Refiners Association, San Antonio, TX, March 25, 1985.

4. Russell to Thomas, October 1985, Milton Russell Special Collection.

5. Daniel Beardsley, memorandum to Milton Russell, "Application of Risk Concepts," September 10, 1984, Milton Russell Special Collection.

6. A special review is initiated by the Office of Pesticides when it suspects that an already-registered product has "unreasonable adverse effects on people or the environment" (https://www.epa.gov/pesticide-reevaluation/reregistration-and-other-review-programs-predating-pesticide-registration, last accessed December 11, 2018). The review involves evaluating existing data, acquiring new information and/or studies, assessing the identified risk, and determining appropriate risk reduction measures; it may result in the cancellation of the registration.

7. A pilot effort had been attempted at the end of 1984, described as a "quick and dirty analysis" of all efforts related to risk reduction across programs. But this was done by the OPPE alone, without cooperation with program offices, and the office focused on cancer risks only (see EPA 1984a).

8. Milton Russell, interview with the author.

9. Alar is a pesticide that some studies had proved carcinogenic at the maximum tolerated dose, and that the EPA had informally placed under special review at the end of the 1970s. In 1980, after negotiations with the manufacturer, the idea of a formal special review was abandoned, and the manufacturer was asked instead to reregister the product. In 1984, however, the CAG produced a "Health and Environmental Effects Profile" for 1,1 Dimethylhydrazine (EPA/600X-84/134), with a quantitative potency estimate, confirming its carcinogenic nature. The Office of Pesticides reinitiated the special review and the cancellation procedure in 1985. It produced a draft position document, reasserting that alar was a probable human carcinogen (EPA 1985a). The SAP, which reviews all rules prepared by the pesticides office, criticized the draft position document. The panel delayed the action by arguing that there was insufficient experimental data available to perform a quantitative risk assessment or potency estimate as CAG did, and that more cancer studies should be performed. (The SAP opinion was critically regarded a couple of years later, when it surfaced that many of the members of the group had worked with the pesticide industry as consultants.) In 1986, the EPA followed the opinion of the SAP, announcing that Uniroyal could continue marketing the product, but the company would have to collect and frequently report monitoring data and chronic toxicity test results. For more on this, see the early and detailed account presented in Jasanoff (1987).

10. Terry Davies, interview with the author.

11. H.R. 2910, Mercury Environmental Risk and Comprehensive Utilization Reduction Initiative, 105th Congress (1997–1998).

12. The White House memo of January 20, 1992, on "Reducing the Burden of Government Regulation," with its ambition "to weed out unnecessary and burdensome government regulations" and issue a 90-day moratorium on all regulatory proposals, testifies to this stance (White House 1992). When Clinton came to office, he asked that those regulations that were on hold (there were several dozen of them, apparently) be returned to agencies to be reviewed by a Clinton appointee, to effectively end the de facto embargo that the OIRA/OMB created on all new pieces of regulation.

13. Anonymous policy analyst, Office of Policy Analysis, interview with the author.

14. Anonymous economist, Office of Policy Analysis, interview with the author.

15. H.R.4306, the Risk Assessment Improvement Act of 1994, 103rd Congress (1993–1994).

16. This was part of H.R.1814, the Environmental Research, Development, and Demonstration Authorization Act of 1995.

Chapter 9

1. The new estimate was that exposure to one-tenth of a pictogram per kilogram of body weight of dioxin every day over seventy years created a one in a million cancer risk. The previous estimate was that such a risk materialized at a much lower dose (in cases of exposure to one six-thousandth of pictogram per kilogram of body weight) (Powell 1999).

2. The debate revolved around the qualification as adverse or harmful of the (not contested) respiratory effects in conditions of low concentration/durable exposure, as opposed to high-concentration/short-term exposure.

3. Interview with the author.

4. PBPK is a method that helps specify the doses at which a given substance may be found in particular organs of the body. It is a modeling method by which the various organs and flux (e.g., blood, air) comprising the human physiological system are described in a set of equations. It helps to make a more precise determination concerning the doses at which a given chemical is found in the body of a human, based on experiments in animals, without applying protective uncertainty or safety factors that the standard methodology of risk assessment uses to err on the safe side.

5. Yosie, the former EPA staff director for the SAB, left the agency in 1988 to join the American Petroleum Institute.

6. The reauthorization of the Clean Air Act had been on the agenda of Congress since the beginning of the 1980s, but it failed on several occasions between 1981 and 1988. However, under the influence of a renewed public concern about environmental protection and the commitment of George H. W. Bush during the presidential election campaign in 1988 to strengthen air pollution control, Republican opposition in Congress to a revision of the Clean Air Act diminished (Oren 1991; Vogel 2012). The adoption of the Clean Air Act amendments in 1990 benefited from these circumstances.

7. These terms evoke those of economist Lester Lave, a key scholar in the professional field of risk analysis in the United States, during a hearing on environmental issues before the Subcommittee on Transportation and Hazardous Materials of the Committee on Energy and Commerce, House of Representatives, in November 1993 (US Congress 1994a, 45).

8. This was a popular comparison in books and articles of that time about the EPA and failed risk-risk trade-offs. See, for instance, Nichols and Zeckhauser (1986). Justice Stephen Breyer drew from these cases and from the OMB report in his 1993 book (Breyer 1993).

9. Throughout the 1980s and 1990s, Abelson published dozens of editorial papers attacking risk assessment as performed, or as appearing to be performed, by the EPA.

10. That is, other than those that qualify as "criteria pollutants" under title I.

11. S.1630—Clean Air Act Amendments of 1990, 101st Congress (1989–1990), 83.

12. Ibid., 83. For more on this point, see Graham 1985, 122, fn148.

13. See chapter 3 on Wilson's participation in an EPA conference on behalf of the industry in order to criticize the linear approach.

14. Thornton also writes: "I might have called this model [the NRC paradigm] the acceptable discharge paradigm, or the pollution control paradigm, or the technocratic paradigm; all of these names refer to essential elements of today's regulatory system" (Thornton 2001).

15. For environmental critics, the adoption of quantitative risk assessment, coupled with cost-benefit analysis, was deregulation in scientific disguise. In their eyes, Ruckelshaus was essentially pursuing the same Reaganite agenda: "EPA needed to find a way to continue the substance of the Gorsuch-Hernandez deregulation efforts but with the appearance of scientific 'objectivity.' William Ruckelshaus, then in his second turn as EPA administrator, took the risk assessment efforts developed by EPA, FDA, and NRC over the previous 15 years, added cost-benefit analysis (under the rubric of risk management), and proclaimed his agency's commitment to protecting public health and the environment in a 'cost-effective way.' The result was an 'objective' methodology, supported by the National Academy of Sciences, that enabled the EPA to set permit levels and clean-up levels for toxic substances. QRA theoretically gives them a method of determining what levels of pollution or pollutants are 'acceptably safe' without having to evaluate available technology, alternative processes, alternative substances, or community concerns" (Ginsburg 1997, 230).

16. In a subsequent paper, Goldstein more explicitly stated that the separation was a myth: "A myth about the early development of risk assessment was that it was believed to be fully independent of risk management. The simple dichotomy between risk assessment as science and risk management as policy has never really existed." He argued that the separation was barred first by the existence of guidelines, and then by practices of risk characterization and risk communication, by which all of the choices, assumptions, and uncertainties pertaining to the assessment of risk were shared by risk managers (Goldstein 2005, 141).

17. In its 1993 report *Researching Health Risk*, the Office of Technology Assessment (OTA) had reached the conclusion that "risk managers should be in contact with risk assessment researchers, the developers of methods and new approaches. The report sees the managers, who must apply the results of risk assessment, as becoming aware of where research might reduce reliance on assumptions, highlight qualitative uncertainties, improve the process, and improve public confidence" (Gough,

cited in US Congress 1994c, 10). The OTA suggested that risk assessment should be more integrated with research from the outset.

18. However, as the reaction of Warner North to Ruckelshaus's June 1983 speech shows (see North 2003, and chapter 5), the authors of RAFG clearly did not see their report as unilaterally advocating such a separation. What one could say is that their categories and overall framework of thinking were so effective that they influenced the way that people in the agency thought and behaved. Hence, those who saw themselves as doing risk assessment started to insulate themselves from others, just by virtue of embracing this identity. Organizational decisions in the agency did the rest.

19. Barnes wrote: "The vigorous developments in risk assessment that took place in the late 1970s and early 1980s shimmered and sparkled but, like so much Jello, lacked a unifying, undergirding structure. It was the NAS paradigm that finally succeeded in nailing much—but not all—of this Jello® to the wall. For example (to borrow from the comparison of the sacred and the secular), the paradigm provides a common, somewhat demystified language which both anointed practitioners (risk assessors) and laypeople (the rest of us) can use when communicating about risk. Both groups can appreciate that all scriptures (risk assessments) should have a common underlying structure (i.e., the four elements in the paradigm). Further, the paradigm lays out the moral equivalent of the Prime Directive (i.e. the separation of church [risk assessment] and state [risk management])" (Barnes 1993, 10).

20. The expression, for instance, is found in the NRC report *Issues in Risk Assessment* and in volume 19 of the *EPA Journal*, both published in 1993. During the 1970s and 1980s, the phrase "Red Book" referred to the FDA manual describing the data and methods applicable to the safety evaluation of food additives (Tardiff and Rodricks 1988).

21. A review of the use of risk management reporting formats across offices also showed that overall, they were accurate and useful. Where the formats were incomplete and insufficient, this was generally at the level of more contentious and sophisticated parts of the risk assessment, such as exposure assessment and quantitative estimation of cancer potency. Most of the time, or so it appeared, the delays were caused not of the review of regulatory aspects and the costs and benefits of anticipated decisions, but of the risk assessment. See Daniel Fiorino, Chief Regulation and Information Management Division Memorandum to Richard Morgenstern, Ron Smith, and Milton Russell, OPPE, "Use of the Risk Management Information Reporting Format," October 24, 1984, Milton Russell Special Collection.

22. The summary of a later workshop sponsored by AIHC et al. (1992, 5) similarly stated that risk analyses had to be "relevant, timely and comprehensible," and had to "provide a variety of risk measures (e.g., both societal and individual risk estimates), as well as a clear statement of uncertainties."

23. In the EPA's 1986 guidelines, the standard of proof of causal association between exposure and cancer was threefold: No bias could explain the association, the possibility of a hidden ("confounding") factor has been ruled out, and, most important, "the association is unlikely to be due to chance." This last item alone constituted the "statistical significance" criterion. In practice, it was expressed by the application of a 95 percent confidence level on all epidemiological studies—a standard of proof that environmental health scientists and environmental groups frequently decried because it excluded critical studies from consideration.

24. Flue-Cured Tobacco Coop. Stabilization Corp. v. EPA, No. 6:93CV00370, 1995 U.S. Dist. LEXIS 7521 (M.D.N.C. May 23, 1995).

25. Flue-Cured Tobacco Co-Op Stabil. vs. U.S. EPA, 4 F. Supp. 2d 435 (M.D.N.C. 1998), 466.

26. Anonymous former administrator chief of staff, interview with the author.

27. Ibid.

28. Anonymous ORD risk assessment scientists, interview with the author.

29. The 1980 CAG document on the calculation of the unit risk estimate for air pollutants, for instance, ends with a section noting that the unit risk estimate is a "rough indication" of the relative potency of the agent compared to other carcinogens. The document states that confidence levels for the estimate are generally not calculated "due to the difficulty of accounting for the uncertainty in the data" (EPA 1980b). The Water Office guidance of 1982 mentions the uncertainty surrounding the mechanisms of carcinogenic action, as well as the evidence of mutagenicity of a substance, to justify employing a conventional toxicity evaluation method (safety factor applied to NOAEL) in parallel with a linear extrapolation model, and presents the results side by side, with some elements to judge the relative weight of each of them (EPA 1982). The 1984 guidelines for deriving numerical national water quality criteria warn only that criteria should be established only if the data are of a good-enough quality, but they do not describe methods to analyze uncertainties in the resulting estimates (Stephan et al. 1983). The 1984 document on probabilistic analysis for water quality treats uncertainty via the application of the classical margins of safety and pleads for caution in altering this margin of safety, which provides a "cushion" for scientific uncertainty (EPA 1984c). The approach crafted by the Office of Pesticides' Hazard Evaluation Division for Ecological Risk Assessment is no more precise about the estimation of uncertainties. Although it mentions quantitative uncertainty analysis in the evaluation of ecological risk scenarios, the document remains close to the conventional elaboration of toxicity estimates, understood as "rough" indications of the potential risks of pesticides to nontarget organisms. At that time, it seemed simply impossible, given the data available, to engage in any sort of quantitative estimation of model uncertainty (Urban and Cook 1986).

30. Gil Omenn, interview with the author.

31. Bernard Goldstein, assistant administrator, ORD, memorandum to Alvin Alm, deputy administrator, and to Milton Russell, assistant administrator, OPPE, "Conservatism," October 28, 1983, Milton Russell Special Collection.

32. In 1991, two reports came out, both drafted under the aegis of the Board of Environmental Studies and Toxicology (BEST) at the NRC, and both were requested by the recently created Agency for Toxic Substances and Disease Registry (ATDSR) to push the development of exposure assessment and epidemiology in risk assessment: *Human Exposure Assessment for Airborne Pollutants: Advances and Opportunities* and *Environmental Epidemiology, Volume 1: Public Health and Hazardous Wastes*. These documents reemphasized the difficulties involved in performing exposure assessment, the lack of data about where substances were found in the environment and at what doses, and the pathways through which people were exposed to these substances, over what period of time, and at what doses. Despite the acknowledged need to construct larger databases for exposure, these assessments continued to rely on typically nonvalidated models.

33. Anonymous senior scientist of the Office of Pesticides, interview with the author.

34. The NRC report was also the source of a compromise in Congress that allowed the endless debate on the application of the Delaney clause finally to be closed.

35. Anonymous former administrator chief of staff, interview with the author.

36. Boland, J. E. Executive Order. August 5, 1992. Philip Morris Incorporated. Bates No. 2022852158–2160 at 2158. Minneapolis: Minnesota Tobacco Document Depository.

37. The staff of the Superfund program argued that they were already distinguishing between risk assessment and risk management, making it clear that baseline risk assessments were just one element for consideration by risk managers. Uncertainty analysis was already covered because the office applied a 95 percent confidence level on the arithmetic mean of site sampling data and used IRIS estimates. The program was slightly more challenged by the requirement to propose measures of high-end exposure and to take into account sensitive populations explicitly. However, the office argued, this was in progress. A new concept of "reasonable maximum exposure" was being articulated (Longest and Diamond 1992).

38. The year 1993 saw a succession of no fewer than three congressional hearings, on environmental issues, on risk assessment and the regulatory process, and on risk assessment research. Those followed hearings on the role of science at EPA in 1992 and on the strengths and limitation of the utilization of risk assessment for policy decisions in 1991 (US Congress 1991, 1992, 1994a, 1994b, 1994c).

39. Anonymous, 1986, "Board on Toxicology and Environmental Health Hazards," 53, NAS-NRC Archives.

40. This is now called BEST. See note 11 in chapter 4.

41. The 1993 CRAM report had three parts. The two other parts concerned ecological risk assessment and the use of the maximum tolerated dose in animal experiments. The workshop and resulting report addressed the questions of whether that dose created imbalances in the organism of the animal under testing, so that the cancer effects that may eventually be observed would not just be linked to the substance itself, but to these imbalances; and what other dose may be used that would reveal carcinogenic effects, without artificially inducing them? The group did not resolve this issue, but concluded that "use of the MTD itself does not predict whether a material will elicit a carcinogenic response in a standard animal bioassay. The basis of the relationship is not clear" (NRC 1993b, 62). It argued that the rationale for the selection of a dose in an animal experiment should always be made clear when reporting results of the study.

42. Anonymous member of the NRC panel, interview with the author.

43. Brauman, Finkel, North, and McClellan had the opportunity to warm up for the discussion in one of the many conferences on risk regulation in the Washington, D.C., area in those days—notably the 1992 conference organized by Resources for the Future (where Finkel was based at the time) on comparative risk assessment and risk ranking.

44. Interview with the author.

45. According to the two authors, this could best be implemented as part of a tiered approach to risk assessment (which also was suggested in the body of the report and defended by the committee as a whole), by which chemicals would be placed in three separate categories. The first covered chemicals "with the least amount of data," for which qualitative characterization was most appropriate. The second included "chemicals with more extensive data," for which generic default options would remain applicable. The third category was that of chemicals for which available data sets were "extensive." For those chemicals, "multiple risk calculations corresponding to alternative models and data sets corresponding to individuals and populations" may be provided (NRC 1994, 635). The two authors considered that the choice to depart from the default was the prerogative of risk managers, even though this claim was inconsistent with the principle that defaults should be overridden when better science was available, for who other than risk assessors could and should take care of appreciating the availability of better scientific knowledge? The two authors themselves wrote that "weighing the plausibility of alternatives is a highly judgmental evaluation that must be carried out by scientists" (ibid., 633).

46. In a footnote, he also took support from a previous NRC report on *Human Exposure Assessment for Airborne Pollutants* which asserted that public health policy

requires that decisions be made despite incomplete evidence, with the aim of protecting public health in the future.

47. Interview with the author.

48. Anonymous employee of the NRC, interview with the author.

49. Anonymous former administrator chief of staff, interview with the author.

50. Clinton, in his presidential campaign, was explicit about the need to be cost-effective in environmental policymaking and to resolve the paralysis and inconsistencies in scientific environmental assessments: "One of our biggest problems is the problem represented by the Superfund and some other areas, where you've got to fix something bad that's already happened, where we spend too much money on lawyers, too much money on consultants, the endless decisions. It's almost impossible to get anybody at the local level to agree what the best solution is.... I am just appalled by the paralysis and the political divisions and the fact that the money is being blown." he declared at the Economic Summit in Little Rock, Arkansas, on December 15, 1992. When she took office, Browner was queried on what she would do to improve the awful situation of those who are "governed by EPA," an agency that takes "two contradictory positions at the same time and puts the heat on that state and the private sector," or "takes one position, then another, and then another" (*EPA Journal* 1993). This was one of the reasons for her initial distancing from making risk assessment the systematic basis of all agency decisions, as Republicans required.

51. This office was formally named the Office of Science, Planning, and Regulatory Evaluation.

52. Other than that, it took time for Browner to deliver on her promise of placing science at the heart of the EPA's work. Recruiting the half-dozen top scientists, as suggested in *Safeguarding Science,* proved difficult, given the budget constraints. Browner could only pledge to recruit more researchers for ORD, thanks to the money saved on external contractors (Stone 1994a).

53. H.R.4306, the Risk Assessment Improvement Act passed by the 103rd Congress (1993–1994), directed the EPA administrator to issue risk assessment guidelines regularly.

54. In a memo from Philip Morris reporting on the policies initiated by Browner, it is indicated that she was releasing her own risk characterization policy because "most EPA risk assessors and risk managers simply placed the 1992 risk characterization policy in the bottom drawer of their desks and did not use it." Boland, J. E. Executive Order. August 5, 1992. Philip Morris Incorporated. Bates No. 2022852158–2160 at 2158. Minneapolis: Minnesota Tobacco Document Depository.

55. Edward Ohanian, interview with the author.

56. Ibid.

57. Various officials of the ORD and of the Office of Policy, interview with the author.

58. Al Meyerhoff, an attorney with the Natural Resources Defense Council, was among those who explained in the media that the increased flexibility would dramatically slow the whole process of regulating hazardous chemicals (Hanson 1996).

59. The EPA was now speaking of the effective dose—namely, the dose associated with a 10 percent biological response. Rather than an objective threshold that would be manifested in biological reactions, an effective dose is a regulatory artifact. The EPA later got in the habit of calling this a point of departure that was deliberately constructed to provide a floor from which the extrapolation to other doses could then be made. If the mode of action led to an expected linear low-dose tumor incidence relationship, then a straight line was drawn from the ED10 to the origin (linear). The molecular approach to cancer risk was presented by the EPA and accepted as a major change. Indeed, it came down to abandoning the default application of the LMS model, in favor of a quasi-avowal that there was a threshold.

60. The new criteria were (1) temporal relationship, (2) consistency, (3) magnitude of the association, (4) biological gradient, (5) specificity of the association, (6) biological plausibility, (7) information completeness, (8) coherence (EPA 1986a).

61. Interview with the author.

Chapter 10

1. The main project affected was the ongoing reflection on the methodology of minimal data values for chemicals under consideration in IRIS: a principle by which no safe dose would be set for chemicals for which studies and data were missing, and uncertainty appeared too large (Inside EPA 2010).

2. Speaking at the 40th anniversary of the EPA, he was quoted as saying, "I was assistant administrator of ORD soon after the Red Book appeared. My standard speech about the then new risk assessment paradigm contained the assertion that it would take ten years before we would know whether risk assessment was of value to the agency. I suspect that a similar time will be needed for sustainability" (Goldstein 2011, 308).

3. A separate initiative in the Air Office echoes the same sort of move toward enlarging the scientific base for decision-making from risk assessment to a broader set of techniques and methods. That initiative consists of producing science integration documents instead of risk assessments in order to set up standards for levels of chemicals in ambient air. Science integration was originally a proposal from the American Petroleum Institute, approved by the EPA administrator in 2006 (Cole 2007).

4. Interview with the author.

5. Ibid.

6. In subsequent discussions in the SAB and in the Science Policy Council on "science integration" (which partly followed from *Science and Decisions*), the language concerns "managers" only.

7. Specifically, this refers to the reports *Phthalates and Cumulative Risk Assessment: The Tasks Ahead* (NRC 2008) and *Toxicity Testing in the 21st Century: A Vision and A Strategy* (NRC 2007). Both offered generic advice on how to perform risk assessment in the EPA. To plan the conference, telephone interviews were conducted with participants, who were questioned about their knowledge and views of the NRC reports, their program's risk assessment activities, their needs for new and/or updated guidance, and their vision for risk assessment and the steps that the agency can take to achieve that vision" (Human Health Risk Assessment Colloquium Summary Report, April 25, 2012).

8. The framework "draws on agency experience" and "takes into account the recommendations" of the Silver Book and "a considerable body of additional expert advice," starting with the Red Book, the 1994 NRC report *Science and Judgment,* and the internal EPA report of 1984, *Risk Assessment and Management: Framework for Decision-Making* (EPA 1984b). It also cites eight other agencywide risk frameworks, from the Risk Assessment Guidance for Superfund Part, to the Framework for Ecological Risk Assessment, to the Guidance on Cumulative Risk Assessment.

9. During the campaign, the contest between Bush and Vice President Al Gore, an environmental champion, forced the former to green his discourse. He pledged to regulate carbon dioxide emissions and greenhouse gases. But as soon as he was in office, he reversed these pledges, overturning an eleventh-hour decision by Clinton to set a standard of 10 ppb for arsenic in water.

10. She later revealed that her departure was triggered by the White House's weakening of EPA regulations on air pollution from aging power plants. According to Whitman, Vice President Dick Cheney pushed hard for a rule that "didn't hamper industry," and the final rule was written "at the direction of the White House," over the objections of Whitman and EPA staff (Becker and Gellman 2007).

11. In the terms of the notice: "The appropriate level of precaution in risk assessment and management is complicated by the need to balance efforts to mitigate these potential risks with countervailing risks that may arise from other sources" (OMB 2003b, 5498).

12. Interview with the author.

13. Several important policy principles are summarized in the document—notably, the definition of what it is to be "conservative" and protective of the environment and of public health when one aims to root decisions in the best possible science.

Since *Science and Judgment* and the protracted debate that took place within the confines of the committee between "plausible conservatism" and "using all the science available" (as discussed in chapter 9), the agency made an effort to clarify its own position. The staff paper recalls this position: namely, uncertainty cannot justify inaction. The agency "seeks to adequately protect public and environmental health by ensuring that risk is not likely to be underestimated" (EPA 2004, 11). The staff paper also recalls the progress and changes made across the agency in the consideration of uncertainties and variabilities.

14. Under the pressure of the OMB and requests for enhanced analysis of uncertainties in risk assessments, the number of substances listed in IRIS declined dramatically in those years (GAO 2008).

15. This substance, a rocket fuel engine, was a bone of contention between the EPA and the US Department of Defense. It had been known to cause cancer since at least the 1950s. In 1999, the agency requested the department to send a list of sites contaminated with perchlorate. By 2003, the latter had still not complied, but the agency went forward anyway, defining benchmark levels for the cleanup of polluted sites under the cross fire of senators Barbara Boxer (D-CA) and James Inhofe (R-OK), acting for and against the establishment of a protective standard, respectively.

16. Graham also detailed OIRA's Nancy Beck to Gray's Office of Science Advisor at the EPA, where she was put in charge of the formulation of the agency's response to the OMB. In the end, the EPA did not really have to respond, as the bulletin initiative was essentially killed by the NRC's response.

17. Gil Omenn and Paul Slovic, also members of the earlier RAC, reviewed the report. Omenn concurred with the NRC's overall severe assessment (Omenn 2006).

18. Under Bush's EPA, the positions of assistant administrator for ORD and science advisor were occupied by the same person.

19. This last review was pushed by a Republican senator, David Vitter of Louisiana. Vitter threatened the EPA that he would oppose the nomination of the proposed new chief for ORD, Paul Anastas, until he came to a compromise with Lisa Jackson, the new administrator appointed by President Obama. Companies with stakes in the production of formaldehyde appear to have funded Vitter (Sapien 2010).

20. In 2012, Congress adopted a provision by which the EPA was directed to apply the recommendations of the NAS on formaldehyde, to report to the House and Senate how it did so, and to contract with the NAS for a set of three more reviews of IRIS assessments.

21. A ruling of the Supreme Court on greenhouse gases in Massachusetts [*Massachusetts v. Environmental Protection Agency*, 549 U.S. 497 (2007)] also took some of the heat out of the debate when it pushed the agency to regulate these gases, saying it should provide a scientific basis to justify not doing so (Greenhouse 2007).

22. Interview with the author.

23. Ibid.

24. Ibid.

25. Ibid.

26. Ibid.

27. He published a separate paper in an academic journal to outline the concept, as well as several other papers to test the concept in case studies (Finkel 2011).

28. Dourson was one of the fathers of the RfD/RfC concepts. He was a longtime staff member of the OHEA (from the early 1980s until 1995). In 2017, he was nominated for the job of director of the OPTS. But his past work as a consultant for the industry, combined with the revelation that he systematically defended less protective standards than the EPA for each of the chemicals on which he was consulted, made him appear too close to the regulated industry, which gradually compromised his candidacy, which he withdrew.

29. That included the Office of Modeling, Monitoring Systems, and Quality Assurance; Office of Environmental Engineering and Technology Demonstration; Office of Environmental Processes and Effects Research; Office of Health Research; and OHEA.

30. The downside of the reorganization, however, was that it made the ORD *more* autonomous from program offices and limited their interaction with the risk management side (Powell 1999). For instance, Huggett launched an exercise in prioritizing subjects for ORD research, ranking the thirty-one risks that the report *Unfinished Business* listed back in 1987. But the managers of program offices were not invited to participate in the exercise (Risk Policy Report 1997).

31. When considering more diffuse cause-effect relationships, as in the case of environmental hazards affecting species or milieus, the logic of risk assessment remains, but the preliminary, qualitative steps for identification of the hazard become more sophisticated. As noted by Suter (2008, 286), "The ORNL ERA framework differed from the Red Book framework in having 3 preliminary steps that differed from the hazard identification step in the Red Book framework. The first, defining the endpoints, was necessitated by the diversity of ecological entities and attributes that might be at risk (Suter 1989). The second, development of source terms, was neglected by the Red Book, which focused on human exposure processes. However, the synthetic fuels case studies clarified the importance of determining where, when, and how pollutants are released to the environment. The third, description of the environment, was necessitated by the recognition that ecological risks are determined by the environmental context."

32. Interview with the author.

33. Ibid.

34. The idea of addressing a variety of risks in an integrated manner also draws on a longer history of codification of the methods of the agency. It belongs specifically to what the agency specialists called *cumulative risk assessment (CRA)*. The EPA produced its first synthetic document about CRA in 1997, approximately ten years after the Committee on Pesticides in the Diets of Infants and Children of the NRC started to sensitize the agency to the problem of multiple exposures. That guidance, drawing from the experience of the Office of Pesticides, produced a new vision of risk assessment, incorporated into a broader dialogue to plan the assessment and future actions stemming from it. It stipulated that risk assessment should not be performed as a matter of habit, on isolated risks or hazards presumed to be the source of the problem. That problem should be appreciated for its complexity: A risk is a risk because the chemical hazard often compounds with such factors as "existing health condition, anxiety, nutritional status, crime, and congestion" (EPA 1997, 2). Rather than risks, these are stressors. In pinpointing the problem and planning for the analytic task, the "needs of risk management," fed back from previous decision points, are an important point of reference. In terms of practical innovation, the report suggested establishing the "analysis plan," or conceptual model that identifies the stressors, the exposed populations, and the end points that will be addressed in the risk assessment, as well as the relationships among them. The report recalls that the analysis plan can simply use and quote existing guidance (Ibid., 3). Again, the practice is borrowed from ecological risk assessment work, where the development of a conceptual model was already an old practice, rooted in the very first formalized plans for ecological risk assessment in the 1980s at the Oak Ridge National Laboratory.

35. The establishment of this risk commission was mandated in the Clean Air amendments of 1990, with the goal of reviewing the NRC report that these same amendments ordered the EPA to obtain in order to review its risk assessment methods. The commission did not start its work until 1994, however, because of the presidential election of 1992 and House and Senate elections of 1994. Gil Omenn, a former member of the OSTP under President Jimmy Carter in the 1970s and a coauthor of RAFG, was nominated by Congress and became chair. Alan Kessler, nominated by the White House, cochaired the group. Bernie Goldstein, a professor of environmental medicine and former head of the ORD during Ruckelshaus's second term as EPA administrator, was nominated by the NAS. Omenn and Goldstein had particular influence on the group and the ideas presented in the report.

36. Interview with the author.

37. The report also distanced itself from RAFG. Volume 1 does not cite RAFG; rather, it starts from the definition of risk management that is given in this report and goes on to broaden it: "Risk management is the process of identifying, evaluating, selecting, and implementing actions to reduce risk to human health and to ecosystems. The goal of risk management is scientifically sound, cost-effective, integrated actions

that reduce or prevent risks while taking into account social, cultural, ethical, political, and legal considerations. Our definition of risk management is broader than the traditional definition, which is restricted to the process of evaluating alternative regulatory actions and selecting among them" (Presidential and Congressional Commission 1997a, 1). Volume 2 of the report, however, mentions RAFG as the source of the "now universally recognized four-step framework for characterizing the likelihood of adverse health effects from particular chemical exposures" (Presidential and Congressional Commission 1997b, 4).

References

Aakhus, Mark. 1999. "Science Court: A Case Study in Designing Discourse to Manage Policy Controversy." *Knowledge, Technology and Policy* 12, no. 2: 20–37.

Abelson, P. H. 1990. "Uncertainties about Health Effects of Radon." *Science* 250, no. 4979 (October 19): 353.

Abt, Eileen, Joseph Rodricks, Jonathan Levy, Lauren Zeise, and Thomas A. Burke. 2010. "Science and Decisions: Advancing Risk Assessment." *Risk Analysis* 30, no. 7: 1029–36.

Alexander, Ernest R. 1982. "Design in the Decision-Making Process." *Policy Sciences* 14, no. 3 (June 1): 279–92.

Alexander, Ernest R. 2005. "Institutional Transformation and Planning: From Institutionalization Theory to Institutional Design." *Planning Theory* 4, no. 3: 209–23.

Allison, Graham T. 1971. *Essence of Decision: Explaining the Cuban Missile Crisis.* Boston: Little Brown.

American Industrial Health Council (AIHC). 1979. "Recommended Framework for Identifying Carcinogens and Regulating Them in Manufacturing Situations." Washington, DC: American Industrial Health Council, October 11.

American Industrial Health Council (AIHC). 1980. "Proposal for a Science Panel: Rough Draft." Washington, DC: American Industrial Health Council, February 21.

American Industrial Health Council (AIHC). 1989. "Presentation of Risk Assessments of Carcinogens: Report of an Ad Hoc Study Group on Risk Assessment Presentation." Washington, DC: American Industrial Health Council.

American Industrial Health Council (AIHC). 1992. "Improving Risk Characterization." Washington, DC: American Industrial Health Council.

American Industrial Health Council (AIHC). 1995. "Advances in Risk Characterization." Washington, DC: American Industrial Health Council.

Ames, Bruce N. 1979. "Identifying Environmental Chemicals Causing Mutations and Cancer." *Science* 204, no. 4393 (May 11): 587–93.

Ames, Bruce N., and L. Swirsky Gold. 1989. "Pesticides, Risk, and Applesauce." *Science* 244, no. 4906: 755.

Ames, Bruce N., Renae Magaw, and Lois Swirsky Gold. 1987. "Ranking Possible Carcinogenic Hazards." *Science* 236, no. 4799: 271–80.

Anastas, Paul. 2010. "ORD: The Path Forward." US Environmental Protection Agency (EPA), Office of Research and Development, March 31.

Anastas, Paul. 2012. "Fundamental Changes to EPA's Research Enterprise: The Path Forward." *Environmental Science and Technology* 46, no. 2 (January 17): 580–86.

Anderson, Fred, Mary-Ann Chirba-Martin, Donald Elliott, Cynthia Farina, Ernest Gellhorn, John Graham, Boyden Gray, Jeffrey Holmstead, Ronald Levin, Lars Noah, Katherine Rhyne and Jonathan Wiener. 2000. "Regulatory Improvement Legislation: Risk Assessment, Cost-Benefit Analysis, and Judicial Review." *Duke Environmental Law & Policy Forum* 11, no. 1: 89–138.

Andrews, Richard N. L. 2011. "The EPA at 40: An Historical Perspective." *Duke Environmental Law and Policy Forum* 21, no. 2: 223–58.

Anonymous. 1976. "Office of Management and Budget Plays Critical Part in Environmental Policymaking, Faces Little External Review." *BNA*, 693–97.

Anonymous. 1978. "EPA Begins a Review of Industrial Pesticide That Contains Dioxin." *Wall Street Journal*. August 7, sec. 1.

Anonymous. 1979. "Dispute Over the Risks of 2,4,5-T Continues Despite a Federal Ban: Use on Rangeland Continues Reluctance on Decision." *New York Times*, May 27, 42.

Anonymous. 1981. "OSHA, EPA: The Heyday Is Over." *New York Times*, January 4, sec. Business Day.

Anonymous. 1982. "Profile: OMB's Jim Tozzi." *Environmental Reporter*, May, 11–12.

Anonymous. 1983a. "Interagency Risk Management Council: Overview and Statement of Purpose." US Environmental Protection Agency (EPA), December 13.

Anonymous. 1983b. "Interview with Bernard Goldstein, EPA Assistant Administrator for Research and Development." *Journal of the Water Control Pollution Federation* 56, no 10: 1064–68.

Anonymous. 1984a. "Administrator's Job Has Shifted in 10 Years to Public Health Problems, Ruckelshaus Says." *Environmental Reporter*, November 11, 1291–92.

Anonymous. 1984b. "E.P.A. Proposes Guidelines to Assess Perils of Pollutants." *New York Times*, November 20 edition, A21.

Anonymous. 1984c. "Federal Agency Mandates Limits On EDB in Grain-Based Products." *New York Times*. April 23 edition, A17.

Anonymous. 1990. "6 Members of Panel on Smoking Have Ties to Tobacco Group." *New York Times*, November 10 edition.

Anonymous. 1992. "Update on Risk Training Activities." *Risk Assessment Review: A Bimonthly Publication of the Office of Research and Development and of the Regional Risk. Assessment Network*, US Environmental Protection Agency, 7–8.

Anonymous. 1993. "Risk Assessors Must Talk to Risk Managers, EPA's Barnes Says." *Pesticide and Toxic Chemical News*, April 14 edition.

Anonymous. 1994. "EPA May Hold Cancer Risk Assessment Workshop." *Pesticide and Toxic Chemical News*, March 2 edition.

Anonymous. 2008. "An Interview with William Ruckelshaus." *Open Spaces. Views from the Northwest*. http://www.open-spaces.com/issue-v9n4.php.

Arellano-Gault, David, David Demortain, Christian Rouillard, and Jean-Claude Thoenig. 2013. "Bringing Public Organization and Organizing Back In." *Organization Studies* 34, no. 2: 145–67.

Aron, Joan Leslie, and Jonathan A. Patz. 2001. *Ecosystem Change and Public Health: A Global Perspective*. Baltimore: John Hopkins University Press.

Assmuth, Timo, Mikael Hildén, and Christina Benighaus. 2010. "Integrated Risk Assessment and Risk Governance as Socio-Political Phenomena: A Synthetic View of the Challenges." *Science of the Total Environment* 408, no. 18: 3943–53.

Baba, Annamaria, Daniel M. Cook, Thomas O. McGarity, and Lisa A. Bero. 2005. "Legislating "Sound Science": The Role of the Tobacco Industry." *American Journal of Public Health* 95, no. S1: S20–S27.

Bailey, Jeff. 1992. "Dueling Studies: How Two Industries Created a Fresh Spin on the Dioxin Debate." *Wall Street Journal*, February 20.

Bales, Ellen. 2009. "An Element of Uncertainty: Radon and Quantification of Risk in 20th-Century America, 1945–2000." PhD dissertation, University of California, Berkeley.

Bardach, Eugene. 1989. "Social Regulation as a Generic Policy Instrument." In *Beyond Privatization: The Tools of Government Action*, edited by Lester M. Salomon and Michael S. Lund, 197–230. Washington, DC: The Urban Institute.

Barnard, Chester I. 1968. *The Functions of the Executive*. Cambridge, MA: Harvard University Press.

Barnes, Donald G. 1993. "Brother, Times Are Tough: Can You Paradigm?' In *EPA-600/R-93/039*. Dayton, OH: US Environmental Protection Agency (EPA).

Barnes, Donald G. 1994. "Times Are Tough-Brother, Can You Paradigm?' *Risk Analysis* 14, no. 3: 219–23.

Barnthouse, Lawrence W. 1992. "The Role of Models in Ecological Risk Assessment: A 1990s Perspective." *Environmental Toxicology and Chemistry* 11, no. 12: 1751–60.

Barnthouse, Lawrence W., and Glen. W. Suter II, eds. 1986. *User's Manual for Ecological Risk Assessment.* Oak Ridge National Laboratory, TN, March 1.

Barzelay, Michael, and Babak J. Armajani. 1992. *Breaking Through Bureaucracy: A New Vision for Managing in Government.* Berkeley: University of California Press.

Baunsgaard, Vibeke V., and Stewart Clegg. 2012. "Dominant Ideological Modes of Rationality: Organizations as Arenas of Struggle over Members' Categorization Devices." In *Research in the Sociology of Organizations*, edited by David Courpasson, Damon Golsorkhi, and Jeffrey J. Sallaz. 34:199–232. Bingley: Emerald Group Publishing Limited.

Beardsley, Daniel. 1987. "Application of Integrated Risk Analysis at EPA." In *Risk Assessment and Management*, edited by Lester B. Lave, 141–46. Advances in Risk Analysis 5. New York: Springer US.

Beck, Ulrich. 1986. *Risikogesellschaft. Auf dem Weg in eine andere Moderne.* Frankfurt: Suhrkamp.

Beck, Ulrich. 1992. *Risk Society: Towards a New Modernity.* New York: Sage.

Becker, Jo, and Barton Gellman. 2007. "Leaving No Tracks." *Washington Post*, June 27, 2007, A1.

Bedford, Tim, and Roger Cooke. 2001. *Probabilistic Risk Analysis: Foundations and Methods.* Cambridge: Cambridge University Press.

Bennett, Michael J. 1991. *The Asbestos Racket: An Environmental Parable.* Punta Gorda: Free Enterprise Press.

Berg, Marc. 1997. *Rationalizing Medical Work: Decision-Support Techniques and Medical Practices.* Cambridge, MA: MIT Press.

Bernstein, Peter L. 1996. *Against the Gods: The Remarkable Story of Risk.* New York: John Wiley and Sons, Inc.

Bier, Vicki M. 2001. "On the State of the Art: Risk Communication to Decision-Makers." *Reliability Engineering and System Safety* 71, no. 2: 151–57.

Bijker, Wiebe E., Roland Bal, and Ruud Hendriks. 2009. *The Paradox of Scientific Authority: The Role of Scientific Advice in Democracies.* Cambridge, MA: MIT Press.

Bittner, Ernest. 1965. "The Concept of Organization." *Social Research* 32, no. 3: 239–55.

Bjögvinsson, Erling, Pelle Ehn, and Per-Anders Hillgren. 2012. "Design Things and Design Thinking: Contemporary Participatory Design Challenges." *Design Issues* 28, no. 3: 101–16.

Blau, Peter M., and Richard Scott. 1962. *Formal Organizations: A Comparative Approach*. Stanford, CA: Stanford University Press.

Bogen, Kenneth T. 1994. "A Note on Compounded Conservatism." *Risk Analysis* 14, no. 4: 379–81.

Borelli, Thomas. 1992. "Executive Order/NAM Meeting." Minneapolis: Minnesota Tobacco Document Depository, June 29.

Borraz, Olivier. 2008. *Les politiques du risque*. Paris: Les Presses de Sciences Po.

Borraz, Olivier. 2011. "From Risk to the Government of Uncertainty: The Case of Mobile Telephony." *Journal of Risk Research* 14, no. 8: 969–82.

Boudia, Soraya. 2014. "Managing Scientific and Political Uncertainty." In *Powerless Science?: Science and Politics in a Toxic World*, edited by Soraya Boudia and Nathalie Jas, 95–112. New York: Berghahn Books.

Boullier, Henri. 2016. "Autoriser pour Interdire. La fabrique des savoirs sur les molécules et leurs risques dans le Règlement européen REACH." Thèse pour le doctorat de sociologie, Université Paris-Est Marne-la-Vallée.

Boullier, Henri, David Demortain, and Maurice Zeeman. 2019. "Inventing Prediction for Regulation: The Development of (Quantitative) Structure-Activity Relationships for the Assessment of Chemicals at the EPA." *Science & Technology Studies*, no. 4, forthcoming.

Bovens, Mark, and Paul Hart. 1996. *Understanding Policy Fiascoes*. New Brunswick, NJ: Transaction Publishers.

Bower, J. L. 1970. "Managing the Resource Allocation Process: A Study of Corporate Planning and Investment." Cambridge, MA: Harvard Business School, Division of Research, October 15.

Bowker, Geoffrey C., and S. L. Star. 1999. *Sorting Things Out: Classification and Its Consequences*. Cambridge, MA: MIT Press.

Breyer, Stephen. 1993. *Breaking the Vicious Circle: Toward Effective Risk Regulation*. Cambridge, MA: Harvard University Press.

Brody, Jane. 1991. "Some Scientists Say Concern over Radon Is Overblown by E.P.A.: They Dispute an Official Estimate of 8 Million Homes at Risk. One Researcher Sees around 100,000 Houses in Danger." *New York Times*. January 8 edition, C4.

Brown, Mark B. 2009. *Science in Democracy: Expertise, Institutions, and Representation*. Cambridge, MA: MIT Press.

Brown, Richard Harvey. 1978. "Bureaucracy as Praxis: Toward a Political Phenomenology of Formal Organizations." *Administrative Science Quarterly* 23, no. 3: 365–82.

Brown, Valerie, and Elizabeth Grossman. 2015. "Why the United States Leaves Deadly Chemicals on the Market." *In These Times*, November 2.

Brownson, Ross C., and Diana B. Petitti. 1998. *Applied Epidemiology: Theory to Practice*. New York: Oxford University Press.

Brubaker, Roger. 1984. *The Limits of Rationality*. London: Routledge.

Brunsson, Nils. 1985. *The Irrational Organization: Irrationality as a Basis for Organizational Action and Change*. Copenhagen: Copenhagen Business School Press.

Brunsson, Nils, and Bengt Jacobsson. 2000. *A World of Standards*. London: Oxford University Press.

Burgess, Adam. "The Making of the Risk-Centred Society and the Limits of Social Risk Research." *Health, Risk and Society* 8, no. 4: 329–42.

Burmaster, David E., and Robert H. Harris. 1993. "The Magnitude of Compounding Conservatisms in Superfund Risk Assessments." *Risk Analysis* 13, no. 2: 131–34.

Cabantous, Laure, Jean-Pascal Gond, and Michael Johnson-Cramer. 2010. "Decision Theory as Practice: Crafting Rationality in Organizations." *Organization Studies* 31, no. 11 (January 11): 1531–66.

Calkins, D. R., R. L. Dixon, C. R. Gerber, D. Zarin, and G. S. Omenn. 1980. "Identification, Characterization, and Control of Potential Human Carcinogens: A Framework for Federal Decision-Making." *Journal of the National Cancer Institute* 64, no. 1: 169–76.

Callon, Michel, Pierre Lascoumes, and Yannick Barthe. 2009. *Acting in an Uncertain World: An Essay on Technical Democracy*. Cambridge, MA: MIT Press.

Calow, Peter P. 2009. *Handbook of Environmental Risk Assessment and Management*. Oxford: John Wiley and Sons.

Cambrosio, Alberto, and Camille Limoges. 1991. "Controversies as Governing Processes in Technology Assessment." *Technology Analysis and Strategic Management* 3, no. 4: 377–96.

Carnegie Corporation. 1992. *Environmental Research and Development: Strengthening the Federal Infrastructure*. New York: Carnegie Commission on Science, Technology, and Government.

Carnegie Corporation. 1993. *Risk and the Environment: Improving Regulatory Decision Making*. New York: Carnegie Commission on Science, Technology, and Government.

Carpenter, Daniel P. 2001. *The Forging of Bureaucratic Autonomy: Reputations, Networks, and Policy Innovation in Executive Agencies, 1862–1928*. Princeton, NJ: Princeton University Press.

Carpenter, Daniel P. 2010. *Reputation and Power: Organizational Image and Pharmaceutical Regulation at the FDA*. Princeton, NJ: Princeton University Press.

Clark, Ray. 1997. "NEPA: The Rational Approach to Change." *Environmental Policy and NEPA: Past, Present, and Future*, 15–23.

Clark, Timothy B. 1979. "At Last, a Battle Plan for the Regulatory War on Cancer." *National Journal*, October 27.

Clarke, Lee. 1999. *Mission Improbable: Using Fantasy Documents to Tame Disaster*. Chicago: University of Chicago Press.

Clarke, Lee, and James F. Short. 1993. "Social Organization and Risk: Some Current Controversies." *Annual Review of Sociology* 19: 375–99.

Clegg, Stewart R. 1975. *Power, Rule and Domination: A Critical and Empirical Understanding of Power in Sociological Theory and Organizational Life*. London: Routledge.

Clegg, Stewart R. 1989. "Radical Revisions: Power, Discipline and Organizations." *Organization Studies* 10, no. 1 (January 1): 97–115.

Cohen, Michael D., James G. March, and Johan P. Olsen. 1972. "A Garbage Can Model of Organizational Choice." *Administrative Science Quarterly* 17, no. 1: 1–25.

Cole, Carol. 2007. "EPA Boss Defends Agency Actions, Including NAAQS Process Revisions." *Octane Week*, February 12.

Collingridge, David, and Colin Reeve. 1986. *Science Speaks to Power: The Role of Experts in Policy Making*. New York: St Martin's Press.

Collins, Harry M. 1981. "Knowledge and Controversy: Studies of Modern Natural Science." *Social Studies of Science* 11, no. 1: 1–158.

Collins, Harry M. 1974. "The TEA Set: Tacit Knowledge and Scientific Networks." *Social Studies of Science* 4, no. 2: 165–85.

Conolly, Rory B. 2002. "The Use of Biologically Based Modeling in Risk Assessment." *Toxicology* 181 (December 27): 275–79.

Conolly, Rory B., and Mel E. Andersen. 1993. "An Approach to Mechanism-Based Cancer Risk Assessment for Formaldehyde." *Environmental Health Perspectives* 101, Suppl. 6 (December): 169–76.

Considine, Mark, Damon Alexander, and Jenny M. Lewis. 2014. "Policy Design as Craft: Teasing out Policy Design Expertise Using a Semi-Experimental Approach." *Policy Sciences* 47, no. 3 (September 1): 209–25.

Cooke, Roger M. 1991. *Experts in Uncertainty. Opinion and Subjective Probability in Science*. Oxford: Oxford University Press.

Cothern, C. Richard. 1995. *Handbook for Environmental Risk Decision Making: Values, Perceptions, and Ethics*. New York: Lewis Publishers.

Covello, Vincent T., and Frederick W. Allen. 1988. "Seven Cardinal Rules of Risk Communication." Washington, DC: Office of Policy Analysis, US Environmental Protection Agency (EPA).

Cozzens, Susan E., and Edward J. Woodhouse. 1995. "Science, Government and the Politics of Knowledge." In *Handbook of Science and Technology Studies*, edited by S. Jasanoff, G. Markle, J. Petersen, and T. Pinch, 533–53. Thousand Oaks, CA: Sage.

Cranor, Carl F. 1997. *Regulating Toxic Substances: A Philosophy of Science and the Law*. New York: Oxford University Press.

Cross, Nigel. 1982. "Designerly Ways of Knowing." *Design Studies* 3, no. 4: 221–27.

Cullen, Alison C. 1994. "Measures of Compounding Conservatism in Probabilistic Risk Assessment." *Risk Analysis* 14, no. 4 (August 1): 389–93.

Cumming, Robert B. 1981 "Reflections on Decisions, Options, and Risks." *Risk Analysis* 1, no. 2 (1981): 97–99.

Cushman, John H. Jr. 2000. "Alvin Alm, 63, a Leader in Environmental Policy." *New York Times*, July 28, A19.

Dalsgaard, Peter. 2014. "Pragmatism and Design Thinking." *International Journal of Design* 8, no. 1 (April).

Davies, Clarence J. 1970. *The Politics of Pollution*. Indianapolis: Bobbs-Merrill.

Davies, Clarence J. 1996. *Comparing Environmental Risks: Tools for Setting Government Priorities*. Washington, DC: Resources for the Future.

Davies, Clarence J. 2003. "The Red Book Committee: Creative Conflict." *Human and Ecological Risk Assessment* 9, no. 5: 1113–18.

Dean, Mitchell. 1999. "Risk, Calculable and Incalculable." In *Risk and Sociocultural Theory: New Directions and Perspectives*, edited by Deborah Lupton, 131–59. Cambridge: Cambridge University Press.

Demortain, David. 2011. *Scientists and the Regulation of Risk: Standardising Control*. Cheltenham, UK: Edward Elgar Publishing.

Demortain, David. 2012. "Enabling Global Principle-Based Regulation: The Case of Risk Analysis in the Codex Alimentarius: Enabling Principle-Based Regulation." *Regulation and Governance* 6, no. 2 (June): 207–24.

Demortain, David. 2016. "The Work of Making Risk Frameworks." In *Riskwork: Essays on the Organizational Life of Risk Management*, edited by M. Power, 26–49. New York: Oxford University Press.

Demortain, David. 2017. "Expertise, Regulatory Science and the Evaluation of Technology and Risk: Introduction to the Special Issue." *Minerva* 55, no. 3 (May 24): 139–59.

Dickson, R. Bruce. 1987. "Risk Assessment and the Law: Evolving Criteria by Which Carcinogenicity Risk Assessments Are Evaluated in the Legal Community." In *Uncertainty in Risk Assessment, Risk Management, and Decision Making*. Advances in Risk Analysis Series, edited by Vincent T. Covello, Lester B. Lave, Alan Moghissi, and Ram Uppuluri, 4:145–57. New York: Springer US.

Diesing, Paul. 1962. *Reason in Society: Five Types of Decisions and Their Social Conditions*. Westport, CT: Greenwood Press.

Dietz, Thomas, and Robert W. Rycroft. 1988. *The Risk Professionals*. New York: Russell Sage Foundation.

Dillon, Lindsey, Christopher Sellers, Vivian Underhill, Nicholas Shapiro, Jennifer Liss Ohayon, Marianne Sullivan, Phil Brown, Jill Harrison, Sara Wylie, and the "EPA Under Siege" Writing Group. 2018. "The Environmental Protection Agency in the Early Trump Administration: Prelude to Regulatory Capture." *American Journal of Public Health* 108, no. S2 (April): S89–S94.

DiMaggio, Paul J., and Walter W. Powell. 1983. "The Iron Cage Revisited: Institutional Isomorphism and Collective Rationality in Organizational Fields." *American Sociological Review* 48, no. 2 (April): 147–60.

Dobbin, Frank. 1994. "Cultural Models of Organization: The Social Construction of Rational Organizing Principles." In *The Sociology of Culture: Emerging Theoretical Perspectives*, edited by Diana Crane, 117–41. Chichester: Wiley-Blackwell

Doern, Bruce, David Castle, and Peter W. B. Phillips. 2016. *Canadian Science, Technology, and Innovation Policy: The Innovation Economy and Society Nexus*. Montreal: McGill-Queen's University Press.

Douglas, Mary. 1990. "Risk as a Forensic Resource." *Daedalus* 119, no. 4: 1–16.

Douglas, Mary. 1992. *Risk and Blame: Essays in Cultural Theory*. London: Routledge.

Douglas, Mary, and Aaron Wildavsky. 1983. *Risk and Culture: An Essay on the Selection of Technical and Environmental Dangers*. San Francisco: University of California Press.

Dowling, John, and Jeffrey Pfeffer. 1975. "Organizational Legitimacy: Social Values and Organizational Behavior." *Pacific Sociological Review* 18, no. 1 (January): 122–36.

Driver, Jeffrey, Scott R. Baker, and David McCallum. 2012. *Residential Exposure Assessment: A Sourcebook*. New York: Springer Science and Business Media.

Drucker, Johanna. 2014. *Graphesis: Visual Forms of Knowledge Production*. Cambridge, MA: Harvard University Press.

Dryzek, John S. 1983. "Don't Toss Coins in Garbage Cans: A Prologue to Policy Design." *Journal of Public Policy* 3, no. 4: 345–67.

Eads, George C., and Michael Fix. 1984. *Relief or Reform? Reagan's Regulatory Dilemma*. Washington, DC: Urban Institute Press.

Edelstein, Michael R., and William J. Makofske. 1998. *Radon's Deadly Daughters: Science, Environmental Policy, and the Politics of Risk*. Lanham: Rowman and Littlefield.

Enthoven, Alain C. 1980. *Health Plan: The Practical Solution to the Soaring Cost of Medical Care*. Washington, DC: Beardbooks.

Environmental Protection Agency (EPA). 1976. *Interim Procedures and Guidelines for Health Risk and Economic Impact Assessments of Suspected Carcinogens*. Washington, DC: US Environmental Protection Agency.

Environmental Protection Agency (EPA). 1977. *Report of the Toxics Integration Work Group*. Washington, DC: US Environmental Protection Agency, March 11.

Environmental Protection Agency (EPA). 1979a. "EPA Chemical Activities: Status Report." Toxics Integration Information Series. Washington, DC: US Environmental Protection Agency.

Environmental Protection Agency (EPA). 1979b. "Policy and Procedures for Identifying, Assessing, and Regulating Airborne Substances Posing a Risk of Cancer—Proposed Rule." Washington, DC: US EPA Office of Air Quality Planning and Standards.

Environmental Protection Agency (EPA). 1980a. "The Carcinogen Assessment Group's Method for Determining the Unit Risk Estimate for Air Pollutants, Prepared by CAG/OHEA for the Office of Air Quality Planning and Standards." Washington, DC: US Environmental Protection Agency (EPA), Carcinogen Assessment Group, August 8.

Environmental Protection Agency (EPA). 1980b. *First Report of the Interagency Toxic Substances Data Committee*. Washington, DC: US Environmental Protection Agency (EPA) and Council on Environmental Quality (CEQ), November.

Environmental Protection Agency (EPA). 1980c. "Public Hearing on Proposed National Emission Standards for Identifying, Assessing and Regulating Airborne Substances Posing a Risk of Cancer and Advanced Notice of Proposed Generic Standards, Apex Reporting, Transcript of Proceedings." Washington, DC: US Environmental Protection Agency, March 19.

Environmental Protection Agency (EPA). 1981a. "Agenda for CMA/EPA Discussion of Exposure and Risk Assessment Issues for DEHP." Washington, DC: US Environmental Protection Agency, September 30.

Environmental Protection Agency (EPA). 1981b. "CMA/EPA Meeting on DI-(2-ethylhexyl) phthalate (DEHP)." Washington, DC: US Environmental Protection Agency, September 29.

Environmental Protection Agency (EPA). 1981c. *Report of the Toxics Integration Project.* 905R81108. Washington, DC: US Environmental Protection Agency, Office for Policy and Resource Management, September.

Environmental Protection Agency (EPA). 1982. "Additional US Environmental Protection Agency Guidance for the Health Assessment of Suspect Carcinogens with Specific Reference to Water Quality Criteria." OHEA-C-088. Washington, DC: US Environmental Protection Agency, June 21.

Environmental Protection Agency (EPA). 1983a. *Guidance for Performing Exposure Assessments.* Washington, DC: US Environmental Protection Agency.

Environmental Protection Agency (EPA). 1983b. *Guidelines for Performing Regulatory Impact Analysis.* Washington, DC: Office of Policy Analysis, US Environmental Protection Agency.

Environmental Protection Agency (EPA). 1983c. *Trends Likely to Affect EPA in the Next Five to Ten Years.* Washington, DC: US EPA Program Evaluation Division, Office of Management Systems and Evaluation.

Environmental Protection Agency (EPA). 1984a. *The Magnitude and Nature of the Air Toxics Problem in the United States. Draft Report. Elaine Haemi Segger, Alan Jones, Bern Steigerwald, Vivian Thomas.* Washington, DC: Office of Air and Radiation, Office of Policy, Planning and Evaluation, US Environmental Protection Agency, September.

Environmental Protection Agency (EPA). 1984b. *Risk Assessment and Management: Framework for Decision Making.* EPA-600-9-85-002. Washington, DC: US Environmental Protection Agency, December.

Environmental Protection Agency (EPA). 1984c. "Use of Probabilistic Information in the Water Quality-Based Approach." EPA-600/D-84-192. Duluth, MN: Environmental Research Laboratory-Duluth, U.S. Environmental Protection Agency.

Environmental Protection Agency (EPA). 1985a. "Daminozide Special Review Position Document 2/3/4, Draft." Washington, DC: US Environmental Protection Agency, September 12.

Environmental Protection Agency (EPA). 1985b. "Overview of EPA Risk/Risk Communication Activities." Washington, DC: US Environmental Protection Agency.

Environmental Protection Agency (EPA). 1985c. "The Use of Risk Assessment in EPA Regional Operations." Atlanta: Region IV, US Environmental Protection Agency.

Environmental Protection Agency (EPA). 1986a. *Guidelines for Carcinogen Risk Assessment.* EPA/630/R-00/004. Washington, DC: Risk Assessment Forum, U.S. Environmental Protection Agency.

Environmental Protection Agency (EPA). 1986b. *Guidelines for Estimating Exposure.* EPA/600/8-87/045. Washington, DC: Office of Health and Environmental Assessment, US Environmental Protection Agency.

Environmental Protection Agency (EPA). 1987. *Unfinished Business: A Comparative Assessment of Environmental Problems—Overview Report.* 000R87901. Washington, DC: Office of Policy Analysis, Office of Policy Planning and Evaluation, US Environmental Protection Agency.

Environmental Protection Agency (EPA). 1988. *Report of the Subcommittee on Risk Reduction: Research Strategies Committee.* SAB-EC-088-40-E. Washington, DC: Office of the Administrator, Science Advisory Board, US Environmental Protection Agency.

Environmental Protection Agency (EPA). 1989. "Comparing Risks and Setting Environmental Priorities: Overview of Three Regional Projects." Washington, DC: US Environmental Protection Agency, Office of Policy, Planning and Evaluation.

Environmental Protection Agency (EPA). 1990a. "The Office of Research and Development." 20460. Washington, DC: US Environmental Protection Agency, September.

Environmental Protection Agency (EPA). 1990b. "Research to Improve Health Risk Assessments (RIHRA) Program." Washington, DC: Office of Research and Development, US Environmental Protection Agency, June.

Environmental Protection Agency (EPA). 1992a. "Framework for Ecological Risk Assessment." EPA/630/R-92/001. Washington, DC: Risk Assessment Forum, US Environmental Protection Agency, February.

Environmental Protection Agency (EPA). 1992b. "Guidelines for Exposure Assessment, Published on May 29, Federal Register 57(104): 22888–938," May.

Environmental Protection Agency (EPA). 1992c. "Implementing the Deputy Administrator's Risk Characterization Memorandum." Washington, DC: US Environmental Protection Agency, Office of Solid Waste and Emergency Response, May 26.

Environmental Protection Agency (EPA). 1992d. "Memorandum from F. Henry Habicht II, Deputy Administrator, to Assistant Administrators and Regional Administrators, Guidance on Risk Characterization for Risk Managers and Risk Assessors." Washington, DC: US Environmental Protection Agency, Office of the Administrator, February 26.

Environmental Protection Agency (EPA). 1992e. *Safeguarding the Future: Credible Science, Credible Decisions: The Report of the Expert Panel of the Role of Science at EPA to William K. Reilly, Administrator, US Environmental Protection Agency.* Washington, DC: US Environmental Protection Agency.

Environmental Protection Agency (EPA). 1994a. *Design for the Environment. Workshop on Identifying a Framework for the Future of Human Health and Environmental Risk Ranking.* EPA-744-S-93-001. Washington, DC: Office of Pollution Prevention and Toxics, US Environmental Protection Agency.

Environmental Protection Agency (EPA). 1994b. "Draft Revisions to the Guidelines for Carcinogen Risk Assessment." EPA/600/BP-92-003. Washington, DC: Risk Assessment Forum, US Environmental Protection Agency, September.

Environmental Protection Agency (EPA). 1994c. *Report on the Workshop on Cancer Risk Assessment Guidelines Issues.* EPA/630/R-94/005a. Washington, DC: Office of Research and Development, U.S. Environmental Protection Agency.

Environmental Protection Agency (EPA). 1994d. *Research, Development, and Technical Services at EPA: A New Beginning. Report to the Administrator.* Washington, DC: Agency-Wide Steering Committee, U.S. Environmental Protection Agency.

Environmental Protection Agency (EPA). 1994e. *An SAB Report: Review of MITRE Corp. Draft Report on the EPA Laboratory Study, Prepared by the Research Strategies Advisory Committee.* Washington DC: Science Advisory Board, US Environmental Protection Agency, May.

Environmental Protection Agency (EPA). 1995a. *Guidance for Risk Characterization.* Washington, DC: Science Policy Council, US Environmental Protection Agency.

Environmental Protection Agency (EPA). 1995b. "William K. Reilly: Oral History Interview." Washington, DC: US Environmental Protection Agency.

Environmental Protection Agency (EPA). 1996a. "Proposed Guidelines for Carcinogen Risk Assessment." EPA/600/P-92/003Ca. Washington, DC: US Environmental Protection Agency.

Environmental Protection Agency (EPA). 1996b. "Strategic Plan for the Office of Research and Development." Washington, DC: Office of Research and Development, US Environmental Protection Agency, May.

Environmental Protection Agency (EPA). 1997. "Guidance on Cumulative Risk Assessment. Part 1. Planning and Scoping." Washington, DC: Science Policy Council, US Environmental Protection Agency, July 3.

Environmental Protection Agency (EPA). 1999. *Residual Risk: Report to Congress.* EPA-453/R-99-001. Washington, DC: Office of Air and Radiation, Office of Air Quality Planning and Standards, US Environmental Protection Agency.

Environmental Protection Agency (EPA). 2000a. "Science Council Policy Handbook: Risk Characterization." Washington, DC: Office of Science Policy, Office of Research and Development, US Environmental Protection Agency.

Environmental Protection Agency (EPA). 2000b. *Taking Toxics out of the Air*. Washington, DC: Office of Air Quality Planning and Standards, US Environmental Protection Agency.

Environmental Protection Agency (EPA). 2004. "An Examination of EPA Risk Assessment Principles and Practices." Staff paper prepared for the US Environmental Protection Agency by members of the Risk Assessment Task Force. EPA/100/B-04/001. Washington, DC: Office of the Science Advisor, US Environmental Protection Agency.

Environmental Protection Agency (EPA). 2012. "Framework for Human Health Risk Assessment to Inform Decision Making. External Review Draft." Washington, DC: Office of the Science Advisor, US Environmental Protection Agency, Risk Assessment Forum, July.

Environmental Protection Agency (EPA). 2013. "Douglas M. Costle: US EPA Oral History Interview." Washington, DC: US Environmental Protection Agency.

Environmental Protection Agency (EPA). 2014. "Framework for Human Health Risk Assessment to Inform Decision Making. External Review Draft." Washington, DC: Office of the Science Advisor, US Environmental Protection Agency, Risk Assessment Forum, April.

EPA Journal. 1993. "An Interview with Carol M. Browner." 19, no 1, January–March, EPA-175-N-93-014, 6-8.

Erickson, Paul, Judy L. Klein, Lorraine Daston, Rebecca Lemov, Thomas Sturm, and Michael D. Gordin. 2013. *How Reason Almost Lost Its Mind: The Strange Career of Cold War Rationality*. Chicago: University of Chicago Press.

Evans, John S., John D. Graham, George M. Gray, and Adrienne Hollis. 1992. "Summary of Workshop to Review an OMB Report on Regulatory Risk Assessment and Management Report." *Risk: Issues in Health and Safety* 3:71–84.

Evans, Karen G. 2000. "Reclaiming John Dewey: Democracy, Inquiry, Pragmatism, and Public Management." *Administration and Society* 32, no. 3: 308–28.

Ewald, François. 1986. *L'Etat Providence*. Paris: Grasset.

Ewald, François. 1991. "Insurance and Risk." In *The Foucault Effect: Studies in Governmentality*, edited by A. Burchell, 197–210. Chicago: University of Chicago Press.

Ezrahi, Yaron. 1990. *The Descent of Icarus: Science and the Transformation of Contemporary Democracy*. Cambridge, MA: Harvard University Press.

Fabiansson, Charlotte, and Stefan Fabiansson. 2016. *Food and the Risk Society: The Power of Risk Perception*. New York: Routledge.

Farmer, F. R. 1967. "Siting Criteria—A New Approach." *Containment and Siting of Nuclear Power Plants. Proceedings of a Symposium on the Containment and Siting of Nuclear Power Plants*. International Atomic Energy Agency, April 3-7, Vienna.

Farr, James, Jacob S. Hacker, and Nicole Kazee. 2006. "The Policy Scientist of Democracy: The Discipline of Harold D. Lasswell." *American Political Science Review* 100, no. 4: 579.

Faustman, Elaine M., and Gilbert S. Omenn. 2010. "Risk Assessment." In *Cassarett and Doull's Essentials of Toxicology*, edited by Curtis Klaassen and John B. Watkins III, 2nd ed., 47–56. New York: McGraw-Hill Medical.

Federal Focus. 1991. "Toward Common Measures: Recommendations for a Presidential Executive Order on Environmental Risk Assessment and Risk Management Policy." Washington, DC: Institute for Regulatory Policy, Federal Focus, Inc.

Federal Register. 1979. "National Emission Standards for Identifying, Assessing and Regulating Airborne Substances Posing a Risk of Cancer." October 10.

Federal Register. 1980a. "Ambient Water Quality Criteria Documents: Notice of Availability." 45: 79317.

Federal Register. 1980b. "Chemical Carcinogens: A Review of the Science and Its Associated Principles ("OSTP Principles")." 50: 10371–442.

Federal Register. 1980c. "Mutagenicity Risk Assessment: Proposed Guidelines," 45: 74984.

Federal Register. 1992. "Draft Report: A Cross-Species Scaling Factor for Carcinogen Risk Assessment Based on Equivalence of mg/kg3/4/day." 57: 24152–173, June 5.

Federal Register. 1993. "Executive Order 12866—Regulatory Planning and Review." 58: 51735-51742, October 4.

Federal Register. 2003. "Draft 2003 Report to Congress on the Costs and Benefits of Federal Regulations." 68: 5492–527, February 3.

Finkel, Adam M. 1990. *Confronting Uncertainty in Risk Management; a Guide for Decision Makers a Report*. Washington, DC: Resources for the Future.

Finkel, Adam M. 1991. "Edifying Presentation of Risk Estimates: Not as Easy as It Seems." *Journal of Policy Analysis and Management* 10, no. 2: 296–303.

Finkel, Adam M. 1994. "The Case for 'Plausible Conservatism' in Choosing and Altering Defaults." In *Science and Judgment in Risk Assessment*, 601–28. Washington, DC: National Academies Press.

Finkel, Adam M. 2011. "'Solution-Focused Risk Assessment": A Proposal for the Fusion of Environmental Analysis and Action." *Human and Ecological Risk Assessment: An International Journal* 17, no. 4: 754–87.

Finkel, Adam M., and John S. Evans. 1987. "Evaluating the Benefits of Uncertainty Reduction in Environmental Health Risk Management." *JAPCA* 37, no. 10 (October 1): 1164–71.

Finkel, Adam M., and Dominic Golding. 1994. *Worst Things First? The Debate over Risk-Based National Environmental Priorities*. Washington, DC: Resources for the Future.

Fiorino, Daniel J. 1995. *Making Environmental Policy*. San Francisco: University of California Press.

Fischer, Frank. 2000. *Citizens, Experts, and the Environment: The Politics of Local Knowledge*. Durham, NC: Duke University Press.

Fischhoff, Baruch. 1977. "Cost Benefit Analysis and the Art of Motorcycle Maintenance." *Policy Sciences* 8, no. 2: 177–202.

Fischhoff, Baruch, Paul Slovic, Sarah Lichtenstein, Stephen Read, and Barbara Combs. 1978. "How Safe Is Safe Enough? A Psychometric Study of Attitudes towards Technological Risks and Benefits." *Policy Sciences* 9, no. 2: 127–52.

Fisher, Elizabeth C. 2007. *Risk Regulation and Administrative Constitutionalism*. Oxford: Hart Publishing.

Fisher, Elizabeth. 2013. "Framing Risk Regulation: A Critical Reflection." *European Journal of Risk Regulation* 2013: 125–32.

Fleming, Bruce. 1994. "Has EPA Gone Mad?' *Pulp and Paper* 68, no. 2: 138.

Forester, John. 1984. "Bounded Rationality and the Politics of Muddling Through." *Public Administration Review* 44, no. 1: 23–31.

Fortun, Mike, and Sylvan S. Schweber. 1993. "Scientists and the Legacy of World War II: The Case of Operations Research (OR)." *Social Studies of Science* 23, no. 4 (November 1): 595–642.

Frantzen, Kurt A. 2016. *Risk-Based Analysis for Environmental Managers*. New York: CRC Press.

Fredrickson, Leif, Christopher Sellers, Lindsey Dillon, Jennifer Liss Ohayon, Nicholas Shapiro, Marianne Sullivan, Stephen Bocking, et al. 2018. "History of US Presidential Assaults on Modern Environmental Health Protection." *American Journal of Public Health* 108, no. S2 (April 2018): S95–S103.

Freeman, Jody, and Adrian Vermeule. 2007. "Massachusetts v EPA: From Politics to Expertise." *Supreme Court Review* 2007, no. 1: 51–110.

Freudenburg, William R., Robert Gramling, and Debra J. Davidson. 2008. "Scientific Certainty Argumentation Methods (SCAMs): Science and the Politics of Doubt." *Sociological Inquiry* 78, no. 1 (1 February 2008): 2–38.

Frickel, Scott. 2004. *Chemical Consequences: Environmental Mutagens, Scientist Activism, and the Rise of Genetic Toxicology*. New Brunswick, NJ: Rutgers University Press.

Friis, Robert H. 2012. *The Praeger Handbook of Environmental Health*. Vol. 1, *Foundations of the Field*. Santa Barbara, CA: Praeger.

Frosdick, Steve. 1997. "The Techniques of Risk Analysis Are Insufficient in Themselves." *Disaster Prevention and Management: An International Journal* 6, no. 3: 165–77.

Fujimura, Joan H. 1987. "Constructing 'Do-Able' Problems in Cancer Research: Articulating Alignment." *Social Studies of Science* 17, no. 2: 257–93.

Fuller, John G. 1975. *We Almost Lost Detroit*. New York: Reader's Digest Press.

Funtowicz, Silvio O., and Jerome R. Ravetz. 1993. "Science for the Post-normal Age." *Futures* 25, no. 7: 739–55.

Furlong, Scott R. 1995. "Reinventing Regulatory Development at the Environmental Protection Agency." *Policy Studies Journal* 23, no. 3 (September): 466–82.

Gardner, Donald E., James D. Crapo, and Roger O. McClellan. 2000. *Toxicology of the Lung*. Philadelphia: Taylor & Francis.

Gauchat, Gordon. 2012. "Politicization of Science in the Public Sphere: A Study of Public Trust in the United States, 1974 to 2010." *American Sociological Review* 77, no. 2: 167–87.

Gay, Paul du, and Thomas Lopdrup-Hjorth, 2016. "Fear of the Formal." *European Journal of Cultural and Political Sociology* 3, no. 1: 6–40.

Gibb, Steven. 2001. "Science and Economics Prominent on EPA Agenda." *Issues in Science and Technology* 17, no. 3: 57–60.

Gieryn, Thomas F. 1983. "Boundary-Work and the Demarcation of Science from Non-Science: Strains and Interests in Professional Ideologies of Scientists." *American Sociological Review* 48, no. 6: 781–95.

Gill, Matthew. 2009. *Accountants' Truth: Knowledge and Ethics in the Financial World*. Oxford: Oxford University Press.

Ginsburg, Robert. 1997. "Quantitative Risk Assessment and the Illusion of Safety." In *Work, Health, and Environment: Old Problems, New Solutions*, edited by Charles Levenstein and John Wooding, 8–15. New York: Guilford Press.

Glicksman, Robert, and Christopher H. Schroeder. 1991. "EPA and the Courts: Twenty Years of Law and Politics." *Law and Contemporary Problems* 54, no. 4: 249–309.

Golden, Marissa Martino. 2013. *What Motivates Bureaucrats? Politics and Administration during the Reagan Years*. New York: Columbia University Press.

Golding, Dominic. 1992. "A Social and Programmatic History of Risk Research." In *Social Theories of Risk*, edited by Dominic Golding and Sheldon Krimsky, 23–52. London: Praeger Publishers.

Goldstein, Bernard D. 1993. "If Risk Management Is Broke, Why Fix Risk Assessment?" *EPA Journal* 19: 37.

Goldstein, Bernard D. 2005. "Advances in Risk Assessment and Communication." *Annual Review of Public Health* 26, no. 1 (April 21): 141–63.

Goldstein, Bernard D. 2011. "EPA at 40: Reflections on the Office of Research and Development." *Duke Environmental Law and Policy Forum* 21, no. 2: 295–308.

Goldstein, Bernard D., and Russellyn S. Carruth. 2003. "Implications of the Precautionary Principle for Environmental Regulation in the United States: Examples from the Control of Hazardous Air Pollutants in the 1990 Clean Air Act Amendments." *Law and Contemporary Problems* 66, no. 4: 247–61.

Goodin, Robert E. 1996. *The Theory of Institutional Design*. Cambridge: Cambridge University Press.

Goodstein, Eban. 1995. "Benefit-Cost Analysis at the EPA." *Journal of Socio-Economics* 24, no. 2 (June 1): 375–89.

Gore, Albert. 1993. *Improving Regulatory Systems: Accompanying Report of the National Performance Review, Office of the Vice President*. Washington, DC: US Government Publishing Office, September.

Gough, Michael, and Steven J. Milloy. 1996. "EPA's Cancer Risk Guidelines: Guidance to Nowhere." *Cato Institute Policy Analysis No. 263*, November 12, 25.

Graham, John D. 1985. "The Failure of Agency-Forcing: The Regulation of Airborne Carcinogens under Section 112 of the Clean Air Act." *Duke Law Journal* 1985, no. 1: 100–150.

Graham, John D., Laura C. Green, and Marc J. Roberts. 1991. *In Search of Safety: Chemicals and Cancer Risk*. Cambridge, MA: Harvard University Press.

Graham, John D., Paul R. Noe, and Elizabeth L. Branch. 2005. "Managing the Regulatory State: The Experience of the Bush Administration." *Fordham Urban Law Journal* 33, no. 4: 101–48.

Graham, John D., and March Sadowitz. 1993. "Risk Analysis and the EPA Cabinet Elevation Bill." *Risk in Perspective-Harvard Center for Risk Analysis* 1, no. 4: 1–2.

Gratt, Lawrence B. 1987. "The Definition of Risk and Associated Terminology for Risk Analysis." In *Risk Assessment in Setting National Priorities*, edited by James J. Bonin and Donald E. Stevenson, 675–80. New York: Springer Science and Business Media.

Gray, George M. 1993. "The Challenge of Risk Characterization." Report to the Office of Technology Assessment, US Congress. National Technical Information Service Publication PB93–218857.

Gray, George M., Joshua T. Cohen, and John D. Graham. 1993. "The Challenge of Risk Characterization: Current Practice and Future Directions." *Environmental Health Perspectives* 101, Supplement 6: Health Effects of Gasoline: 203–8.

Greenhouse, Linda. 2007. "Justices Say EPA Has Power to Act on Harmful Gases." *New York Times*, April 3, sec. Washington.

Greenwood, Royston, Roy Suddaby, and C. R. Hinings. 2002. "Theorizing Change: The Role of Professional Associations in the Transformation of Institutionalized Fields." *Academy of Management Journal* 45, no. 1: 58–80.

Greenwood, Ted. 1984a. *Knowledge and Discretion in Government Regulation*. New York: Praeger.

Greenwood, Ted. 1984b. "The Myth of Scientific Incompetence of Regulatory Agencies." *Science, Technology, and Human Values* 9, no. 1: 83–96.

Gulis, Gabriel, Odile Mekel, Balázs Ádám, and Liliana Cori. 2013. *Assessment of Population Health Risks of Policies*. New York and London: Springer Science and Business Media.

Gutmann, Amy, and Dennis F. Thompson. 1998. *Democracy and Disagreement*. Cambridge, MA: Harvard University Press.

Habermas, Jürgen. 1971. *Toward a Rational Society: Student Protest, Science, and Politics*. Boston: Beacon Press.

Hackett, Edward J., Olga Amsterdamska, Michael Lynch, and Judy Wajcman, eds. 2008. *The Handbook of Science and Technology Studies*. Cambridge, MA: MIT Press.

Hacking, Ian. 1990. *The Taming of Chance*. Vol. 17. Cambridge: Cambridge University Press.

Hacking, Ian. 2003. "Risk and Dirt." In *Risk and Morality*, edited by R. Ericson and A. Doyle, 22–47. Toronto: Toronto University Press.

Hafner, K. 2007. "Ida R. Hoos Is Dead at 94; a Critic of Systems Analysis." *New York Times*, May 5, C10.

Haight, H., L. Haxby, P. Ifland, and A. Norberg. 1980. "Risk-Benefit-Cost Background Report: Issues, Terms, Methodologies and Activities. A Document Developed for Risk-Benefit-Cost Scoping Study Steering Group." CMA 062852. Washington, DC: Chemical Manufacturers Association.

Hakim, Danny, and Eric Lipton. 2018. "Pesticide Studies Won E.P.A.'s Trust, Until Trump's Team Scorned "Secret Science." *New York Times*, August 24, sec. Business.

Halberstam, David. 1988. *The Best and the Brightest*. New York: Random House.

Halffman, Willem. 2005. "Science-Policy Boundaries: National Styles?" *Science and Public Policy* 32, no. 6: 457–67.

Hammitt, James K. 1997. "Improving Comparative Risk Analysis." *Duke Environmental Law and Policy Forum* 8: 81–100.

Hanson, David. 1996. "Cancer Risk of Chemicals: EPA to Assess Broader Range of Data." *Chemical and Engineering Newsselznick* 74, no. 17 (April 22): 6.

Harris, Richard A., and Sidney M. Milkis. 1989. *The Politics of Regulatory Change: A Tale of Two Agencies*. London: Oxford University Press.

Harrison, Kathryn, and George Hoberg. 1994. *Risk, Science, and Politics: Regulating Toxic Substances in Canada and the United States*. Montreal: McGill–Queen's Press.

Hayne, Christie, and Clinton Free. 2014. "Hybridized Professional Groups and Institutional Work: COSO and the Rise of Enterprise Risk Management." *Accounting, Organizations and Society* 39, no. 5 (July): 309–30.

Hayns, M. R. 1999. "The Evolution of Probabilistic Risk Assessment in the Nuclear Industry." *Process Safety and Environmental Protection*, Developments in Process Safety, 77, no. 3 (May): 117–42.

Heifetz, Ronald A. 2009. *Leadership without Easy Answers*. Cambridge, MA: Harvard University Press.

Hepler-Smith, Evans 2019. "Molecular Bureaucracy: Toxicological Information and Environmental Protection." *Environmental History*. doi:10.1093/envhis/emy134.

Hickson, David J. 1987. "Decision-Making at the Top of Organizations." *Annual Review of Sociology* 13, no. 1: 165–92.

Hilgartner, Stephen. 1992. "The Social Construction of Risk Objects." In *Organizations, Uncertainties and Risk*, edited by James Short Jr and Lee Clarke, 39–53. Boulder: Westview Press.

Hilgartner, Stephen. 2000. *Science on Stage: Expert Advice as Public Drama*. Stanford, CA: Stanford University Press.

Hillier, Frederik S., and Gerard J. Lieberman. 1980. *Introduction to Operations Research*. San Francisco: Holden-Day.

Hindess, Barry. 1987. "Rationality and the Characterization of Modem Society." In *Max Weber, Rationality and Modernity*, edited by Scott Lash and Sam Whimster. London: Routledge.

Hippel (Von), Frank. 1977. "Looking Back on the Rasmussen Report." *Bulletin of the Atomic Scientists* 33, no. 2: 42–47.

Hisschemöller, Matthijs, and Rob Hoppe. 1995. "Coping with Intractable Controversies: The Case for Problem Structuring in Policy Design and Analysis." *Knowledge and Policy* 8, no. 4: 40–60.

Hoberg, George. 1992. *Pluralism by Design: Environmental Policy and the American Regulatory State.* New York: Greenwood Publishing Group.

Holgate, Stephen T., Hillel S. Koren, Jonathan M. Samet, and Robert L. Maynard. 1999. *Air Pollution and Health.* London: Academic Press.

Hood, Christopher. 1976. *The Limits of Administration.* London: Wiley.

Hood, Christopher. 1986. *The Tools of Government.* Chatham, NJ: Chatham House Publishers.

Hood, Christopher. 1995. "When Extremes Meet: SPRAT vs SHARK in Public Risk Management." In *Accident and Design: Contemporary Debates in Risk Management,* edited by Christopher Hood and David Jones, 209–27. London: Routledge.

Hood, Christopher, and David K. C. Jones. 1996. *Accident and Design: Contemporary Debates in Risk Management.* London: UCL Press.

Hood, Christopher, Henry Rothstein, and Robert Baldwin. 2001. *The Government of Risk: Understanding Risk Regulation Regimes.* Oxford: Oxford University Press.

Hoos, Ida R. 1969. *Systems Analysis in Social Policy: A Critical Review.* Ann Harbor: Institute of Economic Affairs.

Hoos, Ida R. 1970. "Systems Analysis as a Technique for Solving Social Problems—a Realistic Overview." *Socio-Economic Planning Sciences* 4, no. 1 (March 1): 27–32.

Hoos, Ida R. 1973. "Systems Techniques for Managing Society: A Critique." *Public Administration Review* 33, no. 2 (March): 157.

Hoos, Ida R. 1979. "Societal Aspects of Technology Assessment." *Technological Forecasting and Social Change* 13, no. 3 (April 1): 191–202.

Hoos, Ida R. 1983. *Systems Analysis in Public Policy: A Critique.* Berkeley: University of California Press.

Hoppe, Robert. 2005. "Rethinking the Science-Policy Nexus: From Knowledge Utilization and Science Technology Studies to Types of Boundary Arrangements." *Poiesis and Praxis* 3, no. 3 (May 12): 199–215.

Hornstein, Donald T. 1992. "Reclaiming Environmental Law: A Normative Critique of Comparative Risk Analysis." *Columbia Law Review* 92, no. 3: 562–633.

Howell, William G., and David E. Lewis. 2002. "Agencies by Presidential Design." *Journal of Politics* 64, no. 4: 1095–1114.

Howlett, Michael. 2005. "What Is a Policy Instrument?: Tools, Mixes, and Implementation Styles." *Designing Government: From Instruments to Governance*, edited by Pearl Eliadis, Margaret Hill and Michael Howlett, 31–50. Montreal: McGill-Queens University Press.

Howlett, Michael. 2009. "Governance Modes, Policy Regimes and Operational Plans: A Multi-Level Nested Model of Policy Instrument Choice and Policy Design." *Policy Sciences* 42, no. 1 (February 11): 73–89.

Howlett, Michael, and Raul P. Lejano. 2013. "Tales from the Crypt: The Rise and Fall (and Rebirth?) of Policy Design." *Administration and Society* 45, no. 3 (April 1): 357–81.

Huber, Michael, and Henry Rothstein. 2013. "The Risk Organisation: Or How Organisations Reconcile Themselves to Failure." *Journal of Risk Research* 16, no. 6 (June): 651–75.

Hultman, Nathan E., David M. Hassenzahl, and Steve Rayner. 2010. "Climate Risk." *Annual Review of Environment and Resources* 35: 283–303.

Hume, Brit. 2001. "Political Headlines." *Fox News: Special Report with Brit Hume*, May 17, sec. News; Domestic.

Hutter, Bridget, and Michael Power. 2005. "Organizational Encounters with Risk: An Introduction." In *Organizational Encounters with Risk*, edited by Bridget Hutter and Michael Power, 1–32. Cambridge: Cambridge University Press.

Inside EPA. 2006. "OMB Delay of Risk Definition May Allow Flexible Perchlorate Levels." *Inside EPA Weekly Report*, January 13.

Inside EPA. 2010. "EPA Scraps Staff Panel, Creating Doubts on Fate of Key Research Efforts." *Inside EPA Weekly Report*, October 21.

Institute of Medicine (IOM). 2001. *Clearing the Smoke: Assessing the Science Base for Tobacco Harm Reduction*. Washington, DC: National Academies Press.

Institute of Medicine (IOM). 2013. *Environmental Decisions in the Face of Uncertainty*. Washington, DC: National Academies Press.

Interagency Regulatory Liaison Group (IRLG). 1977. "Draft Work Plans of the Interagency Regulation Liaison Group." *Chemical Regulation Reporter* 1: 1426–32.

Interagency Regulatory Liaison Group (IRLG). 1979. *Scientific Bases for Identification of Potential Carcinogens and Estimation of Risks. Report of the Interagency Regulatory Liaison Group, Work Group on Risk Assessment*. Washington, DC: Executive Office, IRLG. February 17. Republished in *Journal of National Cancer Institute* 63: 241–68.

Irwin, Alan, Henry Rothstein, Steven Yearley, and Elaine McCarthy. 1997. "Regulatory Science—Towards a Sociological Framework." *Futures* 29, no. 1: 17–31.

Jamieson, Dale. 1996. "Scientific Uncertainty and the Political Process." *Annals of the American Academy of Political and Social Science* 545, no. 1 (May 1): 35–43.

Jardine, Cindy, Steve Hrudey, John Shortreed, Lorraine Craig, Daniel Krewski, Chris Furgal, and Stephen McColl. 2003. "Risk Management Frameworks for Human Health and Environmental Risks." *Journal of Toxicology and Environmental Health Part B: Critical Reviews* 6, no. 6: 569–718.

Jardini, David R. 2000. "Out of the Blue Yonder: The Transfer of Systems Thinking from the Pentagon to the Great Society." In *Systems, Experts and Computers*, edited by A. C. Hughes and T. P. Hughes, 311–57. Cambridge, MA: MIT Press.

Jasanoff, Sheila. 1982. "Science and the Limits of Administrative Rule-Making: Lessons from the OSHA Cancer Policy." *Osgoode Hall Law Journal* 20: 536.

Jasanoff, Sheila. 1985. "Peer Review in the Regulatory Process." *Science, Technology, and Human Values* 10, no. 3: 20–32.

Jasanoff, Sheila. 1987a. "Contested Boundaries in Policy-Relevant Science." *Social Studies of Science* 17, no. 2: 195–230.

Jasanoff, Sheila. 1987b. "EPA's Regulation of Daminozide: Unscrambling the Messages of Risk." *Science, Technology, and Human Values* 12, no. 3/4: 116–24.

Jasanoff, Sheila. 1990. *The Fifth Branch: Science Advisers as Policymakers*. Cambridge, MA: Harvard University Press.

Jasanoff, Sheila. 1992. "Science, Politics, and the Renegotiation of Expertise at EPA." *Osiris* 7: 194–217.

Jasanoff, Sheila. 1993a. "Bridging the Two Cultures of Risk Analysis." *Risk Analysis* 13, no. 2: 123–29.

Jasanoff, Sheila. 1993b. "Relating Risk Assessment and Risk Management." *EPA Journal* 19. (Jan/Feb/Mar): 35–37.

Jasanoff, Sheila. 1995. "Procedural Choices in Regulatory Science." *Technology in Society* 17, no. 3: 279–93.

Jasanoff, Sheila. 1998. "Science and Judgment in Environmental Standard Setting." *Applied Measurement in Education* 11, no. 1: 107–20.

Jasanoff, Sheila. 1999. "The Songlines of Risk." *Environmental Values* 8, no. 2: 135–52.

Jayjock, Michael A., Jeremiah Lynch, and Deborah Imel Nelson. 2000. *Risk Assessment Principles for the Industrial Hygienist*. Fairfax: American Industrial Hygiene Association.

Johnson, Barry L., and James J. Reisa. 2003. "Essays in Commemoration of the 20th Anniversary of the National Research Council's Risk Assessment in the Federal Government: Managing the Process." *Human and Ecological Risk Assessment* 9, no. 5: 1093–99.

Junginger, Sabine. 2014. "Towards Policy-Making as Designing: Policy-Making beyond Problem-Solving and Decision-Making." In *Design for Policy*, edited by Christian Bason, 57–69. London: Routledge.

Kalberg, Stephen. 1980. "Max Weber's Types of Rationality: Cornerstones for the Analysis of Rationalization Processes in History." *American Journal of Sociology* 85, no. 5: 1145–79.

Kantrowitz, Arthur. 1967. "Proposal for an Institution for Scientific Judgment." *Science* 156, no. 3776: 763–64.

Kantrowitz, Arthur. 1975. "Controlling Technology Democratically." *American Scientist* 63, no. 5: 505–9.

Kassim, Tarek A., and Kenneth J. Williamson. 2005. *Environmental Impact Assessment of Recycled Wastes on Surface and Ground Waters: Concepts; Methodology and Chemical Analysis*. New York: Springer Science and Business Media.

Kelly, Kathryn E. 1995. "Cleaning Up EPA's Dioxin Mess: Why Regulate." *Wall Street Journal*. June 29.

Kittel, Charles. 1947. "The Nature and Development of Operations Research." *Science* 105, no. 2719 (February 7): 150–53.

Klein, Julie Thompson. 1996. *Crossing Boundaries: Knowledge, Disciplinarities, and Interdisciplinarities*. Charlottesville: University of Virginia Press.

Kramer, Fred A. 1975. "Policy Analysis as Ideology." *Public Administration Review* 35, no. 5: 509–17.

Krimsky, Sheldon, and Dominic Golding. 1992. *Social Theories of Risk*. Westport, CT: Praeger.

Kuehn, Robert R. 1996. "The Environmental Justice Implications of Quantitative Risk Assessment." *University of Illinois Law Review*, no. 1: 103–72.

Kuzmack, Arnold M., and Robert E. McGaughy. 1975. *Quantitative Risk Assessment for Community Exposure to Vinyl Chloride*. US Environmental Protection Agency (EPA), Office of Planning and Management and Office of Health and Ecological Effects.

Kysar, Douglas A. 2010. *Regulating from Nowhere: Environmental Law and the Search for Objectivity*. New Haven, CT: Yale University Press.

Lallement, Michel. 2013. *Tensions majeures: Max Weber, l'économie, l'érotisme*. Paris: Editions Gallimard.

Landau, Martin. 1992. "Catastrophic Errors and Bureaucracy." In *The State of Public Bureaucracy*, edited by Larry B. Hill, 217–22. Armonk, NY: Sharpe, Inc.

Landy, Marc K., Marc J. Roberts, and Stephen Thomas. 1994. *The Environmental Protection Agency: Asking the Wrong Questions from Nixon to Clinton*. New York: Oxford University Press.

Langley, Ann. 1989. "In Search of Rationality: The Purposes Behind the Use of Formal Analysis in Organizations." *Administrative Science Quarterly* 34, no. 4: 598–631.

Langley, Ann. 1990. "Patterns in the Use of Formal Analysis in Strategic Decisions." *Organization Studies* 11, no. 1 (January): 17–45.

Langley, Ann. 1991. "Formal Analysis and Strategic Decision Making." *Omega* 19, no. 2 (January 1): 79–99.

Laroche, Hervé. 1995. "From Decision to Action in Organizations: Decision-Making as a Social Representation." *Organization Science* 6, no. 1 (February 1): 62–75.

Lascoumes, Pierre, and Patrick Le Gales. 2007. "Introduction: Understanding Public Policy through Its Instruments. From the Nature of Instruments to the Sociology of Public Policy Instrumentation." *Governance* 20, no. 1: 1–21.

Lash, Scott, Bronislaw Szerszynski, and Brian Wynne. 1996. *Risk, Environment and Modernity: Towards a New Ecology*. London: SAGE.

Lasswell, Harold. 1942. "The Developing Science of Democracy." In *The Future of Government in the United States: Essays in Honor of Charles E. Merriam*, edited by L. D. White, 25–48. Chicago: University of Chicago Press.

Lasswell, Harold. 1951. "The Policy Orientation." In *The Policy Sciences: Recent Developments in Scope and Method*, edited by Daniel Lerner and Harold Lasswell, 3–15. Stanford, CA: Stanford University Press.

Lave, Lester B. 1982. *Quantitative Risk Assessment in Regulation*. Washington, DC: Brookings Institution.

Law, John. 1994. *Organizing Modernity*. Oxford: Blackwell.

Layzer, Judith A. 2012. *Open for Business: Conservatives' Opposition to Environmental Regulation*. Cambridge, MA: MIT Press.

Lehman, Arnold J., Wilbur I. Patterson, Bernard Davidow, and Ernest C. Hagan. 1955. "Procedures for the Appraisal of the Toxicity of Chemicals in Foods, Drugs and Cosmetics." *Food, Drug and Cosmetics Law Journal* 10: 679.

Leiss, William. 2000. "Between Expertise and Bureaucracy: Risk Management Trapped at the Science-Policy Interface." In *Risky Business: Canada's Changing Science-based Policy and Regulatory Regime*, edited by Bruce *Doern and Ted Reed*, 49–74. Toronto: University of Toronto Press.

Leung, Hon-Wing, and Dennis J. Paustenbach. 1995. "Physiologically Based Pharmacokinetic and Pharmacodynamic Modeling in Health Risk Assessment and

Characterization of Hazardous Substances." *Toxicology Letters*, Decision Substances Methodologies for Human Health Risk Assessment of Toxic Substances, 79, no. 1 (September 1): 55–65.

Levenstein, Charles, and John Wooding. 1997. *Work, Health, and Environment: Old Problems, New Solutions*. New York: Guilford Press.

Lezaun, Javier. 2006. "Creating a New Object of Government: Making Genetically Modified Organisms Traceable." *Social Studies of Science* 36, no. 4: 499–531.

Lindblom, Charles E. 1959. "The Science of 'Muddling Through.'" *Public Administration Review* 19, no. 2: 79–88.

Lindblom, Charles E., and David K. Cohen. 1979. *Usable Knowledge: Social Science and Social Problem Solving*. Vol. 21. New Haven, CT: Yale University Press.

Linder, Stephen H., and B. Guy Peters. 1984. "From Social Theory to Policy Design." *Journal of Public Policy* 4, no. 3: 237–59.

Linder, Stephen H., and B. Guy Peters. 1995. "The Two Traditions of Institutional Designing: Dialogue versus Decision?" In *Institutional Design*, edited by David Weimer, 133–60. Boston: Kluwer Academic Publishers.

Linders, Jan B. H. J. 2012. *Modelling of Environmental Chemical Exposure and Risk*. Dordrecht: Kluwer Academic Publishers.

Lippincott, Don. 1985. "Environmental Protection Agency: Ruckelshaus Returns." Kennedy School of Government Case Program, Harvard University, Cambridge, MA.

Lodge, Martin, and Kai Wegrich. 2016. "The Rationality Paradox of Nudge: Rational Tools of Government in a World of Bounded Rationality." *Law and Policy* 38, no. 3 (July 1): 250–67.

Longest, Henry L. II, and Bruce Diamond. 1992. "Implementing the Deputy Director's Risk Characterization Memorandum." US EPA Office of Solid Waste and Emergency Response, May 26.

Lowrance, William W. 1976. *Of Acceptable Risk: Science and the Determination of Safety*. Los Altos, CA: William Kaufman.

Lynch, Jeremiah. 1996. "Risk Reform." *American Industrial Hygiene Association Journal* 57, no. 9: 798.

Macey, Jonathan R. 1992. "Organizational Design and Political Control of Administrative Agencies." *Journal of Law, Economics, and Organization* 8, no. 1: 93–110.

Main, Jeremy. 1991. "The Big Clean-up Gets It Wrong." *Fortune*, May 20.

Majone, Giandomenico. 1992. *Evidence, Argument, and Persuasion in the Policy Process*. New Haven, CT: Yale University Press.

Majone, Giandomenico, and Edward S. Quade, eds. 1980. *Pitfalls of Analysis*. Chichester, UK: International Institute for Applied Systems Analysis and John Wiley and Sons.

March, James G. 1982. "Theories of Choice and Making Decisions." *Society* 20, no. 1: 29–39.

March, James G. 2006. "Rationality, Foolishness, and Adaptive Intelligence." *Strategic Management Journal* 27, no. 3: 201–14.

March, James G., and Herbert A. Simon. 1958. *Organizations*. New York: John Wiley & Sons.

Marcus, Alfred Allen. 1980. *Promise and Performance: Choosing and Implementing an Environmental Policy*. Vol. 39. Westport, CT: Greenwood Press.

Martin, Brian, and Eveleen Richards. 1995. "Scientific Knowledge, Controversy, and Public Decision Making." In *Handbook of Science and Technology Studies*, edited by S. Jasanoff, G. Markle, J. Petersen, and T. Pinch, 506–25. Thousand Oaks, CA: SAGE Publications.

Mayo, Deborah G., and Rachelle D. Hollander. 1991. *Acceptable Evidence: Science and Values in Risk Management*. New York: Oxford University Press.

Mazmanian, Melissa, and Christine M. Beckman. 2018. "'Making' Your Numbers: Engendering Organizational Control through a Ritual of Quantification." *Organization Science* 29, no. 3 (April 13): 357–79.

Mazur, Allan. 1973. "Disputes between Experts." *Minerva* 11, no. 2: 243–62.

Mazur, Allan. 1977. "Science Courts." *Minerva* 15, no. 1: 1–14.

McCarthy, Gina, and Janet McGabe. 2018. "Scott Pruitt's Attack on Science Would Paralyze the E.P.A." *New York Times*, March 26 edition, sec. Opinion.

McClellan, Roger O. 1994. "Science and Judgment in Risk Assessment—A Commentary on the NRC Report." *Regulatory Toxicology and Pharmacology* 20, no. 3, Part 2 (December): S142–S168.

McClellan, Roger O. 2003. "Risk Assessment: Replacing Default Options with Specific Science." *Human and Ecological Risk Assessment* 9, no. 1: 421–38.

McClellan, Roger, and Warner North. 1994. "Making Full Use of Scientific Information in Risk Assessment." In *Science and Judgment in Risk Assessment*, 629–40. Washington, DC: National Academies Press.

McCray, Lawrence. 1983. "An Anatomy of Risk Assessment: Scientific and Extra-Scientific Components in the Assessment of Scientific Data on Cancer Risks." Washington, DC: National Research Council.

McCubbins, M. D. 1985. "The Legislative Design of Regulatory Structure." *American Journal of Political Science* 29, no. 4: 721–48.

McGarity, Thomas O. 1978–1979. "Substantive and Procedural Discretion in Administrative Resolution of Science Policy Questions: Regulating Carcinogens in EPA and OSHA." *Georgetown Law Journal* 67: 729–810.

McGarity, Thomas O. 1983. "OSHA's Generic Carcinogen Policy: Rule Making under Scientific and Legal Uncertainty." *Law and Science in Collaboration: Resolving Regulatory Issues in Science and Technology* 78: 55–104.

McGarity, Thomas O. 1986. "Regulatory Reform in the Reagan Era." *Maryland Law Review* 45, no. 2: 253–73.

McGarity, Thomas O. 1991a. "The Internal Structure of EPA Rulemaking." *Law and Contemporary Problems* 54, no. 4: 57–111.

McGarity, Thomas O. 1991b. *Reinventing Rationality: The Role of Regulatory Analysis in the Federal Bureaucracy*. Cambridge: Cambridge University Press.

McGarity, Thomas O. 1996. "The Expanded Debate over the Future of the Regulatory State." *University of Chicago Law Review* 63, no. 4: 1463–1532.

McGarity, Thomas O. 1998. "A Cost-Benefit State." *Administrative Law Review* 50: 7–80.

McGarity, Thomas O. 2001. "Deflecting the Assault: How EPA Survived a 'Disorganized Revolution' by 'Reinventing' Itself a Bit." *Environmental Law Reporter* 31, no. 11249: 11259–60.

McGarity, Thomas O. 2003a. "On the Prospect of 'Daubertizing' Judicial Review of Risk Assessment." *Law and Contemporary Problems* 66, no. 4: 155–225.

McGarity, Thomas O. 2003b. "Our Science Is Sound Science and Their Science Is Junk Science: Science-Based Strategies for Avoiding Accountability and Responsibility for Risk-Producing Products and Activities." *University of Kansas Law Review* 52: 897–937.

McGarity, Thomas O. 2012. "EPA at Helm's Deep: Surviving the Fourth Attack on Environmental Law." *Fordham Environmental Law Review* 24, no. 2: 205–41.

McGarity, Thomas O., and Wendy Elizabeth Wagner. 2008. *Bending Science: How Special Interests Corrupt Public Health Research*. Cambridge, MA: Harvard University Press.

McKone, Thomas E., and Kenneth T. Bogen. 1991. "Predicting the Uncertainties in Risk Assessment." *Environmental Science and Technology* 25, no. 10: 1674–81.

McQueen, Charlene. 2010. *Comprehensive Toxicology*. Amsterdam: Elsevier.

Meltsner, Arnold J. 1976. *Policy Analysts in the Bureaucracy*. San Francisco: University of California Press.

Meyer, Alan D. 1984. "Mingling Decision Making Metaphors." *Academy of Management Review* 9, no. 1: 6–17.

Meyer, John, and Brian Rowan. 1977. "Institutionalized Organizations: Formal Structure as Myth and Ceremony." *American Journal of Sociology* 83, no. 2: 340–63.

Michaels, David, ed. 2008. *Doubt Is Their Product: How Industry's Assault on Science Threatens Your Health*. Oxford: Oxford University Press.

Miller, Clark. 2017. "It's Not a War on Science." *Issues in Science and Technology*, Spring.

Mintzberg, Henry. 1980. "Beyond Implementation: An Analysis of The Resistance to Policy Analysis." *INFOR: Information Systems and Operational Research* 18, no. 2 (May 1): 100–134.

Mirer, Franklin E. 2003. "Distortions of the "Mis-Read" Book: Adding Procedural Botox to Paralysis by Analysis." *Human and Ecological Risk Assessment: An International Journal* 9, no. 5 (September 1, 2003): 1129–43.

Mirowski, Philip. 1999. "Cyborg Agonistes: Economics Meets Operations Research in Mid-Century." *Social Studies of Science* 29, no. 5: 685–718.

Mirowski, Philip. 2002. *Machine Dreams: Economics Becomes a Cyborg Science*. Cambridge: Cambridge University Press.

MITRE. 1994. *Assessment of the Scientific and Technical Laboratories and Facilities of the U.S. Environmental Protection Agency*. MTR 94W0000082V1. McLean, VA, and Washington, DC: Center for Environment, Resources and Space.

Moe, Terry M. 1989. "The Politics of Bureaucratic Structure." In *Can the Government Govern?*, edited by John E. Chubb and Paul E. Peterson, 285–323. Washington, DC: Brookings Institution.

Moffitt, Susan L. 2014. *Making Policy Public: Participatory Bureaucracy in American Democracy*. Cambridge: Cambridge University Press.

Mooney, Chris. 2004. "Paralysis by Analysis." *Environmental Forum* 36, no. 5 (May 1): 38–42.

Morgan, Granger, Max Henrion, and Mitchell Small. 1992. *Uncertainty: A Guide to Dealing with Uncertainty in Quantitative Risk and Policy Analysis*. Cambridge: Cambridge University Press.

Morgenstern, Richard, and Stuart Sessions. 1988. "EPA's Unfinished Business." *Environment* 30, no. 6 (1988): 14.

Mosher, Lawrence. 1983. "'Acceptable Risk'—Can the Government Decide Whether to Be Safe or Sorry?" *National Journal*, December 3.

Muggli, Monique E., Richard D. Hurt, and James Repace. 2004. "The Tobacco Industry's Political Efforts to Derail the EPA Report on ETS." *American Journal of Preventive Medicine* 26, no. 2 (February): 167–77.

Nakashima, Ellen. 2002. "For Bush's Regulatory "Czar," The Equation Is Persuasion." *Washington Post*, March 10, 35.

Nash, Linda. 2017. "From Safety to Risk: The Cold War Contexts of American Environmental Policy." *Journal of Policy History* 29, no. 1 (January): 1–33.

National Academy of Sciences (NAS). 1991. "Report on an Intergovernmental Public Meeting on Risk Assessment in the Federal Government: Asking the Right Questions." Washington, DC: National Academy of Sciences, November 19.

National Academy of Sciences (NAS). 2006. "EPA Assessment of Dioxin Understates Uncertainty about Health Risks and May Overstate Human Cancer Risk," July 11.

National Research Council (NRC). 1969. "Guidelines for Estimating Toxicologically Insignificant Levels of Chemicals in Food." In *Evaluating the Safety of Food Chemicals*. Washington, DC: National Academies Press.

National Research Council (NRC). 1977. *Decision Making in the Environmental Protection Agency: A Report to the U.S. Environmental Protection Agency*. Washington, DC: National Academies Press.

National Research Council (NRC). 1981. *The Handling of Risk Assessments in NRC Reports. A Report to the Governing Board, National Research Council, by the Governing Board Committee on the Assessment of Risk*. Washington, DC: National Academies Press.

National Research Council (NRC). 1982. *Risk and Decision Making: Perspectives and Research*. Washington, DC: National Academies Press.

National Research Council (NRC). 1983. *Risk Assessment in the Federal Government: Managing the Process*. Washington, DC: National Academies Press.

National Research Council (NRC). 1989. *Improving Risk Communication*. Washington, DC: National Academies Press.

National Research Council (NRC). 1991. *Human Exposure Assessment for Airborne Pollutants: Advances and Opportunities*. Washington, DC: National Academies Press.

National Research Council (NRC). 1993a. *Issues in Risk Assessment*. Washington, DC: National Academies Press.

National Research Council (NRC). 1993b. *Pesticides in the Diets of Infants and Children*. Washington, DC: National Academies Press.

National Research Council (NRC). 1993c. *Research to Protect, Restore, and Manage the Environment*. Washington, DC: National Academies Press.

National Research Council (NRC). 1994. *Science and Judgment in Risk Assessment.* Washington, DC: National Academies Press.

National Research Council (NRC). 1996. *Understanding Risk: Informing Decisions in a Democratic Society*. Washington, DC: National Academies Press.

National Research Council (NRC). 1997. *Building a Foundation for Sound Environmental Decisions*. Washington, DC: National Academies Press.

National Research Council (NRC). 2000. *Strengthening Science at the U.S. Environmental Protection Agency: Research-Management and Peer-Review Practices*. Washington, DC: National Academies Press.

National Research Council (NRC). 2006a. *Assessing the Human Health Risks of Trichloroethylene: Key Scientific Issues*. Washington, DC: National Academies Press.

National Research Council (NRC). 2006b *Health Risks from Dioxin and Related Compounds: Evaluation of the EPA Reassessment.* Washington, DC: National Academies Press.

National Research Council (NRC). 2007. *Toxicity Testing in the 21st Century: A Vision and a Strategy*. Washington, DC: National Academies Press.

National Research Council (NRC). 2008. *Phthalates and Cumulative Risk Assessment: The Tasks Ahead*. Washington, DC: National Academies Press.

National Research Council (NRC). 2009. *Science and Decisions: Advancing Risk Assessment.* Washington, DC: National Academies Press.

National Research Council (NRC). 2010. *Review of the Environmental Protection Agency's Draft IRIS Assessment of Tetrachloroethylene*. Washington, DC: National Academies Press.

National Research Council (NRC). 2011a. *Review of the Environmental Protection Agency's Draft IRIS Assessment of Formaldehyde*. Washington, DC: National Academies Press.

National Research Council (NRC). 2011b. *Sustainability and the U.S. EPA*. Washington, DC: National Academies Press.

Nelkin, Dorothy. 1971a. *Nuclear Power and Its Critics: The Cayuga Lake Controversy*. Ithaca, NY: Cornell University Press.

Nelkin, Dorothy. 1971b. "Scientists in an Environmental Controversy." *Social Studies of Science* 1, no. 3–4 (January 10): 245–61.

Nelkin, Dorothy. 1975. "The Political Impact of Technical Expertise." *Social Studies of Science* 5, no. 1: 35–54.

Nelkin, Dorothy. 1984. *Controversy: The Politics of Technical Decision*. Beverly Hills, CA: SAGE.

Nelkin, Dorothy. 1985. *The Language of Risk: Conflicting Perspectives on Occupational Health*. London: SAGE Publications.

Nelkin, Dorothy. 1995. "Science Controversies the Dynamics of Public Disputes in the United States." In *Handbook of Science and Technology Studies*, edited by Sheila Jasanoff, Gerald Markle, James Peterson, and Trevor Pinch, 444–56. Thousand Oaks, CA: SAGE Publications, Inc.

Newman, Michael C. 1994. *Quantitative Methods in Aquatic Ecotoxicology*. Boca Raton, FL: Lewis Publishers.

Newman, Michael C. 2014. *Fundamentals of Ecotoxicology: The Science of Pollution, Fourth Edition*. Boca Raton, FL: Lewis Publishers.

Newman, Michael, and David Rosenberg. 1985. "Systems Analysts and the Politics of Organizational Control." *Omega* 13, no. 5 (January 1): 393–406.

Nichols, Albert L., and Richard J. Zeckhauser. 1986. "Perils of Prudence: How Conservative Risk Assessments Distort Regulation, The Regulating Risk." *Regulation* 10: 13–24.

Nicolaidis, Nicholas G. 1960. "Policy-Decision and Organization Theory." PhD dissertation, University of Southern California.

North, Warner D. 2003. "Reflections on the Red/Mis-Read Book, 20 Years After." *Human and Ecological Risk Assessment* 9, no. 5: 1145–54.

Norton, Bryan G. 2005. *Sustainability: A Philosophy of Adaptive Ecosystem Management*. Chicago: University of Chicago Press.

O'Brien, Mary. 2000. *Making Better Environmental Decisions: An Alternative to Risk Assessment*. Cambridge, MA: MIT Press.

Office of Management and Budget (OMB). 1990. "Appendix C: Current Regulatory Issues in Risk Assessment and Risk Management," Regulatory Program of the United States Government April 1, 1990—March 31, 1991, 13–26. Washington, DC: US Government Printing Office.

Office of Management and Budget (OMB). 2003a. "Draft 2003 Report to Congress on the Costs and Benefits of Federal Regulations." Washington, DC: Office of Management and Budget, Office of Information and Regulatory Affairs. February 3.

Office of Management and Budget (OMB). 2003b. *Informing Regulatory Decisions: 2003 Report to Congress on the Costs and Benefits of Federal Regulations and Unfunded Mandates on State, Local, and Tribal Entities*. Washington, DC: Office of Management and Budget, Office of Information and Regulatory Affairs.

Office of Management and Budget (OMB). 2006. *Proposed Risk Assessment Bulletin.* Washington, DC: Office of Management and Budget, Office of Information and Regulatory Affairs.

Office of Management and Budget (OMB). 2007. "Memorandum M-07–24, Updated Principles for Risk Analysis." Washington, DC: Office of Management and Budget, Office of Information and Regulatory Affairs, September.

Office of Technology Assessment (OTA). 1981. "Assessment of Technologies for Determining Cancer Risks from the Environment." Washington, DC: US Government Printing Office.

Office of Technology Assessment (OTA). 1993. *Researching Health Risks.* Office of Technology Assessment, US Congress. Washington, DC: DIANE Publishing.

O'Leary, Rosemary. 1995. *Environmental Change. Federal Courts and the EPA.* Philadelphia: Temple University Press.

Oliver, Christine. 1991. "Strategic Responses to Institutional Processes." *Academy of Management Review* 16, no. 1 (January 1): 145–79.

O'Malley, Pat 2004. *Risk, Uncertainty and Government.* London: GlassHouse Press.

Omenn, Gilbert S. 2006. "Public Comment on OMB Draft Risk Assessment Bulletin." *Regulatory Analysis—AEI-Brookings Joint Center for Regulatory Studies* 6, no. 4 (June): 1–10.

Omenn, Gilbert S. 2011. "Physician-Scientist: Linking Science, Medicine, and Public Policy." In *Medicine Science and Dreams: The Making of Physician-Scientists*, edited by David A. Schwartz, 269–87. New York: Springer.

Oren, Craig N. 1991. "The Clean Air Act Amendments of 1990: A Bridge to the Future." *Environmental Law* 21, no. 4: 1817–42.

Ostrom, Elinor, Thomas Dietz, Nives Dolsak, Paul C. Stern, Susan Stonich, and Elke U. Weber. 2002. *The Drama of the Commons.* Washington, DC: National Academies Press.

Parsons, Talcott. 1956. "Suggestions for a Sociological Approach to the Theory of Organizations-I." *Administrative Science Quarterly* 1, no. 1: 63–85.

Pasztor, Andy. 1983. "EPA Changes Stand On Formaldehyde For Health Reasons." *Wall Street Journal.* November 18, sec. 1.

Pasztor, Andy. 1984. "EPA Decides to Crack Down on the Use Of Formaldehyde in Certain Industries." *Wall Street Journal.* May 18.

Pasztor, Andy. 1986. "EPA Issues Rules on Arsenic Emissions, Easing Limits It Had Proposed Earlier." *Wall Street Journal.* July 8.

Patton, Dorothy E., and Robert J. Huggett. 2003. "The Risk Assessment Paradigm as a Blueprint for Environmental Research." *Human and Ecological Risk Assessment* 9, no. 5: 1337–48.

Pellizzoni, Luigi. 2009. "Revolution or Passing Fashion? Reassessing the Precautionary Principle." *International Journal of Risk Assessment and Management* 12, no. 1: 14–34.

Percival, Robert V. 2011. "Who's in Charge-Does the President Have Directive Authority over Agency Regulatory Decisions." *Fordham Law Review* 79, no. 6: 2487–2540.

Perrow, Charles. 1982. "Not Risk but Power." *Contemporary Sociology* 11, no. 3: 298–300.

Perrow, Charles. 1984. *Normal Accidents: Living with High-Risk Technologies.* New York: Basic Books.

Pfiffner, John M. 1960. "Administrative Rationality." *Public Administration Review* 20, no. 3: 125.

Porter, Theodore M. 1992. "Quantification and the accounting ideal in science." *Social Studies of Science* 22, no. 4: 633–51.

Porter, Theodore M. 1995. *Trust in Numbers: The Pursuit of Objectivity in Science and Public Life*. Princeton, NJ: Princeton University Press.

Powell, Mark R. 1999. *Science at EPA: Information in the Regulatory Process.* Washington, DC: Resources for the Future.

Power, Michael. 1997. *The Audit Society: Rituals of Verification.* Oxford: Oxford University Press.

Power, Michael. 2004. *The Risk Management of Everything: Rethinking the Politics of Uncertainty.* London: Demos.

Power, Michael. 2007. *Organized Uncertainty: Designing a World of Risk Management.* Oxford: Oxford University Press.

Power, Michael. 2014. "Risk, Social Theories, and Organizations." In *Handbook of Sociology, Social Theory, and Organization Studies: Contemporary Currents*, edited by Paul Adler, Paul du Gay, Glenn Morgan, and Michael Reed, 370–92. Oxford: Oxford University Press.

Power, Michael. 2016. "Introduction—Riskwork: The Organizational Life of Risk Management." In *Riskwork: Essays on the Organizational Life of Risk Management*, edited by Michael Power, 1–25. Oxford: Oxford University Press.

Power, Michael, and Lynn S. McCarty. 1998. "A Comparative Analysis of Environmental Risk Assessment/Risk Management Frameworks." *Environmental Science and Technology* 32, no. 9: 224A–231A.

Power, Michael, and Lynn S. McCarty. 2002. "Trends in the Development of Ecological Risk Assessment and Management Frameworks." *Human and Ecological Risk Assessment* 8, no. 1: 7–18.

Power, Michael, Tobias Scheytt, Kim Soin, and Kerstin Sahlin. 2009. "Reputational Risk as a Logic of Organizing in Late Modernity." *Organization Studies* 30, no. 2–3 (February 1): 301–24.

Presidential and Congressional Commission on Risk Assessment and Risk Management (PCCRARM). 1997a. *Framework for Environmental Health Risk Management: Final Report, Volume 1*. Washington, DC.

Presidential and Congressional Commission on Risk Assessment and Risk Management (PCCRARM). 1997b. *Framework for Environmental Health Risk Management: Risk Assessment and Risk Management in Regulatory Decision-Making: Final Report; Volume 2*. Washington DC.

Preuss, Peter W., and William H. Farland. 1993. "A Flagship Risk Assessment." *EPA Journal* 19, no. 1: 24–25.

Primack, Joel. 1975. "An Introduction to the Issues." *Bulletin of the Atomic Scientists* 31, no. 7 (September 1): 15–19.

Proctor, Robert. 1995. *Cancer Wars: How Politics Shapes What We Know and Don't Know about Cancer*. New York: Basic Books.

Quarles, John, and William H. Lewis. 1991. "The New Clean Air Act." *American Ceramic Society Bulletin* 70, no. 3: 405–8.

Raiffa, Howard. 1982. *The Art and Science of Negotiation*. Cambridge, MA: Harvard University Press.

Rampton, Sheldon. 2001. "Industry Ally John Graham Is Wrong Choice to Be Nation's Regulatory Gatekeeper." *Center for Media and Democracy's PR Watch* (blog), May 16. http://www.prwatch.org/spin/2001/05/429/industry-ally-john-gra.

Rasmussen, Norman. 1975. "An Assessment of Accident Risks in US Commercial Nuclear Power Plants." WASH-1400, NU-REG 75/014. Washington, DC: US Nuclear Regulatory Commission.

Rayner, Steve. 2003. "Who's in Charge?: The Worldwide Displacement of Democratic Judgment in an Age of Expert Assessment." *Economic and Political Weekly* 38, no. 48: 5113–19.

Rayner, Steve. 2007. "The Rise of Risk and the Decline of Politics." *Environmental Hazards* 7, no. 5: 165–72.

Reed, Michael. 2005. "Beyond the Iron Cage? Bureaucracy and Democracy in the Knowledge Economy and Society." In *The Values of Bureaucracy*, edited by Paul du Gay, 115–40. Oxford: Oxford University Press.

Reich, Robert B. 1985. "Public Administration and Public Deliberation: An Interpretive Essay." *Yale Law Journal* 94, no. 7: 1617–41.

Renn, Ortwin. 2005. *White Paper No. 1. Risk Governance: Towards an Integrative Approach*. Geneva: International Risk Governance Council.

Renn, Ortwin. 2008. *Risk Governance: Coping with Uncertainty in a Complex World.* London: Earthscan.

Renn, Ortwin, and Katherine Walker. 2008. *Lessons Learned: A Re-Assessment of the IRGC Framework on Risk Governance*. New York: Springer.

Rest, Kathleen M., and Michael H. Halpern. 2007. "Politics and the Erosion of Federal Scientific Capacity: Restoring Scientific Integrity to Public Health Science." *American Journal of Public Health* 97, no. 11: 1939–44.

Rich, Robert F., and Kelly R. Merrick. 2007. "Use and Misuse of Science: Global Climate Change and the Bush Administration." *Virginia Journal of Social Policy and Law* 14, no. 3: 223–52.

Riles, Annelise. 2004. "Real Time: Unwinding Technocratic and Anthropological Knowledge." *American Ethnologist* 31, no. 3: 392–405.

Rip, Arie. 1986. "The Mutual Dependence of Risk Research and Political Context." *Science and Technology Studies* 4, no. 3: 3–15.

Risk Policy Report. 1997. "ORD Creates "Composite" Agency-Wide Risk-Based Research Priorities." *Risk Policy Report,* March 21.

Risk Policy Report. 2010a. "Anastas' Priorities Seek To Boost Sustainability, Make ORD More Relevant." *Risk Policy Report,* April 20.

Risk Policy Report. 2010b. "NAS Panel Sees Hurdle in Bid to Craft Novel Sustainable Approach for EPA." *Inside EPA Weekly Report,* December 3.

Rittel, Horst W. J., and Melvin M. Webber. 1973. "Dilemmas in a General Theory of Planning." *Policy Sciences* 4, no. 2: 155–69.

Roberts, Leslie. 1991. "Dioxin Risks Revisited." *Science* 251, no. 4994: 624–26.

Rodricks, Joseph V. 2007. *Calculated Risks: Understanding the Toxicity and Human Health Risks of Chemicals in Our Environment.* 2nd ed. Cambridge: Cambridge University Press.

Rosenbaum, Walter A. 1995. "The Bureaucracy and Environmental Policy." In *Environmental Politics and Policy: Theories and Evidence,* edited by James P. Lester, 212–37. Durham, NC: Duke University Press.

Rothstein, Henry. 2006. "The Institutional Origins of Risk: A New Agenda for Risk Research." *Health, Risk and Society* 8, no. 3: 215–21.

Rothstein, Henry, and John Downer. 2012. "'Renewing DEFRA': Exploring the Emergence of Risk-Based Policy-Making in UK Central Government." *Public Administration* 90, no. 3: 781–99.

Rothstein, Henry, Michael Huber, and George Gaskell. 2006. "A Theory of Risk Colonization: The Spiralling Regulatory Logics of Societal and Institutional Risk." *Economy and Society* 35, no. 1 (February): 91–112.

Rowe, William D. 1975. *An "Anatomy" of Risk*. Washington, DC: US Environmental Protection Agency (EPA), March.

Rowe, William D. 1977. *An Anatomy of Risk*. New York: Wiley.

Ruckelshaus, William D. 1983a. "How E.P.A Faces the Arsenic Risk." *New York Times*. July 23, 22.

Ruckelshaus, William D. 1983b. "Our Challenge: Remarks at the Semi-annual Meeting of the Chemical Manufacturers Association." New York City. November 8.

Ruckelshaus, William D. 1983c. "Science, Risk and Public Policy." *Science* 221, no. 4615: 1026–28.

Ruckelshaus, William D. 1984. "Risk in a Free Society." *Risk Analysis* 4, no. 3 (September 1): 157–62.

Ruckelshaus, William D. 2017. "A Lesson Trump and the E.P.A. Should Heed." *New York Times*, March 7 edition, sec. Opinion.

Rushefsky, Mark E. 1986. *Making Cancer Policy*. Albany, NY: SUNY Press.

Russell, Milton. 1984a. "Communicating with the Public on Issues of Environmental Risk: Issues and Options. Memorandum to Al Alm." Washington, DC 20460: Office of Policy, Planning and Evaluation, US Environmental Protection Agency (EPA), January 17.

Russell, Milton. 1984b. "Equity and the Choice of Risk Measures. Memo to the Administrator." Office of Policy, Planning and Evaluation, US Environmental Protection Agency (EPA), June 7.

Saint-Martin, Denis, and Christine Rothmayr Allison. 2011. "Rationalism and Public Policy: Mode of Analysis or Symbolic Politics?" *Policy and Society* 30, no. 1 (February): 19–27.

Salter, Liora. 1988. *Mandated Science: Science and Scientists in the Making of Standards*. Dordrecht, Netherlands: Kluwer Academic Publishers.

Samet, Jonathan M., and Thomas A. Burke. 2001. "Turning Science into Junk: The Tobacco Industry and Passive Smoking." *American Journal of Public Health* 91, no. 11: 1742–44.

Sanderson, Ian. 2009. "Intelligent Policy Making for a Complex World: Pragmatism, Evidence, and Learning." *Political Studies* 57, no. 4 (December): 699–719.

Sapien, Joaquin. 2010. "How Senator Vitter Battled the EPA over Formaldehyde's Link to Cancer." *ProPublica* (blog), April 15.

Sarewitz, Daniel. 2004. "How Science Makes Environmental Controversies Worse." *Environmental Science and Policy* 7, no. 5: 385–403.

Schön, Donald A., and Martin Rein. 1994. *Frame Reflection: Toward the Resolution of Intractable Policy Controversies*. New York: Basic Books.

Schot, Johan, and Arie Rip. 1997. "The Past and Future of Constructive Technology Assessment." *Technological Forecasting and Social Change* 54, no. 2–3: 251–68.

Schwarz, Michiel, and Michael Thompson. 1990. *Divided We Stand: Re-Defining Politics, Technology and Social Choice*. Philadelphia: University of Pennsylvania Press.

Schwing, Richard C., and Walter A. Albers. 1980. *Societal Risk Assessment: How Safe Is Safe Enough?* New York: Springer.

Science Advisory Board (SAB). 1977. *Report on the Research, Development, Monitoring, and Technical Suppport System of the US EPA*. Washington, DC: United States Environmental Protection Agency, Science Advisory Board, Environmental Measurements Advisory Committee. August.

Science Advisory Board (SAB). 1983. *Review of Reorganization Proposals for Office of Research and Development Laboratories and Headquarters' Staff Offices*. Washington DC: Report of the Science Advisory Board.

Science Advisory Board (SAB). 1987. *EPA Report of the Director of the Science Advisory Board for Fiscal Year 1987*. Washington, DC: United States Environmental Protection Agency, Science Advisory Board.

Science Advisory Board (SAB). 2000. *Toward Integrated Environmental Decision-Making*. Washington, DC: United States Environmental Protection Agency, Science Advisory Board.

Scott, Richard W. 1987. "The Adolescence of Institutional Theory." *Administrative Science Quarterly* 32, no. 4: 493–511.

Scott, Richard W. 2002. "Formalization: Reflection or Construction?" *Contemporary Sociology* 31, no. 6 (November 6): 635.

Scott, Richard W. 2008. "Lords of the Dance: Professionals as Institutional Agents." *Organization Studies* 29, no. 2 (February 1): 219–38.

Sedman, Richard M., and Paul W. Hadley. 1992. "Risk Assessment and Risk Management: Mending the Schism." *Risk: Issues in Health and Safety* 3: 189–94.

Sellers, Christopher C. 1999. *Hazards of the Job: From Industrial Disease to Environmental Health Science: From Industrial Disease to Environmental Health Science*. Chapel Hill: University of North Carolina Press.

Sellers, Christopher. 2018. "How Republicans Came to Embrace Anti-Environmentalism." *Vox*, July 6.

Selznick, Philip. 1949. *TVA and the Grass Roots: A Study in the Sociology of Formal Organization*. San Francisco: University of California Press.

Selznick, Philip. 1996. "'Institutionalism' 'Old' and 'New.'" *Administrative Science Quarterly* 41, no. 2: 270–77.

Severo, Richard. 1976. "Chemical Flowing Illegally into Niagara: Toxic Chemical Flowing Illegally into the Niagara, Other Discharge Allowed a Flame-Retardant Dismay Expressed." *New York Times*, sec. New Jersey Pages.

Sewell, William H. 1992. "A Theory of Structure: Duality, Agency, and Transformation." *American Journal of Sociology* 98, no. 1 (July 1): 1–29.

Sexton, Ken. 1995. "Science and Policy in Regulatory Decision Making: Getting the Facts Right about Hazardous Air Pollutants." *Environmental Health Perspectives* 103, Suppl 6: 213.

Sexton, Ken, and Stephen H. Linder. 2014. "Integrated Assessment of Risk and Sustainability in the Context of Regulatory Decision Making." *Environmental Science and Technology* 48, no. 3: 1409–18.

Shabecoff, Philip. 1983a. "Agency Reopens Formaldehyde Regulation Issue." *New York Times*, November 18, sec. U.S.

Shabecoff, Philip. 1983b. "E.P.A. Chief Urges Uniform Pollution Assessment." *New York Times*, June 23.

Shabecoff, Philip. 1983c. "E.P.A.'s Recent Past May Overshadow Its Future." *New York Times*. March 27, sec. Week in Review.

Shabecoff, Philip. 1983d. "E.P.A. Says It Maps New Top Priority: Aide Says Toxic Wastes Will Become Main Task Under 1985 Fiscal Year Plan." *New York Times*. November 21.

Shabecoff, Philip. 1983e. "Forecast for EPA Was Stormy from the Start." *New York Times*, February 20, sec. Week in Review.

Shabecoff, Philip. 1983f. "Higher Risks Seen in New EPA Rules: Agency Aides Say 2 Proposals Exceed Past Acceptability." *New York Times*. September 18, 1.

Shabecoff, Philip. 1983g. "Threat Posed by Dioxin Subject of Growing Fear." *New York Times*. February 12.

Shabecoff, Philip. 1986. "Pesticides Finally Top the Problem List at E.P.A." *New York Times*. March 6, B12.

Shabecoff, Philip. 2003. *A Fierce Green Fire: The American Environmental Movement*. Washington, DC: Island Press.

Shapin, Steven, and Simon Schaffer. 1985. *Leviathan and the Air Pump: Hobbes, Boyle and the Politics of Experiment*. Princeton, NJ: Princeton University Press.

Shapiro, Sidney A. 2007. "OMB and the Politicization of Risk Assessment." *Environmental Law* 37, no. 4: 1083–106.

Shapiro, Sidney A. 2009. "Political Science: Regulatory Science after the Bush Administration." *Duke Journal of Constitutional Law and Public Policy* 4, no. 32: 31–43.

Shapiro, Sidney A. 2011. "Cost-Benefit Analysis: An Organizational Design Perspective." *New York University Environmental Law Journal* 19, no. 1: 194–224.

Shapiro, Sidney, and Robert Glicksman. 2003. *Risk Regulation at Risk: Restoring a Pragmatic Approach*. Stanford, CA: Stanford University Press.

Sharlin, Harold I. 1985. "EDB: A Case Study in the Communication of Health Risk." Washington, DC: US Environmental Protection Agency, January 9.

Sharlin, Harold I. 1987. "Macro-Risks, Micro-Risks, and the Media: The EDB Case." In *The Social and Cultural Construction of Risk: Essays on Risk Selection and Perception*, edited by Vincent T. Covello and Branden B. Johnson, 183–98. Dordrecht, Netherlands: D. Reidel Publishing Company.

Short, James F., Jr. 1984. "The Social Fabric at Risk: Toward the Social Transformation of Risk Analysis." *American Sociological Review* 49, no. 6: 711–25.

Shostak, Sara. 2013. *Exposed Science: Genes, the Environment, and the Politics of Population Health*. Berkeley: University of California Press.

Shrader-Frechette, Kristin Sharon. 1985. *Risk Analysis and Scientific Method*. New York: Springer.

Shrader-Frechette, Kristin Sharon. 1991. *Risk and Rationality: Philosophical Foundations for Populist Reforms*. San Francisco: University of California Press.

Shrader-Frechette, Kristin Sharon. 1995. "Evaluating the Expertise of Experts." *Risk: Health, Safety and Environment* 6, no. 2: 115–26.

Shulman, Seth. 2008. *Undermining Science: Suppression and Distortion in the Bush Administration*. Berkeley: University of California Press.

Sigel, Astrid, Helmut Sigel, and Roland K. O. Sigel. 2011. *Metal Ions in Toxicology: Effects, Interactions, Interdependencies*. Berlin: de Gruyter.

Silbergeld, Ellen. 1991. "Risk Assessment and Risk Management: An Uneasy Divorce." In *Acceptable Evidence: Science and Values in Risk Management*, edited by Deborah G. Mayo and Rachelle D. Hollander, 99–114. New York: Oxford University Press.

Silbergeld, Ellen. 1993. "Risk Assessment: The Perspective and Experience of US Environmentalists." *Environmental Health Perspectives* 101, no. 2: 101–4.

Silbergeld, Ellen. 1994. "Testimony before the Subcommittee on Technology, Environment and Aviation on Strengthening Risk Assessment within EPA." Environmental Defense Fund, March 9.

Simon, Herbert Alexander. 1969. *The Sciences of the Artificial*. Cambridge, MA: MIT Press.

Simon, Ted. 2014. *Environmental Risk Assessment: A Toxicological Approach*. New York: CRC Press.

Slovic, Paul. 1999. "Trust, Emotion, Sex, Politics, and Science: Surveying the Risk-Assessment Battlefield." *Risk Analysis* 19, no. 4: 689–701.

Slovic, Paul. 2003. "Going beyond the Red Book: The Sociopolitics of Risk." *Human and Ecological Risk Assessment* 9, no. 5: 1181–90.

Spear, Robert C., Frédéric Y. Bois, Tracey Woodruff, David Auslander, Jennifer Parker, and Steve Selvin. 1991. "Modeling Benzene Pharmacokinetics across Three Sets of Animal Data: Parametric Sensitivity and Risk Implications." *Risk Analysis* 11, no. 4: 641–54.

Spenlehauer, Vincent. 2013. "L'ancrage de l'évaluation des Politiques Publiques dans les Sciences Sociales aux Etats-Unis: Reconstruction d'un Linéament Intellectuel." *Revue Française d'Administration Publique* 148: 877–93.

Stadler, Linda. 1993. "Corrosion Proof Fittings V. EPA: Asbestos in the Fifth Circuit—A Battle of Unreasonableness." *Tulane Environmental Law Journal* 6: 423.

Star, Susan Leigh. 1992. "The Trojan Door: Organizations, Work, and the 'Open Black Box.'" *Systemic Practice and Action Research* 5, no. 4: 395–410.

Star, Susan Leigh. 1995. "The Politics of Formal Representations: Wizards, Gurus, and Organizational Complexity." In *Ecologies of Knowledge: Work and Politics in Science and Technology*, edited by Susan Leigh Star, 88–118. Albany: State University of New York Press.

Star, Susan Leigh, and James R. Griesemer. 1989. "Institutional Ecology, Translation and Boundary Objects: Amateurs and Professionals in Berkeley's Museum of Vertebrate Zoology." *Social Studies of Science* 19: 387–420.

Star, Susan Leigh, and Karen Ruhleder. 1996. "Steps toward an Ecology of Infrastructure: Design and Access for Large Information Spaces." *Information Systems Research* 7, no. 1: 111–34.

Stephan, Charles E., Donald I. Mount, David J. Hansen, John R. Gentile, Gary A. Chapman, and William A. Brungs. 1983. *Guidelines for Deriving Numerical National Water Quality Criteria for the Protection of Aquatic Organisms and Their Uses*. Washington, DC: US Environmental Protection Agency (EPA), Office of Water Regulations and Standards Criteria, PB85–227049.

Stern, Paul. 2009. "Generic Lessons about Societal Responses to Emerging Technologies as Involving Risks," http://www.wilsoncenter.org/sites/default/files/Generic_Lessons_about_Risks.pdf.

Stigler, G. J. 1971. "The Theory of Economic Regulation." *Bell Journal of Economics and Management Science* 2, no. 1: 3–21.

Stinchcombe, Arthur L. 1997. "On the Virtues of the Old Institutionalism." *Annual Review of Sociology* 23: 1–18.

Stinchcombe, Arthur L. 2001. *When Formality Works: Authority and Abstraction in Law and Organizations*. Chicago: University of Chicago Press.

Stone, Deborah. 1989. "Causal Stories and the Formation of Policy Agendas." *Political Science Quarterly* 104, no. 2: 281–300.

Stone, Deborah. 2012. *Policy Paradox: The Art of Political Decision Making*. New York: W. W. Norton and Company.

Stone, Richard. 1994a. "Can Carol Browner Reform EPA?" *Science*, January 21, 312–15.

Stone, Richard. 1994b. "Strict Ethical Rules Confound EPA Science." *Science* 263, no. 5145 (January 21): 313.

Storni, Cristiano. 2012. "Unpacking Design Practices The Notion of Thing in the Making of Artifacts." *Science, Technology and Human Values* 37, no. 1 (January 1): 88–123.

Strauss, Anselm. 1988. "The Articulation of Project Work: An Organizational Process." *Sociological Quarterly* 29, no. 2: 163–78.

Suchman, Mark C. 1995. "Managing Legitimacy: Strategic and Institutional Approaches." *Academy of Management Review* 20, no. 3: 571–610.

Suddaby, Roy, and Royston Greenwood. 2001. "Colonizing Knowledge: Commodification as a Dynamic of Jurisdictional Expansion in Professional Service Firms." *Human Relations* 54, no. 7: 933–53.

Susskind, Lawrence, Ravi Jain, and Andrew Martyniuk. 2001. *Better Environmental Policy Studies: How to Design and Conduct More Effective Analyses*. Washington, DC: Island Press.

Suter, Glenn W., II. 1990. "Uncertainty in Environmental Risk Assessment." In *Acting Under Uncertainty: Multidisciplinary Conceptions*, 203–30. New York: Springer.

Suter, Glenn W., II. 2008. "Ecological Risk Assessment in the United States Environmental Protection Agency: A Historical Overview." *Integrated Environmental Assessment and Management* 4, no. 3: 285–89.

Swidler, Ann. 1973. "The Concept of Rationality in the Work of Max Weber." *Sociological Inquiry* 43, no. 1 (January 1): 35–42.

Tardiff, Robert G., and Joseph V. Rodricks. 1988. "Comprehensive Risk Assessment." In *Toxic Substances and Human Risk*, 391–434. New York: Springer.

Taylor, Serge. 1984. *Making Bureaucracies Think: The Environmental Impact Statement Strategy of Administrative Reform*. Stanford, CA: Stanford University Press.

Tesh, Sylvia Noble. 2000. *Uncertain Hazards: Environmental Activists and Scientific Proof*. Ithaca, NY: Cornell University Press.

Thacher, David, and Martin Rein. 2004. "Managing Value Conflict in Public Policy." *Governance* 17, no. 4: 457–86.

Thomas, William. 2015. *Rational Action: The Sciences of Policy in Britain and America, 1940–1960*. Cambridge, MA: MIT Press.

Thompson, Kimberly M., Paul F. Deisler Jr., and Richard Schwing. 2005. "Interdisciplinary Vision: The First 25 Years of the Society for Risk Analysis (SRA), 1980–2005." *Risk Analysis* 25, no. 6: 1333–86.

Thornton, Joe. 2001. *Pandora's Poison: Chlorine, Health, and a New Environmental Strategy*. Cambridge, MA: MIT Press.

Thornton, Patricia H., William Ocasio, and Michael Lounsbury. 2012. *The Institutional Logics Perspective: A New Approach to Culture, Structure, and Process*. Oxford: Oxford University Press.

Tierney, Kathleen. 2014. *The Social Roots of Risk: Producing Disasters, Promoting Resilience*. Stanford, CA: Stanford University Press.

Tomkin, Shelley Lynne. 2016. *Inside OMB: Politics and Process in the President's Budget Office*. London: Routledge.

Townley, Barbara. 2008. *Reason's Neglect: Rationality and Organizing*. Oxford: Oxford University Press.

Train, Russell E. *Politics, Pollution, and Pandas: An Environmental Memoir.* Washington, DC: Island Press, 2003.

Tribe, Laurence H. 1972. "Policy Science: Analysis or Ideology?" *Philosophy and Public Affairs* 2, no. 1: 66–110.

Union of Concerned Scientists (UCS). 2008. *Interference at the EPA. Science and Politics at the US Environmental Protection Agency.* Union of Concerned Scientists, Scientific Integrity Program. Cambridge, MA: UCS Publications.

Urban, Douglas J., and Norman J. Cook. 1986. *Hazard Evaluation Division—Standard Evaluation Procedure—Ecological Risk Assessment.* Washington, DC: US Environmental Protection Agency (EPA), Office of Pesticides Program.

US Congress. 1982a. *Environmental Protection Agency, Private Meetings and Water Protection Programs Hearings before a Subcommittee of the Committee on Government Operations, House of Representatives, Ninety-Seventh Congress, First Session, October 21 and November 4, 1981.* Washington, DC: US Government Printing Office.

US Congress. 1982b. *Environmental Protection Agency Research and Development Posture Hearing: Hearing before the Subcommittee on Natural Resources, Agricultural Research, and Environment of the Committee on Science and Technology, U.S. House of Representatives, Ninety-Seventh Congress, First Session, October 22, 1981.* Washington, DC: US Government Printing Office.

US Congress. 1983a. *Control of Carcinogens in the Environment: Hearing before the Subcommittee on Commerce, Transportation, and Tourism of the Committee on Energy and Commerce, House of Representatives, Ninety-Eighth Congress, First Session, March 17, 1983.* Washington, DC: US Government Printing Office.

US Congress. 1983b. *Formaldehyde: Review of Scientific Basis of EPA's Carcinogenic Risk Assessment. Hearing before the Subcommittee on Investigations and Oversight of the Committee on Science and Technology, U.S. House of Representatives, Ninety-Seventh Congress, Second Session, May 20, 1982.* Washington, DC: US Government Printing Office.

US Congress. 1983c. *Regulation of Pesticides: Hearings before the Subcommittee on Department Operations, Research, and Foreign Agriculture of the Committee on Agriculture, House of Representatives, Ninety-Eighth Congress, First Session.* Washington, DC: US Government Printing Office.

US Congress. 1984. *Dioxin—the Impact on Human Health: Hearings Before the Subcommittee on Natural Resources, Agriculture Research, and Environment of the Committee on Science and Technology, House of Representatives, Ninety-Eighth Congress, First Session, June 30, July 13, 28, 1983.* Washington, DC: US Government Printing Office.

US Congress. 1985. *H.R. 4192, the Risk Assessment Research and Demonstration Act of 1983: Hearings before the Subcommittee on Natural Resources, Agriculture Research, and Environment of the Committee on Science and Technology, U.S. House of Representatives,*

Ninety-Eighth Congress, Second Session, May 17, June 5, 1984. Washington, DC: US Government Printing Office.

US Congress. 1991. *Risk Assessment: Strengths and Limitations of Utilization for Policy Decisions: Hearing before the Subcommittee on Environment of the Committee on Science, Space, and Technology, House of Representatives, One Hundred Second Congress, First Session, May 21, 1991*. Washington, DC: US Government Publishing Office.

US Congress. 1992. *The Role of Science at EPA and Fiscal Year 1993 Budget Authorization for EPA's Office of Research and Development: Hearing before the Committee on Science, Space, and Technology, U.S. House of Representatives, One Hundred Second Congress, Second Session, March 19, 1992*. Washington, DC: US Government Printing Office.

US Congress. 1994a. *Environmental Issues: Hearings before the Subcommittee on Transportation and Hazardous Materials of the Committee on Energy and Commerce, House of Representatives, One Hundred Third Congress, First Session, Including H.R. 1924, H.R. 1925, and H.R. 2105, November 17, 1993—Risk Assessment, November 18, 1993—Environmental Justice*. Washington, DC: US Government Publishing Office.

US Congress. 1994b. *How Safe Is Safe Enough?: Risk Assessment and the Regulatory Process : Hearing before the Subcommittee on Investigations and Oversight of the Committee on Science, Space, and Technology, U.S. House of Representatives, One Hundred Third Congress, First Session, July 27, 1993*. Washington, DC: US Government Publishing Office.

US Congress. 1994c. *Risk Assessment Research: Hearing before the Subcommittee on Technology, Environment, and Aviation of the Committee on Science, Space, and Technology, U.S. House of Representatives, One Hundred Third Congress, First Session, November 16, 1993*. Washington, DC: US Government Publishing Office.

US Congress. 1994d. *Risk Based Decisionmaking at the Environmental Protection Agency: Joint Hearing Before the Environment, Energy, and Natural Resources Subcommittee and Legislation and National Security Subcommittee of the Committee on Government Operations, House of Representatives, One Hundred Third Congress, Second Session, February 1, 1994*. Washington, DC: US Government Printing Office.

US Congress. 1994e. *Strengthening Risk Assessment within EPA: Hearing before the Subcommittee on Technology, Environment, and Aviation of the Committee on Science, Space, and Technology, U.S. House of Representatives, One Hundred Third Congress, Second Session, March 9, 1994*. Washington, DC: US Government Printing Office.

US Congress. 1995. *Risk Assessment and Cost/Benefit Analysis for New Regulations: Joint Hearings before the Subcommittee on Commerce, Trade, and Hazardous Materials and the Subcommittee on Health and Environment of the Committee on Commerce, House of Representatives, One Hundred Fourth Congress, First Session, on Title III—H.R. 9, the Job Creation and Wage Enhancement Act of 1995, February 1 and 2, 1995*. Washington, DC: US Government Publishing Office.

US Congress. 2006. *EPA's Fiscal Year 2007 Science and Technology Budget Proposal: Hearing before the Subcommittee on Environment, Technology, and Standards, Committee on Science, House of Representatives, One Hundred Ninth Congress, Second Session, March 16, 2006*. Washington, DC: US Government Printing Office.

US Government Accountability Office (GAO). 2008. *Toxic Chemicals. EPA's New Assessment Process Will Increase Challenges EPA Faces in Evaluating and Regulating Chemicals*. Washington, DC: US Government Accountability Office (GAO), March.

US House of Representatives. 2003. "Politics and Science in the Bush Administration. Prepared for Rep. Henry A. Waxman." Washington, DC: Special Investigations Division, Committee of Government Reform, House of Representatives. November.

Vallero, Daniel. 2003. *Engineering the Risks of Hazardous Wastes*. Amsterdam: Butterworth-Heinemann.

Van Zwanenberg, P., and E. Millstone. 2005. *BSE: Risk, Science and Governance*. Oxford: Oxford University Press.

Vinsel, Lee. 2012. "From R&D to Technology Assessment: Regulatory Agencies, Strategies, and the Changing Nature of Knowledge-Production in US Automotive Emissions Control, 1960–1980." Paper for the workshop "Regulatory Agencies as Sites of Regulatory Knowledge." Institut Francilien Recherche Innovation Société. Paris, June.

Vogel, David. 1988. "The 'New' Social Regulation in Historical and Comparative Perspective." In *American Law and the Constitutional Order: Historical Perspectives*, edited by Lawrence M. Friedman and Harry N. Scheiber, 431–48. Cambridge, MA: Harvard University Press.

Vogel, David. 2012. *The Politics of Precaution: Regulating Health, Safety, and Environmental Risks in Europe and the United States*. Princeton, NJ: Princeton University Press.

Vogel, Sarah A. 2013. *Is It Safe? BPA and the Struggle to Define the Safety of Chemicals*. Berkeley: University of California Press.

Wagner, Wendy E. 1995. "The Science Charade in Toxic Risk Regulation." *Columbia Law Review* 95, no. 7: 1613–723.

Wagner, Wendy E., and Rena Steinzor. 2018. "Deconstructing Regulatory Science." 2018. *Regulatory Review*, June 19.

Wagner, Wendy, Elizabeth Fisher, and Pasky Pascual. 2018. "Whose Science?: A New Era in Regulatory 'Science Wars.'" *Science* 362, no. 1415: 636–39.

Wartenberg, Daniel, and Caron Chess. 1993. "The Risk Wars: Assessing Risk Assessment." *New Solutions: A Journal of Environmental and Occupational Health Policy* 3, no. 2 (August 1): 16–25.

Weatherwax, Robert K. 1975. "Virtues and Limitations of Risk Analysis." *Bulletin of the Atomic Scientists* 31, no. 7 (September 1): 29–32.

Weber, Max, Guenther Roth, and Claus Wittich. 1978. *Economy and Society: An Outline of Interpretive Sociology*. Berkeley: University of California Press.

Weick, Karl. 1993. "Organizational Redesign as Improvisation." In *Organizational Change and Redesign: Ideas and Insights for Improving Performance*, edited by George P. Huber and William H. Glick, 346–79. Oxford: Oxford University Press.

Weil, C. S., and D. D. McCollister. 1963. "Safety Evaluation of Chemicals, Relationship between Short-and Long-Term Feeding Studies in Designing an Effective Toxicity Test." *Journal of Agricultural and Food Chemistry* 11, no. 6 (1 November): 486–91.

Weimer, David L. 1993. "The Current State of the Design Craft: Borrowing, Tinkering, and Problem-Solving." *Public Administration Review* 53, no. 2: 110–20.

Weimer, David Leo, ed. 1995. *Institutional Design*. Boston: Kluwer Academic Publishers.

Weinberg, Alvin M. 1972. "Science and Trans-Science." *Minerva* 10, no. 2: 209–22.

Weinberg, Alvin M. 1981. "Reflections on Risk Assessment." *Risk Analysis* 1, no. 1 (1981): 5–7.

Weinberg, Philip, and Kevin A. Reilly. 2013. *Understanding Environmental Law,* 3rd edition. New Providence: Lexis Nexis.

Weingart, Peter. 1999. "Scientific Expertise and Political Accountability: Paradoxes of Science in Politics." *Science and Public Policy* 26, no. 3: 151–61.

Wennersten, Ronald, and Jan Fidler. 2007. "Methods for Risk Assessment within the Framework of Sustainable Development." Working Paper. Department of Industrial Ecology School of Industrial Engineering and Management, Royal Institute of Technology (KTH), Stockholm.

Whelan, Elizabeth. 1993. *Toxic Terror: The Truth behind the Cancer Scares*. Amherst: Prometheus Books.

Whelan, Elizabeth. 1994. "Mouse Terrorism: Pseudoscience and the Regulation of Environmental Chemicals." *Commonsense*. 1, no. 3: 30-47.

White House. 1992. "Reducing the Burden of Government Regulation." Memorandum for Certain Department and Agency Heads, January 28.

Wildavsky, Aaron B. 1988. *Searching for Safety*. New Brunswick: Transaction Publishers.

Wilson, James D. 1990. "Afterword: Some Thoughts on What Was Learned and Some Science Policy Issues." In *Scientific Issues in Quantitative Cancer Risk Assessment,* edited by Suresh H. Moolgavkar, 306–12. Boston: Birkhauser.

Wilson, Richard, and William Clark. 1991. "Risk Assessment and Risk Management: Their Separation Should Not Mean Divorce." In *Risk Analysis: Prospects and Opportunities*, edited by Constantine Zervos, Kathleen Knox, Lee Abramson, and Rob Coppock, 187–95. Boston: Springer.

Winickoff, David E., and Douglas M. Bushey. 2010. "Science and Power in Global Food Regulation: The Rise of the Codex Alimentarius." *Science, Technology and Human Values* 35, no. 3: 356.

Wood, Dan B., and John Bothe. 2004. "Political Transaction Costs and the Politics of Administrative Design." *Journal of Politics* 66, no. 1: 176–202.

Woodruff, Tracey J., and Frédéric Bois. 1993. "Optimization Issues in Physiological Toxicokinetic Modeling. A Case Study with Benzene." *Toxicology Letters* 69, no. 2 (August): 181–96.

Woodruff, Tracey J., Frédéric Y. Bois, David Auslander, and Robert C. Spear. 1992. "Structure and Parameterization of Pharmacokinetic Models: Their Impact on Model Predictions." *Risk Analysis* 12, no. 2: 189–201.

World Health Organization (WHO). 2005. *Principles of Characterizing and Applying Human Exposure Models*. Geneva: World Health Organization.

World Health Organization (WHO). 2009. "Principles for Modeling Dose-Response for the Risk Assessment of Chemicals." Geneva: World Health Organization/ILOP/UNEP.

Wynne, Brian. 1975. "The Rhetoric of Consensus Politics: A Critical Review of Technology Assessment." *Research Policy* 4, no. 2 (May): 108–58.

Wynne, Brian. 1982. "Institutional Mythologies and Dual Societies in the Management of Risk." In *The Risk Analysis Controversy*, edited by Howard C. Kunreuther and Eryl V. Ley, 127–43. Berlin, Germany: Springer Berlin Heidelberg.

Wynne, Brian. 1987. "Uncertainty—Technical and Social." In *Science for Public Policy*, edited by Harvey Brooks and Chester Cooper, 95–115. Oxford: Pergamon Press.

Wynne, Brian. 1992. "Carving out Science (and Politics) in the Regulatory Jungle." *Social Studies of Science* 22, no. 4: 745–58.

Wynne, Brian. 1995. "Technology Assessment and Reflexive Social Learning: Observations from the Risk Field." In *Managing Technology in Society: The Approach of Constructive Technology Assessment*, edited by Arie Rip, Johan Schot, and Thomas Misa, 19–36. London: Pinter.

Wynne, Brian. 1996. "May the Sheep Safely Graze? A Reflexive View of the Expert-Lay Knowledge Divide." In *Risk, Environment and Modernity: Towards a New Ecology*, edited by Scott Lash, Bronislaw Szerszynski, and Brian Wynne, 44–83. London: SAGE Publications.

Wynne, Brian. 2002. "Risk and Environment as Legitimatory Discourses of Technology: Reflexivity inside Out?" *Current Sociology* 50, no. 3: 459–77.

Wynne, Brian. 2005. "Risk as Globalizing 'Democratic' Discourse? Framing Subjects and Citizens." In *Science and Citizens: Globalization and the Challenge of Engagement*, edited by Melissa Leach, Ian Scoones, and Brian Wynne, 66–82. London: Zed Books.

Wynne, Brian. 2013. *Rationality and Ritual: Participation and Exclusion in Nuclear Decision-Making*. London: Routledge.

Yellin, Joel. 1983. "Science, Technology, and Administrative Government: Institutional Designs for Environmental Decisionmaking." *Yale Law Journal* 92, no. 7 (June): 1300–333.

Yosie, Terry F. 1989. "Science and Sociology: The Transition to a Post-Conservative Risk Assessment Era." In *Risk Assessment in Setting National Priorities*, edited by James Bonin and Donald Stevenson, 1–11. New York: Springer.

Zeeman, Maurice, and J. Gilford. 1993. "Ecological Hazard Evaluation and Risk Assessment under EPA's Toxic Substances Control Act (TSCA): An Introduction." *In Environmental Toxicology and Risk Assessment.* Vol. 1, *ASTM STP 1179*, edited by Wayne Landis, John Hughes, and Michael Lewis, 7–21. Philadelphia: American Society for Testing and Materials.

Zucker, Lynne G. 1983. "Organizations as Institutions." In *Research in the Sociology of Organizations*, edited by Samuel Bacharach, 2:1–47. Greenwich, CT: JAI Press.

Index

2-4-5-T, 90
2,3,7,8-tetrachlorodibenzo-p-dioxin (TCDD), 234

Acceptable daily intake (ADI), 57, 123, 353n18
  ADI policy group, 193–194
  safety factor, 173, 194
Administrative Procedure Act, 76
Agency for Toxic Substances and Disease Registry (ATSDR), 361n32
Air Force, 32
Air Office, 64, 67–68, 82, 187, 203–204, 239, 247–248
Air Quality Planning and Standards Office, 187
Alar, 215, 217–220, 238, 306, 308
ALARA (as low as reasonably achievable), 64, 331n9
Albert, Roy, 61, 64, 70, 311
Aldrin, 55, 57
Allen, Derry, 159
Alliance for Risk Assessment, 294
Allied Chemical, 58–59
Alm, Alvin, 141, 153, 160, 305, 312, 148n9. *See also* Options selection and tracking
  cancer assessment guideline, 52, 59–61, 65
  decision-making system, 161, 166–179, 191–197, 200
  EPA's top policy analyst, 59
  management philosophy, 166
  tools for risk assessment and risk management, 171–174
American Association for the Advancement of Science, 110
American Chemistry Council (ACC), 286, 316
American Industrial Health Council (AIHC)
  action on risk assessment guidelines and Interagency Regulatory Liaison Group (IRLG), 71, 73–74
  amendment of the Clean Air Act, 240
  campaign on risk characterization, 245–246, 252
  science panel proposal, 106–116
American Lung Association, 235
American Petroleum Institute, 108, 316
American Public Health Association, 176
Ames, Bruce, 219–220, 238–239
Anastas, Paul, 275, 298
Anderson, Elizabeth (Betty)
  Cancer Assessment Group and development of quantitative risk assessment, 61, 70, 74
  cancer classification, 201, 205, 207
  countering bill on risk assessment, 151

Arbitrary and capricious test, 76, 220, 249
Arsaco, 155–156
Arsenic, 85, 94, 174, 189, 195, 199, 207
  inorganic arsenic risk controversy, 236–237
  uncertainty factor, 306
Asbestos, 189, 232, 235–236
Ash Council, 52, 80, 87, 102
Assumption
  assumptions vs. values, 127–129
  cascading of assumptions, 238
  default assumptions, defaults, 237, 242–243, 263, 271, 278
  no-threshold hypothesis, 63, 67, 206, 245
Atomic Energy Commission, 46, 53, 61
Audiences, 8
  conflict among audiences, 27, 232, 257, 258
  diversity of agency audiences, 17, 27
  expectations of audiences and agency legitimacy, 27, 138, 318
  risk audiences and constituencies, 148, 155
  scientific audiences, 142, 143, 146

Bailar, John, 262
Banbury Conference, 233–234. *See also* Dioxin
Barnes, Donald, 5, 140, 190, 223, 243, 359n19
Beardsley, Dan, 92, 107, 178. *See also* Comparative risk assessment; Toxics integration
Beck, Ulrich, 10, 12, 40–42
Benzene, 174, 187, 278, 306
  benzene case, 75
Berg, Marc, 326n8
Biologically based dose-response (BBDR), 149, 233, 254, 261, 278
Bis(2-ethylhexyl) phthalate (DEHP), 120, 279, 290
Bliley, Thomas J., Jr., 248, 249
Blue Book. *See Science and Judgment* (report)
Board of Environmental Studies and Toxicology (BEST), 361n32, 362n40
Board on Toxicology and Environmental Health Hazards (BOTEHH), 115, 117, 362n39
Boudia, Soraya, 13, 42, 44
Breyer, Stephen, 224, 357n8
Brown, Mark, 3
Browner, Carol, 8, 275, 276, 307, 309, 310
Bureaucratic design
  algorithmic and nonalgorithmic, 25, 26, 281
  commensurative, 26, 105, 211, 244, 273. *See also* Risk ranking
  and configurations of controversy, 27–29, 102–103, 136, 148, 160, 237, 303
  decisionistic, 26, 30, 74, 136, 138, 174–175, 273, 303, 312
  deliberative, 27, 29, 138, 152–159. *See also* Risk communication
  design science, 25, 28, 49, 307, 313
  as dialectical process, 23, 326n9
  and formal organizational structure, 12
  material and symbolic dimensions, 138
  networks, 25, 52, 160, 307, 310–314
  as organizational assemblage, 26, 27, 93, 98, 109
  scientistic, 26–27, 73, 106, 110, 232, 274, 316
Bureaucratic screen, 22, 66, 77, 226, 241, 243, 268, 317
Bureaucratic technologies, 25, 74
  algorithmic and nonalgorithmic, 25, 26, 281
  problem formulation as, 229, 274
  risk assessment guideline, quantitative risk assessment as, 51, 58, 205, 295

risk ranking and comparative risk assessment as, 103
Burke, Thomas, 292, 311, 312
Bush, George H. W., 147, 220, 237
Bush, George W., 2, 8, 284

Cancer classification, 201–206
Cancer principles, 55–61
Cancer risk assessment
guideline, 60–66
harmonization of cancer and noncancer risk assessment, 173
low-dose extrapolation, 63
no-threshold hypothesis, 63, 67, 206, 245
Carcinogen Assessment Group (CAG), 61–66, 99, 205
Carnegie Foundation, 224, 296, 316
Carpenter, Daniel, 8, 15, 17, 22, 56, 81
Carter, Jimmy, 72, 88, 91, 95, 139, 166, 314
Centers for Disease Control and Prevention (CDC), 234
Chemical industry. *See also* American Industrial Health Council (AIHC)
amendment to Clean Air Act, 239
sound science campaign, 249
worries towards crisis of the EPA, 119
Chemical Industry Institute of Toxicology (CIIT), 120, 262, 315
Chemical Manufacturers Association (CMA), 108, 220
Chemical revolution, 87–89, 103
Cheney, Dick, 284, 365n10
Chlorine Institute, 234
Chlorpyrifos, 1
Clay, Don, 163, 176, 178
Clean Air Act (CAA), 3, 84–89, 147, 237–240, 261, 264
Clean Air Scientific Advisory Committee (CASAC), 260, 262
Clean Water Act (CWA), 3, 84–86, 88, 91
Climate change, 275
Clinton, Bill, 8, 224, 225, 229, 266, 289
Cloyna, Ernest, 186
Collins, Harry, 10, 24, 42
Committee on Improving the Risk Analysis Approaches Used by the US EPA, 278. See also *Science and Decisions* (report)
Committee on Public Engineering Policy, 44
Committee on Risk and Decision-Making (CORADM), 110–112, 115, 116, 141, 243
Committee on Risk Assessment Methodology (CRAM), 260
Committee on the Institutional Means for Assessment of Risks to Public Health. *See* Risk Assessment Committee (RAC)
Comparative risk assessment, 44–45, 212–214, 224–226
as bureaucratic technology and design, 223–228
definition of environmental priorities, 224
Unfinished Business (report), 212–216
Competitiveness Council, 237, 251, 256
Comprehensive Environmental Response, Compensation, and Liability Act (CERCLA), 3, 85, 86
Congress
as agency audience, 8, 9
bills on agencies and risk assessment, 113, 150–152, 223, 260, 267
as de facto risk manager, 86, 240
hearings and EPA representation, 119, 216, 224, 226, 258–260
Conservation Foundation, 221
Conservatism, 64, 67, 71, 206. *See also* Assumption
compounded conservatism, 238
industry action against conservatism, 73–74, 106–108, 232

Conservatism (cont.)
plausible conservatism, 264–265
postconservatism, 232–233, 237, 271, 282–286
Consumer Product Safety Commission (CPSC), 70, 75, 88, 89, 176
Controversy. *See also* Fact and value; Uncertainty
interpretive flexibility, 42, 62
public dispute and disagreement about risk, 10–13, 23, 41
value conflicts, 22–23
wicked, unstructured problems, 23
Corn, Morton, 115, 117, 122
Cost-benefit analysis. *See also* Regulatory analysis
as agency tradition, 284
and asbestos, 236
decline under Carol Browner, 225
guideline for regulatory impact analysis, 172–173
institutionalization of cost-benefit analysis through risk management, 126–127, 147, 171–174
as policy science, 33, 44
requirement in statutes and executive orders, 44, 267 (*see also* Executive order 12291; Executive order 12866)
and risk assessment, 65, 172–173
Costle, Douglas, 52, 80, 139, 213, 221, 335n4
chemical revolution, 87–89
ranking scheme, 91–92
regulatory council and regulatory reform, 95–96
Council of Economic Advisers, 95, 153
Council on Environmental Quality, 52, 55, 95, 287
Covello, Vincent, 158, 159
Cross, Nigel, 22, 25
Crump, Kenny, 64, 117, 122, 206, 260
Cumulative risk assessment, 278, 294, 307, 368n34
Daubert vs. Merrel Down Pharmaceuticals, 246
Davies, Clarence (Terry)
comparative risk assessment, 212, 221, 224
creation and design of the EPA, 52, 80, 102
and Risk Assessment Committee, 116, 117, 122, 123
De minimis risk, 75
Decision-making
agency as decision-making system, 165–171, 198
bureaucratic-pluralistic style of, 170–171
consistency problem, 80–87, 193, 287
decision as a representation of agency's action, 22
improving consistency, 99, 162–163, 178–179, 196–197, 204, 214
nosology, 164, 197, 199
options selection and tracking system, 165–171
probabilistic decision-making, 51, 60, 66
two-step process, 65, 73, 108, 116
Deisler, Paul, 127
Delaney amendment, 52–53, 55–56, 75, 270
Democratic Party, Democrats, 29, 52, 152, 212, 224, 290, 313
Department of Agriculture (USDA), 53, 54, 68, 84, 88
Department of Defense, 52, 69, 80, 287
Department of Energy, 69, 166, 218, 287
Department of Health and Human Services, 53, 84, 85, 190
Department of the Interior, 53, 84, 85, 88
Dichlorodiphenyltrichloroethane (DDT), 55, 58
Dieldrin, 55, 57
Di Maggio, Paul, 14, 18, 317

Dioxin, 90–91, 120, 174–175, 232–234, 315
Dobbin, Arthur, 17–18, 326n7
Dole, Vincent, 117, 122
Donn, Viviani W., 193
Dose-response assessment
  assumptions, 120, 263
  component of risk assessment, 7, 129, 131, 164, 153, 197–198, 256
  dose extrapolation, curve-drawing, 56, 57, 63–64, 75, 93, 123
  dose-response models, 250, 259–261, 278, 294
  vs. effect assessment, 62, 175
  harmonization of cancer and noncancer, 173
  vs. health assessment, 123
  point of departure, 286, 364n59
  response assessment 294
Douglas, Mary, 10, 23, 38–39
Dourson, Michael, 294, 367n28
Dowling, John, 21, 22, 319

Eagleton, Thomas, 113
Ecological risk assessment, 254, 274, 279, 291, 296, 298–301
Ecological system, 274, 298–301
  ecological issues vs. health issues, 214–216, 221–223
  ecological restoration, 29, 92, 100, 282, 303
  effects and threats to, 214, 223, 301
Ehrlich, Alan, 193, 353n17
Emergency Planning and Community Right-to-Know Act, 4
Endangered Species Act, 4
Energy Independence and Security Act, 4
Energy Policy Act, 4
Enthoven, Alain, 33, 80, 102
Environmental Defense Fund, 175, 233, 242, 259
Environmental impact assessment, 44
Environmental justice, 29, 225, 275
Environmental management
  comprehensive action on toxic environment, 88
  comprehensive environmental management, 57, 88
  cross-media perspective, 79, 80–82, 212–214
Environmental Protection Agency (EPA)
  autonomy, 3, 15, 17, 20, 75, 114, 150, 240, 267, 287, 291, 308–315
  balkanized structure, 90
  credibility, 2, 7, 9, 15, 22, 23
  cross-office conflict and coordination, 68, 168, 177–179, 192–194, 299, 317
  functional vs. programmatic structure, 80–83, 168–169, 177, 208, 221
  integrated agenda, 14, 28, 29, 79–105, 127, 138, 211–229
  leadership and steering of, 16, 69–70, 90, 169–170, 223, 295, 306
  legitimacy and controversy, 18, 19, 20–22, 23, 28, 81
  organizational identity and image, 3–4, 7–8, 81, 87, 103, 143, 273–274, 295, 308
  organizational integration, 7, 29, 77, 80–100, 161, 167, 172, 178–179, 193, 252
  program offices as siloes, fiefdoms or policy towers, 28, 54, 70, 81, 90, 180, 194, 212
  whole-of-agency policy, 101, 153, 210, 273, 309
Environmental tobacco smoke (ETS), 247–249
Epidemiology, 42, 115, 128, 209, 219, 260
  epidemiology and toxicology, 72, 259
Ethylene dibromide (EDB), 158, 174–176, 220, 238
Executive Order 12291, 44, 97, 168, 172, 227, 237. *See also* Cost-benefit analysis

Executive Order 12866, 227. *See also* Cost-benefit analysis
Exposure assessment
  exposure assessment office, 186
  guideline, 199–201
  maximally exposed individual, 200, 238, 240, 261

Fact and value, 22, 45. *See also* Risk evaluation
Fault-tree analysis, 47, 48, 99
Federal Focus Inc., 251
Federal Insecticide, Fungicide, and Rodenticide Act (FIFRA), 3, 59, 84, 86, 87, 89, 256
Federal Radiation Council, 53
Finkel, Adam, 224, 254, 257, 258, 312. *See also* Conservatism, plausible conservatism; Risk assessment, solution focused
Fiorino, Daniel, 59, 96, 225, 237
Fisher, Elizabeth, 14
Food and Drug Administration, 15, 56, 113, 114, 117
  autonomy and power, 15
  coordination with other agencies, 88, 70
  implementation of Delaney amendment, 75
  and risk assessment, 56
Food Quality Protection Act, 270
Formaldehyde, 76, 120, 170, 174, 176, 187, 207, 278, 290, 306

Gale, Roger, 154, 347n32
Gallo, Michael, 233–234, 260–261
Gingrich, Newt, 267
Glickman, Roger, 3, 11
Goldman, Lynn, 226
Goldstein, Bernie, 152, 165, 180, 311
  assistant administrator for EPA ORD, 186–189
  cancer risk assessment and extrapolation, 206
  as chair of Committee on Risk Assessment Methodology (CRAM), 260
  ozone risk assessment and CASAC, 235
  reorganization of research and science at the EPA, 186–189
  separation of risk assessment and risk management, 243
  sustainability framework, 311, 364n2
Gorsuch, Ann, 4, 119, 137, 141
Graham, John, 246, 285–289
Gray, George, 246, 289–294
Greenhouse gas, 1
Greenwood, Ted, 116, 117, 123, 151, 339n14
Guideline
  1986 revision of risk assessment guideline, 196–197
  1996 revision of risk assessment guideline, 268–271
  as bureaucratic technology, 51, 58, 60, 255, 295
  cancer assessment guideline, 61–67
  exposure assessment guideline, 199–201
  as instrument of a probabilistic organization, 60

Habermas, Jurgen, 35
Habicht, Hank, 236, 248
  memorandum on risk characterization, 250–258
Hard look doctrine, 76, 249
Harvard Center for Risk Analysis, 246, 247, 255, 267, 285–292, 307
Hazard characterization. *See* Dose-response assessment
Hazardous air pollutants, 3, 86, 90, 239–240
Health and Environmental Research Laboratory (HERL), 236

Health assessment, 62
asbestos, 235
cancer assessment group, 196–197, 201, 207
dioxin, 175, 233
inconsistency in, 92
vs. risk assessment, 100–101, 121, 123, 126, 317
Heptachlor, 55
Hernandez, John W., 143, 358n15
assessment of benzene, formaldehyde by science courts, 119–121, 176
Hilgartner, Stephen, 13, 99
Hill, Richard, 164
Hisschemöller, Matthijs, 23
Hitch, Charles J., 32, 49
Hoel, David, 333n18
Hood, Christopher, 9, 16, 43
Hoos, Ida, 31, 32–35
Hoppe, Robert, 23, 327n4
Houk, Vernon, 234
Hugget, Robert, 296, 367n30

Inference options, 132–133, 238, 244, 254, 268–269. *See also* Assumption; Dose-response assessment
Information Quality Act, 288
Institute for Regulatory Policy, 251
Institute of Medicine (IOM), 117, 291, 325n1
Institution
cultural view of rationalization, 17–19
institutional innovation, 15, 317–318
isomorphism, 17–19
new institutionalism, 17–19
Integrated Environmental Management Division (IEMD), 97
Integrated Risk Information System (IRIS), 195, 236, 253, 287–295
Interagency Regulatory Liaison Group (IRLG), 70–75, 117, 133, 149, 250, 311
Interagency Risk Management Council (IRMC), 143, 148–150, 178, 250
Interagency Work Group on Risk Management, 287
Interagency Working Group on Risk Assessment, 250
International Agency for Research on Cancer (IARC), 203–206, 235. *See also* Cancer classification
Irwin, Alan, 10

Jaeschke, Wayne, 73
Jasanoff, Sheila
boundary work and scientific authority, 11, 122
cancer principles, 58
formaldehyde, 120
regulatory science, 10
risk assessment-risk management separation, 242, 243
risk as organizing concept and songlines, 14, 26
Science Advisory Board (SAB), 69
Science and Judgment (report), 261, 314
science policy, 129, 145, 242, 243
Johnson, Stephen, 284

Kantrowitz, Arthur, 109–110
Kates, Robert, 114
Katzen, Sally, 228, 288, 289
Kennedy, Ted, 113
Knowledge representation, 24–25, 106, 111
and assemblage of expertise, 7, 21, 23, 273, 308
and design, 25, 28, 49, 307, 313
guidelines as, 197–199
reorganization of EPA, 161, 164
Risk Assessment in the Federal Government (RAFG) as, 135, 138
Kysar, Douglas, 24

Landau, Martin, 25
Landy, Mark, 52, 54, 70–72, 88, 90, 221, 229

Laroche, Hervé, 22
Lave, Lester, 38, 123
Law, John, 24
Lazen, Al, 134, 135, 339n22
Leavitt, Michael, 284
Lindblom, Charles, 16, 111–112, 116
Linearized Multistage (LMS) model, 64, 67, 76, 206–208, 260
Lippman, Morton, 235
Love Canal, 86, 91, 220, 224
Lowest adverse effect level (LOAEL), 194
Lowrance, William, 45, 107, 123, 142

Majone, Giandomenico, 27, 322n4
March, James, 16, 17, 21
Marcus, Alfred, 54, 81–82
Marine Protection, Research, and Sanctuaries Act, 4
Massachusetts Institute of Technology (MIT), 47, 117
McClellan, Roger, 262–265, 269, 315
McCray, Larry, 118, 135, 138, 151
  anatomy of risk assessment, 121–123, 129, 131, 133
McGarity, Thomas
  Daubert judgment, 246
  differences between EPA program offices, 54
  dioxin, 234
  George H. W. Bush Presidency, 237
  mandate of Carol Browner, 267
  regulatory analysis at the EPA, 177
  Republicans attack on the EPA, 3, 284
  risk assessment guidelines and court review, 61–62, 76
McNamara, Robert, 32, 80
Mechanical objectivity, 24
Merrill, Richard, 115, 117, 118, 123, 131, 135, 138, 224
Meyer, John, 18, 20
Miller, Clark, 3
Milvy, Paul, 203, 204
Mirer, Franklin, 11, 118, 260
Mirex, 55, 58–59
Mirowski, Philip, 33
Moffett, Toby, 119, 340n23
Moore, John (Jack)
  alar, 217
  arsenic, 236
  asbestos, 235
  consistency in decision-making, 179
  options selection and tracking, 170, 174
  and *Risk Assessment in the Federal Government* (RAFG), 163–165
Morgenstern, Richard, 97, 162–163, 215–216, 312
Moses, Lincoln, 262
Moynihan, Daniel, 223, 258
Mrak Commission, 55
Muskie, Edmund, 52

Nash, Linda, 51
National Academy of Engineering, 44, 325n1
National Academy of Sciences (NAS). *See also* Risk assessment, risk assessment board; Risk Assessment Committee (RAC); *Risk Assessment in the Federal Government* (RAFG)
  lobbying of AIHC, 113, 240, 260
  response to Congress on risk assessment, 151, 152, 260, 290
  review of EPA's assessments, 256, 292, 319
  and risk research, 114
  speech of Ruckelshaus at, 4, 141–143
National Aeronautics and Space Administration (NASA), 287
National Ambient Air Quality Standards (NAAQS), 3
National Center for Environmental Assessment (NCEA), 252, 276
National Emissions Standards for Hazardous Air Pollutants (NESHAPS), 3

National Environmental Policy Act, 3, 44, 52
National Institute for Environmental Health Sciences, 139, 163, 187, 264
National Institute for Occupational Safety and Health, 234
National Research Council (NRC). *See* National Academy of Sciences (NAS)
National Science Foundation (NSF), 158, 328n14
National Wildlife Fund (NWF), 233
Natural Resources Defense Council, 176, 217, 218, 364n58
Nelkin, Dorothy, 10, 11, 13, 23, 43
Nixon, Richard, 3, 52–53, 95–96, 102, 117, 221
No Observed Adverse Effect Level (NOAEL), 56, 173, 333n18

O'Malley, Pat, 14
Oak Ridge National Laboratory, 299, 300
Occupational Safety and Health Administration (OSHA), 15, 148, 152, 261
 and *Risk Assessment in the Federal Government* (RAFG), 114, 117, 118
 risk assessment guideline, 70, 72, 73
Office for Indoor Air, 247
Office for Policy, Economics, and Innovations, 284
Office of Air Quality Planning and Standards (OAQPS), 187
Office of Air and Radiation, 83, 86, 94
Office of Drinking Water, 101, 236
Office of Enforcement, 82
Office of Health and Environmental Assessment (OHEA)
 1986 revision of risk assessment guideline, 196–197
 cancer classification, 201–206
 coordination with program offices, 68, 186, 187, 188
 definition of risk assessment, 123
 dioxin, 234
 as mega-laboratory, 70
 ozone, 235
 perchloroethylene (PCE), 201–206
 risk assessment guideline, 57–67
Office of the General Counsel, 50, 64, 168, 205
Office of Information and Regulatory Affairs (OIRA), 97, 285–289. *See also* Graham, John; Information Quality Act; Tozzi, Jim
Office of Management and Budget (OMB). *See also* Office of Information and Regulatory Affairs (OIRA)
 autonomous coordination of agencies, 149. *See also* Interagency Risk Management Council (IRMC)
 and cost-benefit analysis, 168, 225
 criticism of EPA's guidelines and conservatism, 237–239, 256
 review of EPA rules, 53, 96, 101, 102, 173, 218, 237–238
 risk assessment bulletin, 287
 and toxics integration, 95, 97
Office of Management Systems and Evaluation, 168
Office of Pesticides
 2-4-5-T, 90
 and EPA general counsel, 54, 57, 59
 capacities for science and risk assessment, 68, 179
 dioxin, 91
 relationship with EPA ORD, 68, 179
 transfer from USDA to EPA, 84
Office of Pesticides and Toxic Substances (OPTS), 86
 alar, 217
 asbestos, 235
 coordination with other program offices, 95
 ethylene dibromide (EDB), 175
 multimedia assessment, 99

Office of Pesticides and Toxic Substances (OPTS) (cont.)
vs. OPPE, 179
participation in guideline revision, 197
reorganization following RAFG, 179
risk assessment training, 181
terminological confusion, 197
Office of Planning, 97
Office of Policy Analysis, 159, 162, 173, 214, 215
Office of Policy and Resources Management (OPRM), 330n3
Office of Policy, Planning, and Evaluation (OPPE)
as EPA main analytical office, 169
application of the risk assessment-risk management framework and RAFG, 155 167–168, 213
comparative risk assessment, 213–215
coordination of program offices, 80, 213, 223, 244, 167–171
interface with OMB, 96
as mini-OMB, 96, 97, 174, 237
rise and fall in hierarchy of offices, 169, 225, 244
risk communication, 152–158
Office of Pollution Prevention and Toxics (OPPT), 225, 226
Office of Regulatory and Information Policy, 97
Office of Research and Development (ORD). *See also* Carcinogen Assessment Group (CAG); National Center for Environmental Assessment (NCEA); Office of Health and Environmental Assessment (OEHA)
budget and capacities, 1, 69, 141, 143, 218, 226
Office of Technology Transfer and Regulatory Support, 188
regulatory oversight group, 244
relationship with Office of Toxic Substances and Pesticides, 193
relationship with program offices, 7, 68, 69, 99, 168, 188, 193
reorganization and laboratory structure, 69–70, 186–188, 189, 233, 275–276, 295–297
research on technology and engineering, 146
research programmation, 93, 226
risk assessment training, 180–181
sustainability, 274–276
Office of Science Advisor, 281, 366n16
Office of Science and Technology Policy (OSTP)
1984 risk assessment guideline, 208
cancer guideline (Calkins et al. 1980), 72–73, 107, 129, 140
coordination of agency risk assessment, 113, 289
Regulatory Analysis Review Group (RARG), 95
Office of Solid Waste, 258
Office of Technology Assessment (OTA), 128, 226, 316, 327n7
Office of Toxics Integration, 94, 336n15
Office of Water, 86, 101, 152
Omenn, Gilbert, 72, 314. *See also* Office of Science and Technology Policy (OSTP)
Carnegie report on Risk and the Environment, 224
member of Risk Assessment Committee (RAC), 116–117
Presidential/Congressional Commission on Risk Assessment and Risk management, 300–302, 368n35
and Todhunter, John, 163
Operations research. *See* Systems analysis, systems approach
Options selection and tracking, 165–171
Ozone, 234–235, 214, 306

Parsons, Talcott, 21
Particulate matter 10, 84, 177, 284

Passive smoking. *See* Environmental tobacco smoke (ETS)
Patton, Dorothy, 195
Perchlorate, 288, 289, 366n15
Perchloroethylene (PCE), 201–206
Perrow, Charles, 37–38
Pfeffer, Jeffrey, 21, 22, 319
Pfiffner, John, 16
Pharmacokinetics, 259, 261, 263
Philip Morris, 248, 249
Phthalates, 290, 294
Physiologically based pharmaco-kinetics (PBPK), 236, 259, 271, 278, 315
Planning Programming Budgeting System, 32
Policy analysis, 32, 59, 111, 228, 327n4
Planning and scoping, 274–281, 293–295, 303
Policy sciences, 24, 31–37
Polychlorinated biphenyl (PCB), 220
Porter, Theodore, 24
Post-normal science, 10
Powell, Mark, 54, 94, 207, 236, 244, 258, 296, 311
Powell, Walter W., 14, 18, 317
Power, Michael, 6, 9, 11, 19, 22, 90
Prediction, 10, 14, 73, 259. *See also* Physiologically based pharmaco-kinetics (PBPK)
Presidential Advisory Group on Anticipated Advances in Science and Technology, 110
Presidential/Congressional Commission on Risk Assessment and Risk Management (PCCRARM), 300–302, 368n35
Press, Frank, 115, 116, 134, 138, 139, 151
Preuss, Peter
  1996 revision of risk assessment guideline, 270
  dioxin risk assessment, 234
  regulatory oversight group, 267
  memorandum on risk characterization, 252
  Office of Technology Transfer and Regulatory Support, 188
  Science and Decisions (report), 291, 292, 311
  uncertainty analysis at NCEA, 290
Price-Anderson Nuclear Industries Indemnity Act of 1957 (or Price-Anderson Act), 46
Problem formulation, 30, 274, 280–284
Pruitt, Scott, 1
Public participation, 6, 35, 157. *See also* Stakeholder engagement or involvement
Pure Food and Drug Act, 55

Quantification, 11, 33
  as educated guess, 34, 39
  quantomania, 34
Quantitative risk assessment, 33, 43, 46–47, 48, 61–67, 198
Quarles, John, 91
Quayle, Dan, 237, 250, 252, 257, 288

Radon, 211, 216, 218–222
Raiffa, Howard, 37–38, 111–112
RAND Corporation, 32–33
Rasmussen, Norman, 47–48
Rationalization, 15–20
  cultural and strategic view of rationalization, 15–20
  instrumental rationality, 15–17, 47–48
  material rationality, 17, 26, 315. *See also* Weber, Max
  narrative of rationalization, 22, 310
  political history of rationalization, 28
Reactor Safety Study (WASH-1400) (NUREG 75/014), 47–48
Reagan, Ronald. *See also* Regulatory policy
  appointment of Ann Gorsuch, 96, 119
  appointment of William Ruckelshaus, 4, 137–140, 187

Rebuttable Presumption Against Registration (RPAR), 175
Red Book, 240–247, 325n2. *See also Risk Assessment in the Federal Government* (RAFG)
Reference dose, 194–195, 236, 253, 291–293
Regulatory analysis, 92, 95–97
  development under Ruckelshaus and Alm, 166, 177
  format and content, 172–173, 350n28
  Regulatory Analysis Review Group (RARG), 95, 96
  Regulatory Impact Analysis (RIA), 95, 172–173
  and risk management, 129, 147
Regulatory Council, 53, 71, 75, 95, 335n9
Regulatory policy
  deregulation, 1, 103, 141, 307, 358n15
  regulatory reform, 95–97
  regulatory relief, 96
Regulatory science, 10
Reilly, William (Bill), 7, 12, 148, 211
  dioxin risk assessment, 234
  environmental tobacco smoke (ETS), 248
  ozone, 235
  relative risk reduction, 220–225
Rein, Martin, 22, 26, 27
Reisa, James, 5, 279
Republican Party, Republicans, 2, 150, 267, 277, 307, 319
Resource Conservation and Recovery Act (RCRA), 3, 89
Resources for the Future, 117, 153, 224, 254, 312, 316
Riles, Annelise, 23
Rip, Arie, 31, 42–46
Risk
  acceptability, 44, 45, 48, 79, 150, 194, 306
  controversy, 10–11
  definition, 34, 326n6
  health risk vs. ecological risk, 212–216
  individual vs. population risk, 101, 252, 255
  as organizing concept, 26
  vs. safety, 56–57
  as set of design rationales, 26
  as songlines, 14
Risk analysis
  as comparative risk assessment and synoptical view of risks, 44, 97–98, 214, 221
  interdiscipline, 42, 43, 46, 107
  as policy science, 33–49
  vs. risk assessment, 98–102
  Society for Risk Analysis, 42–46, 245, 285,
Risk Analysis (journal), 329n14
Risk analysis principles, 228, 267, 289
Risk assessment. *See also* Dose-response assessment; Exposure assessment; Risk assessment and risk management; Risk characterization; Science policy
  accuracy of, 145, 190–191, 196, 209, 214, 229, 259.
  anatomy of, 121–123
  codification, 122–126
  elements, 5, 122–126, 133, 140, 181
  vs. health assessment, effect assessment, 62, 100–101
  risk assessment board, 133, 151, 339n22, 342n35
  risk assessment policy, 123, 132–133, 140, 145, 208, 255
  scientization, 232, 242, 254, 303
  solution-focussed, 293–294
  statistical confidence levels, 71, 360n29
  weight of evidence (WOE), 61, 73, 120, 173, 197, 205

Risk assessment and risk management. *See also* Risk assessment; Risk management.
- creation of organizational units for, 7
- distinction, functional separation, 129, 132, 134, 145
- as framework and assemblage, 25, 60–62, 106, 136, 161–164, 171–172, 187, 207, 209
- interaction between, 129, 130, 243, 266, 278
- organizational discipline, 171
- as "paradigm," 4, 6
- schism, divorce between, 242–243
- as science and policy, 14, 129
- separation, 2, 9, 134, 145, 148, 240–244, 306
- as steps or modules of decision-making, 121, 127 129, 181
- utility of risk assessment for risk management, 279–282

Risk Assessment Committee (RAC), 115–130, 145, 192, 221, 260, 265, 314

Risk Assessment in the Federal Government: Managing the Process (RAFG)
- ambiguity, 129, 133, 145
- building on EPA work, 122–123, 164
- change to the paradigm after RAFG, 6–7, 223
- codification of risk decision-making, 122–126
- dissemination of the report, 138–140
- as "paradigm," 4–6, 325n3
- reception and interpretation, 4, 7, 130–135, 142–148, 243
- reception in EPA, 163–165
- sequels, 260, 261–266, 278–279
- writing of the report, 115–121

Risk Assessment Forum (RAF), 7, 185–190, 195, 226

Risk Assessment Review, 182, 244

Risk-based budgeting, 224

Risk characterization
- AIHC campaign, 245–246, 252
- as component of the risk framework, 5, 124–126, 129–131, 133, 164, 197
- Habicht memorandum on risk characterization, 250–258
- as narrative interpretation of risk, 126
- presentation of risk to decision-makers, 245
- uncertainties and assumptions in risk assessment, 256–258

Risk communication, 6, 14, 138, 152–159, 218, 224, 226
- communication of risk to decision-makers, 245
- seven cardinal rules of, 159

Risk evaluation
- ambiguous step in decision-making, 111–112, 150
- Committee on Risk and Decision-Making (CORADM), 111, 112, 113,
- deletion by Risk Assessment Committee, 127–128
- value judgment, 46, 107, 115
- "value side" in Risk Assessment Committee (RAC), 122
- values in decision-making, 111, 112, 125

Risk management. *See also* Interagency Risk Management Council (IRMC); *Risk Assessment in the Federal Government* (RAFG); Risk assessment and risk management
- as including cost-benefit analysis, 126–127, 147, 171–174
- invention of the risk management category, 126, 129, 142
- as management of risk assessment and of science policies, 6, 148, 190, 254, 279–281, 293–294, 296–297
- as mode of communication and engagement with publics, 153–156

Risk Management Council (RMC), 7, 192, 266
risk management information reporting format, 173
as space for decision-maker and policy judgment, 14, 142, 148
as weighing of policy alternatives or options, 16, 111–112, 126–127, 130, 243
Risk object, 89–90, 131
Risk perception, 11, 118, 157–158
as branch of risk research, 38–39, 43, 45
Risk professionals, 38–39, 45, 139, 329n17
professional domination, 37–39
Risk-ranking, 28–29, 80, 93–94, 103, 106, 225–228
Risk reduction, 100, 147, 225–227, 274, 286
relative risk reduction, 222–223
Risk regulation, 3, 9, 14, 139, 227, 286
Risk research, 42–49
Rittel, Horst, 22
Ritter, Don, 113, 150–151, 220, 258–260
RJR Nabisco, 249
Robbins, Fred, 116
Rodricks, Joseph, 5, 56, 288, 311
audit of EPA risk assessment practices, 191, 192
member of Risk Assessment Committee (RAC), 115–118, 131, 132, 135
Science and Decisions (report), 278, 292
Rothstein, Henry, 18, 19, 26
Rowe, William, 45–46, 48, 206
Ruckelshaus, William
activist suit strategy, 54
appointment by Ronald Reagan, 140–142
balance between economic development and environmental protection, 147
coordination of federal agencies, 148, 150, 250 (*see also* Interagency Risk Management Council [IRMC])
and cost-benefit analysis, 147, 358n15
EPA structure, 81–82
reputation, 141
stress on health issues, 144, 145, 146, 147
use of *Risk Assessment in the Federal Government* (RAFG), 4, 7, 142–148
Rushefsky, Mark, 60, 63, 71
Russell, Milton, 152–153
application of *Risk Assessment in the Federal Government* (RAFG) in the EPA, 166–167
assistant administrator for OPPE and adviser to EPA administrator, 212–214
memorandum on risk communication, 152–154
pushing risk analysis across EPA, 171, 178–180

Safe Drinking Water Act (SDWA), 3, 89, 19
Safety evaluation, 57
Saffioti, Umberto, 57, 70
Sarewitz, Daniel, 10
Schaffer, Simon, 24
Schatzow, Steven, 215
Scheuer, James, 152, 258
Schön, Donald, 22, 26
Science
authority of science, 40–41
language of science and proof, 35, 41, 62, 241
mimicking of science, 22, 307
scientific method, 16, 32–34
scientization, 232, 242, 254, 303
secondary scientization, 40–42
Science Advisory Board (SAB), 1
effective use of, 189
organization of research at the EPA, 69, 186–187, 296

relative risk reduction, 221–222
review of EPA standards, 96, 119, 189, 207, 234, 236, 240
SAB review vs. National Academy of Sciences review of EPA standards, 290–291
Science and Decisions (report), 279, 292–294
Science and Judgment (report), 262–266
Science and Technology Policy Council, 275
Science and Technology Studies, 24, 242, 261
Science court, 109, 120, 152
Science panel, 106–109, 112, 113. *See also* American Industrial Health Council (AIHC)
Science policy, 129, 165, 192
element of the risk framework, 188, 194, 195, 207, 241, 253
flexibility and clarification of science policies, 231–232, 292
vs. risk assessment policy, 129, 208, 263
Science Policy Council, 7, 266, 268, 275
Scott, Richard, 18–19, 21
Secondhand, passive smoking. *See* Environmental tobacco smoke (ETS)
Sellers, Christopher, 2, 45
Selznick, Philip, 19, 22, 319
Shabecoff, Philip, 11, 91, 143, 177, 215
Shapin, Steven, 24
Shapiro, Sidney, 2, 3, 11, 284, 289
Sharlin, Harold, 158
Shrader-Frechette, Kristin, 314, 328n12, 341n32
Silbergeld, Ellen, 11, 241–242, 259–260
Silver Book. *See Science and Decisions* (report)
Simon, Herbert, 16, 20, 21
Slovic, Paul, 11, 48, 118, 123, 157, 314
Smith, Philip, 139
Social regulation, 55, 71, 96, 119
Sound science campaign, 249
Stakeholder engagement or involvement, 279–283, 295, 301–302
Stallones, Reuel, 115, 116, 122, 133–134, 139, 311
Star, Susan Leigh, 24, 84
Starr, Chauncey, 44, 215
Stauffer Chemicals, 44, 215
Stevens, John Paul (Justice), 75
Stinchcombe, Arthur, 19, 21
Stone, Deborah, 16, 23
Strauss, Anselm, 25
Superfund, 3, 68, 85–86, 137, 146
Supreme Court, 60, 75–76, 246
Sustainability, 274–276
sustainability framework, 276–278, 364n2
Suter, Glenn, 254, 299–300
Systems analysis, systems approach, 32–35

Tacoma hearing, 155–157
Tardiff, Robert (Bob), 115, 339n13
Technocracy, 37
Technology
Maximum Available Control Technology (MACT), 239
technology assessment, 31, 33–36
Technology Assessment and Risk Analysis (TARA), 158
technology-based statutes, 86, 96, 233
technology first then risk approach, 239
technology research at the EPA, 146–147
Tetrachlorodibenzo-p-dioxin (TCDD), 91, 234, 290
Thacher, David, 27
Thomas, Lee, 146, 158, 163, 192, 211, 225
health vs. environmental priorities, 212–216

Thomas, William, 32, 33
Thorton, Joe, 242, 358n14
Thorton, Patricia, 19
Times Beach, 86, 91, 234
Todhunter, John, 120–121, 163, 174, 176
Toxic chemicals
  carcinogenicity, 57–58, 60, 62, 201–206
  chemical of the month syndrome, 68, 216
  developmental toxicity, 270
  genotoxicity, 192, 206
  high visibility chemicals, 88, 145
  mode of action, 278, 292
  toxics and conventional pollutants, 145, 149
Toxicological Excellence in Risk Assessment (TERA), 294
Toxicology, 43–49, 259–260
Toxics integration, 91–99, 169, 221
  Integrated Toxics Strategy Work Group, 91
  Toxics Integration Task Force (TITF), 162
Toxic Substances Control Act (TSCA), 3, 85–89, 91, 97, 176–179, 235–236
Tozzi, Jim, 97, 139, 251, 285, 288
Train, Russell, 59–61, 65, 88, 108
Trichloroethylene (TCE), 290

Uncertainty
  data gaps, 33, 34, 255, 268
  measurement uncertainty, 255
  Science and Decisions (report), 294–295
  structural uncertainty and controversy, 10, 12, 26
  uncertainty analysis, 48, 250, 254–255, 271, 281
  uncertainty analysis under George Gray at ORD, 287, 289, 292
  uncertainty factor, 193–194, 236, 278, 286, 306
  variability, 247, 263, 278–279, 281, 293
Utidjian, William, 118

Vinyl chloride, 91

Wagner, Elizabeth, 2, 13, 234, 249, 315
Walker, Bailus, 266, 292
Wampler, William, 113
Water Office, 123, 179, 197
  2,3,7,8-tetrachlorodibenzo-p-dioxin (TCDD), 91
  arsenic, 236
  and Cancer Assessment Group (CAG), 68, 99, 101
  Perchloroethylene (PCE), 201–204
  separation of science and regulatory work, 244
Waxman, Henry, 248, 258
Weber, Max, 15, 17, 26
Weinberg, Alvin, 3
Weyerhaeuser, 343n5
Wheeler, Andrew, 1
White House Council on Government Reorganization, 52
White, Gilbert, 114
Whitman, Christine Todd, 8, 284
Wildavsky, Aaron, 10, 23, 38–40
Wilson, Richard, 74, 107–108, 241, 243, 244
Wynne, Brian, 10, 25, 35–37

Yosie, Terry, 77, 188–189, 207, 237

Zeise, Lauren, 292

Inside Technology Series

Edited by Wiebe E. Bijker, W. Bernard Carlson, and Trevor Pinch

David Demortain, *The Science of Bureaucracy: Risk Decision-Making and the US Environmental Protection Agency*

Joeri Bruyninckx, *Listening in the Field: Recording and the Science of Birdsong*

Edward Jones-Imhotep, *The Unreliable Nation: Hostile Nature and Technological Failure in the Cold War*

Jennifer L. Lieberman, *Power Lines: Electricity in American Life and Letters, 1882–1952*

Jess Bier, *Mapping Israel, Mapping Palestine: Occupied Landscapes of International Technoscience*

Benoît Godin, *Models of Innovation: The History of an Idea*

Stephen Hilgartner, *Reordering Life: Knowledge and Control in the Genomics Revolution*

Brice Laurent, *Democratic Experiments: Problematizing Nanotechnology and Democracy in Europe and the United States*

Cyrus C. M. Mody, *The Long Arm of Moore's Law: Microelectronics and American Science*

Tiago Saraiva, *Fascist Pigs: Technoscientific Organisms and the History of Fascism*

Teun Zuiderent-Jerak, *Situated Interventions: Sociological Experiments in Healthcare*

Basile Zimmermann, *Technology and Cultural Difference: Electronic Music Devices, Social Networking Sites, and Computer Encodings in Contemporary China*

Andrew J. Nelson, *The Sound of Innovation: Stanford and the Computer Music Revolution*

Sonja D. Schmid, *Producing Power: The Pre-Chernobyl History of the Soviet Nuclear Industry*

Casey O'Donnell, *Developer's Dilemma: The Secret World of Videogame Creators*

Christina Dunbar-Hester, *Low Power to the People: Pirates, Protest, and Politics in FM Radio Activism*

Eden Medina, Ivan da Costa Marques, and Christina Holmes, editors, *Beyond Imported Magic: Essays on Science, Technology, and Society in Latin America*

Anique Hommels, Jessica Mesman, and Wiebe E. Bijker, editors, *Vulnerability in Technological Cultures: New Directions in Research and Governance*

Amit Prasad, *Imperial Technoscience: Transnational Histories of MRI in the United States, Britain, and India*

Charis Thompson, *Good Science: The Ethical Choreography of Stem Cell Research*

Tarleton Gillespie, Pablo J. Boczkowski, and Kirsten A. Foot, editors, *Media Technologies: Essays on Communication, Materiality, and Society*

Catelijne Coopmans, Janet Vertesi, Michael Lynch, and Steve Woolgar, editors, *Representation in Scientific Practice Revisited*

Rebecca Slayton, *Arguments that Count: Physics, Computing, and Missile Defense, 1949–2012*

Stathis Arapostathis and Graeme Gooday, *Patently Contestable: Electrical Technologies and Inventor Identities on Trial in Britain*

Jens Lachmund, *Greening Berlin: The Co-Production of Science, Politics, and Urban Nature*

Chikako Takeshita, *The Global Biopolitics of the IUD: How Science Constructs Contraceptive Users and Women's Bodies*

Cyrus C. M. Mody, *Instrumental Community: Probe Microscopy and the Path to Nanotechnology*

Morana Alač, *Handling Digital Brains: A Laboratory Study of Multimodal Semiotic Interaction in the Age of Computers*

Gabrielle Hecht, editor, *Entangled Geographies: Empire and Technopolitics in the Global Cold War*

Michael E. Gorman, editor, *Trading Zones and Interactional Expertise: Creating New Kinds of Collaboration*

Matthias Gross, *Ignorance and Surprise: Science, Society, and Ecological Design*

Andrew Feenberg, *Between Reason and Experience: Essays in Technology and Modernity*

Wiebe E. Bijker, Roland Bal, and Ruud Hendricks, *The Paradox of Scientific Authority: The Role of Scientific Advice in Democracies*

Park Doing, *Velvet Revolution at the Synchrotron: Biology, Physics, and Change in Science*

Gabrielle Hecht, *The Radiance of France: Nuclear Power and National Identity after World War II*

Richard Rottenburg, *Far-Fetched Facts: A Parable of Development Aid*

Michel Callon, Pierre Lascoumes, and Yannick Barthe, *Acting in an Uncertain World: An Essay on Technical Democracy*

Ruth Oldenziel and Karin Zachmann, editors, *Cold War Kitchen: Americanization, Technology, and European Users*

Deborah G. Johnson and Jameson W. Wetmore, editors, *Technology and Society: Building Our Sociotechnical Future*

Trevor Pinch and Richard Swedberg, editors, *Living in a Material World: Economic Sociology Meets Science and Technology Studies*

Christopher R. Henke, *Cultivating Science, Harvesting Power: Science and Industrial Agriculture in California*

Helga Nowotny, *Insatiable Curiosity: Innovation in a Fragile Future*

Karin Bijsterveld, *Mechanical Sound: Technology, Culture, and Public Problems of Noise in the Twentieth Century*

Peter D. Norton, *Fighting Traffic: The Dawn of the Motor Age in the American City*

Joshua M. Greenberg, *From Betamax to Blockbuster: Video Stores tand the Invention of Movies on Video*

Mikael Hård and Thomas J. Misa, editors, *Urban Machinery: Inside Modern European Cities*

Christine Hine, *Systematics as Cyberscience: Computers, Change, and Continuity in Science*

Wesley Shrum, Joel Genuth, and Ivan Chompalov, *Structures of Scientific Collaboration*

Shobita Parthasarathy, *Building Genetic Medicine: Breast Cancer, Technology, and the Comparative Politics of Health Care*

Kristen Haring, *Ham Radio's Technical Culture*

Atsushi Akera, *Calculating a Natural World: Scientists, Engineers and Computers during the Rise of U.S. Cold War Research*

Donald MacKenzie, *An Engine, Not a Camera: How Financial Models Shape Markets*

Geoffrey C. Bowker, *Memory Practices in the Sciences*

Christophe Lécuyer, *Making Silicon Valley: Innovation and the Growth of High Tech, 1930–1970*

Anique Hommels, *Unbuilding Cities: Obduracy in Urban Sociotechnical Change*

David Kaiser, editor, *Pedagogy and the Practice of Science: Historical and Contemporary Perspectives*

Charis Thompson, *Making Parents: The Ontological Choreography of Reproductive Technology*

Pablo J. Boczkowski, *Digitizing the News: Innovation in Online Newspapers*

Dominique Vinck, editor, *Everyday Engineering: An Ethnography of Design and Innovation*

Nelly Oudshoorn and Trevor Pinch, editors, *How Users Matter: The Co-Construction of Users and Technology*

Peter Keating and Alberto Cambrosio, *Biomedical Platforms: Realigning the Normal and the Pathological in Late-Twentieth-Century Medicine*

Paul Rosen, *Framing Production: Technology, Culture, and Change in the British Bicycle Industry*

Maggie Mort, *Building the Trident Network: A Study of the Enrollment of People, Knowledge, and Machines*

Donald MacKenzie, *Mechanizing Proof: Computing, Risk, and Trust*

Geoffrey C. Bowker and Susan Leigh Star, *Sorting Things Out: Classification and Its Consequences*

Charles Bazerman, *The Languages of Edison's Light*

Janet Abbate, *Inventing the Internet*

Herbert Gottweis, *Governing Molecules: The Discursive Politics of Genetic Engineering in Europe and the United States*

Kathryn Henderson, *On Line and On Paper: Visual Representation, Visual Culture, and Computer Graphics in Design Engineering*

Susanne K. Schmidt and Raymund Werle, *Coordinating Technology: Studies in the International Standardization of Telecommunications*

Marc Berg, *Rationalizing Medical Work: Decision Support Techniques and Medical Practices*

Eda Kranakis, *Constructing a Bridge: An Exploration of Engineering Culture, Design, and Research in Nineteenth-Century France and America*

Paul N. Edwards, *The Closed World: Computers and the Politics of Discourse in Cold War America*

Donald MacKenzie, *Knowing Machines: Essays on Technical Change*

Wiebe E. Bijker, *Of Bicycles, Bakelites, and Bulbs: Toward a Theory of Sociotechnical Change*

Louis L. Bucciarelli, *Designing Engineers*

Geoffrey C. Bowker, *Science on the Run: Information Management and Industrial Geophysics at Schlumberger, 1920–1940*

Wiebe E. Bijker and John Law, editors, *Shaping Technology / Building Society: Studies in Sociotechnical Change*

Stuart Blume, *Insight and Industry: On the Dynamics of Technological Change in Medicine*

Donald MacKenzie, *Inventing Accuracy: A Historical Sociology of Nuclear Missile Guidance*

Pamela E. Mack, *Viewing the Earth: The Social Construction of the Landsat Satellite System*

H. M. Collins, *Artificial Experts: Social Knowledge and Intelligent Machines*

http://mitpress.mit.edu/books/series/inside-technology